MOLECULAR DYNAMICS IN BIOSYSTEMS

MOLECULAR DYNAMICS IN BIOSYSTEMS

The Kinetics of Tracers in Intact Organisms

by

KENNETH H. NORWICH

University of Toronto, Toronto, Ontario, Canada

PERGAMON PRESS

OXFORD · NEW YORK · TORONTO · SYDNEY · PARIS · FRANKFURT

U.K.	Pergamon Press Ltd., Headington Hill Hall, Oxford OX3 0BW, England
U.S.A.	Pergamon Press Inc., Maxwell House, Fairview Park, Elmsford, New York 10523, U.S.A.
CANADA	Pergamon of Canada Ltd., 75 The East Mall, Toronto, Ontario, Canada
AUSTRALIA	Pergamon Press (Aust.) Pty. Ltd., 19a Boundary Street, Rushcutters Bay, N.S.W. 2011, Australia
FRANCE	Pergamon Press SARL, 24 Rue des Ecoles, 75240 Paris, Cedex 05, France
WEST GERMANY	Pergamon Press GmbH, 6242 Kronberg-Taunus, Pferdstrasse 1, Frankfurt-am-Main, West Germany

First edition 1977

Library of Congress Cataloging in Publication Data

Norwich, Kenneth H
Molecular dynamics in biosystems.
1. Molecular biology. 2. Biological transport.
I. Title.
QH506.N67 1977   574.8'8   76-44524
ISBN 0-08-020420-1

In order to make this volume available as economically and rapidly as possible the author's typescript has been reproduced in its original form. This method unfortunately has its typographical limitations but it is hoped that they in no way distract the reader.

QH
506
.N67
1977

Printed in Great Britain by A. Wheaton & Co., Exeter

This book is dedicated to
my dear parents,
JACK AND ETTA NORWICH

CONTENTS

viii   Contents

FOREWORD

This book deals with the measurement of the transport and turnover of molecules in an intact biological organism. Since the primary tool used for such
studies is the tracer, a large part of the book deals with tracer kinetics.
Emphasis is placed throughout on rates of appearance and disappearance of
substances (turnover) within a distributed or non-compartmented system.
I have tried to answer such questions as:

How rapidly does the material "turn over"?

How fast does the fluid flow?

How large is the space through which the material is distributed?

How is our understanding of biology enhanced by the above information?
It is essentially a mathematical study. Fundamental equations are derived
from physical and chemical principles; experimental data are then adduced to
verify and demonstrate the applicability of the equations. Emphasis is always
placed on the applications of the equations in the analysis of experimental
or clinical data.

At the end of most sections of the book are inserted short summaries giving
the applications of the theory. It is hoped that people who are primarily
concerned with experimental studies can obtain the factual information they
require from these *Speaking Practically* summaries, referring to the relevant
portions of the main text when they require more knowledge about some subject.

A feature of this book is its demonstration that biokinetics can be developed
without recourse to the concept of the "compartment". While the theory of
compartments is developed, and even followed historically, it is shown that a
model of the organism as a distributed system achieves all the utility of a
compartmental model and much more besides, often with a good deal less mathematical complexity. The distributed model, which admits the use of vector
analysis and the application of the theory of linear systems, places biokinetics on firmer footing in physics and chemistry. No pseudo-physical concept such as "instantaneous, uniform intermixing within non-localizable

compartments" is required.

A great deal of importance is laid upon the calculation of the rates (in both
the steady and the non-steady state) at which substances appear and disappear
in intact systems. It is stressed that studying changes in the concentration
of metabolites is not sufficient. Any attempt to understand the causes of
changes in the concentrations of metabolites without knowledge of the rates,
or mass fluxes, is like trying to understand the causes of temperature changes
without invoking the concept of caloric flux.

Another characteristic of this treatment of biokinetics is the stress which it
places upon the experimental validation of mathematical results. Wherever
possible, experimental evidence is provided to demonstrate the validity and
limitations of the mathematical equations. That theory and experiment must
advance together *pari passu* was a cardinal principle practised by my teacher,
Dr. Gerald A. Wrenshall, one of the pioneers in the field of biokinetics.

The material presented here forms part of a course on biokinetics offered by
the University of Toronto. This course was designed for senior undergraduates
and graduate students in both the biological and physical sciences, and re-
gistrants have come from physiology, pharmacology, physics and electrical and
mechanical engineering. I hope that this book may serve both as a text to
students such as these, and as a reference for those engaged in research.

I have assumed that my readers have some familiarity with the biological
sciences, and with the principles of physiology in particular. A basic knowledge of mathematics, at the freshman or sophomore level, is necessary. I
have tried to provide theorems in certain specialized areas such as vector
analysis, hydrodynamics and linear systems analysis, so that readers with a
good general knowledge of differential and integral calculus should have no
difficulty proceeding. For the reader less advanced in mathematics, appen-
dices have been provided to introduce some basic properties of vectors, and
to give an elementary method for the numerical solution of differential
equations.

Appendix B has been devoted to a brief discussion of double exponential curve
fitting, and an entire program in Fortran is provided. Since this topic is

now considered passé, the inclusion of such a program may seem superfluous.
I can only say that despite the presence of a good library tape subroutine at
the Computer Centre of the University of Toronto, medical scientists are
usually not able to proceed independently. The program of Appendix B, however, is short, easy to use, and virtually never hangs up.

A number of problems have been provided for the reader to solve. Many of
these problems have been given out as assignments to my students, and many of
the "bugs" in the problems have been ferreted out by these students. However,
some problems are new and I should be grateful to receive any suggestions for
their improvement.

I have, quite deliberately, attempted to be eclectic rather than encyclo-
paedic. The nomenclature adopted, the type of equations developed, the
physiological *Weltanschauung* espoused, are those preferred by myself and my
colleagues. While the discussion is mathematical, only a very small portion
is mathematical theory for its own sake. The material is intended to be
applied.

I wish to acknowledge the aid of the Medical Research Council of Canada, which
has supported a number of the studies reported in this book, through both its
awards and its grants programs. I am most indebted to my colleagues,
Professor Geza Hetenyi Jr. and Professor Samuel Zelin for their critical reviews of the manuscript, and many of their suggestions have been incorporated.
Of course, I retain full responsibility for the accuracy of the material as
it has finally been presented. My sincere thanks go to Professor R.S.C.
Cobbold, Director of the Institute of Biomedical Engineering, who provided me
with the licence to use so many of the Institute's resources in the preparation of this book, and to Mrs. Carolyn Nish, who typed the manuscript in the
form you now see it. If the reader has been involved with the preparation of
camera-ready copy, he can appreciate the enormous demands made upon the typist.
To all these, Carolyn Nish was equal, and it is her indefatigable constitution
and methodical attention to detail which rendered this book (I hope) readable.
My thanks also to Mr. Heinz Loth, who prepared the drawings and to my wife,
Barbara, who proof-read much of the manuscript.

K.H. Norwich
University of Toronto
November 1976

CHAPTER I
INTRODUCTION

SECTION 1
NOMENCLATURE

A substance is secreted by a gland into the blood stream and is subsequently disseminated throughout the entire organism. The theory of this transport process and the experimental evidence for its validation form the nucleus of this book. A metabolite is produced endogenously at a certain *appearance rate* (moles/minute), and is concurrently degraded at other sites within the organism at a certain *disappearance rate*. These rates of appearance and of disappearance may be constant or varying in time, and the theory underlying methods for their measurement will be explored here in some depth. When the rates of appearance have been maintained equal and constant for some time, then they are taken to define the steady state *turnover* of the metabolite. The study of transport and turnover in vivo cannot really be separated one from the other. Both involve the use of isotopic *tracers*, and so quite a lot will be said about tracers. Finally, we shall deal here with the flow of fluids in vivo. Some discussion of flow is mandatory in the sense that convection currents are an important mode of transport of biological materials. Moreover, it transpires that when flow is measured using *indicators*, which are a type of tracer, the various formulae for computing the rates of flow are analogues of the formulae for computing rates of appearance. So the matters we shall be treating in the following pages are the rates of appearance, rates of disappearance, and rates of flow in biological systems. The emphasis will always be on the mathematical theory and the means for its validation.

1

The use of the terms *tracee* and *tracer* to refer to the unlabelled and labelled substances respectively has been recommended by a task force on nomenclature (Brownell et al., 1968) and so these have been adopted here, although alternative terms have been proposed.

SECTION 2
TRACEE, TRACER, ISOTOPE AND ISOTOPE EFFECT

A metabolite or similar substance of interest is present in a biological system of macroscopic proportions in concentrations of the order of magnitude of micrograms or milligrams per millilitre. Since this metabolite is the one whose transport and turnover properties we shall trace, it will be termed the *tracee* and it will be referred to as the substance S. If one or more atoms of a molecule of tracee are replaced by isotopes of these atoms, there results a molecule of *tracer* which will be designated by S*. Generally, tracer is present in the biological system in concentrations about 10^6 times smaller than that of the tracee. Since there are often more than two isotopes of the same element, there may be more than one tracer for any tracee, and these tracers might be designated as S_1*, S_2*....etc. We can also produce multiple tracers for a given tracee by replacing more than one kind of atom in the tracee molecule by its isotope. Thus, for example, glucose may be doubly *labelled* replacing ^{12}C by ^{14}C and ^1H by ^3H. The various isotopes of a given element differ in the number of neutrons within the nucleus of the atom. Isotopes may be stable or radioactive. Stable isotopes are detected gravimetrically using a mass spectrometer; that is, the isotopes are distinguished by differences in their masses. This process is still cumbersome with present technology and so the use of stable isotopes and non-radioactive tracers in metabolic studies is not in vogue. Radio-isotopes which emit beta particles and gamma rays are more readily detectable and hence are the ones used almost exclusively in the fashioning of isotopic tracers. *Tracer* may therefore usually be taken to imply *radioisotopic tracer*.

When it is stated in this book that a quantity of tracer is added to a
system, this will indicate that pure S* rather than a mixture of S and S*
is added.

Isotopes of an element differ in their mass numbers but not in their atomic
numbers. Therefore the chemical properties of the various isotopes are
identical. But the physical properties of the isotopes are functions of
their atomic weights, and hence may differ, and these differences in behaviour have been termed *isotope effects*. Since usually only one atom of a
molecule of interest differs between tracer and tracee, it would not be
expected that much difference could be observed between the behaviour of
tracer and tracee due to isotope effect. In other words, it is generally
expected that isotope effects are negligible in tracer studies, and usually
this is true. However, problems may arise when dealing with small molecules.
For example, if tritium were used as a label for hydrogen (one atom of
hydrogen being replaced by its heavier isotope), then trouble might ensue.
The molecular weight of the tracee, H_2, is 2, while that of the tracer,
H-Tritium, is 4. Thus the molecular weight of the tracer is twice that of
the tracee which could lead to a serious disparity in the behaviour of the
two species S and S*. Conversely, the use of ^{14}C-labelled protein as a
tracer for ^{12}C-protein might produce a mass difference of only 2 parts in,
say, 20,000 — which is inconsequential. The possible perturbations generated
by isotope effects should be borne in mind although they can usually be
neglected.

In the development of tracer theory we posit the existence of perfect tracers
— an ideal we know cannot be fully achieved. We postulate that S and S* are
in every way indistinguishable *to the biological system*, but that we, the
investigators, can by various means distinguish them. Thus we deal with a
type of distinguishable indistinguishability; distinguishable to us but not
to the system. ' This may be an oversimplification. Quantum mechanics has
taught us to beware of making overly simple statments about "indistinguishable" particles. The Maxwell-Boltzmann statistics of the 19th century has
given way to the less intuitive Fermi-Dirac and Bose-Einstein statistics of
the 20th century. No such revolution seems imminent in the tracer world,
but one should be prepared.

SECTION 3
UNITS OF MEASUREMENT

The mass, M, and concentration, C, of tracee in biological studies are
usually measured in units of milligrams (occasionally micrograms) and milligrams (micrograms) per millilitre respectively, although no hard and fast
rule exists. Ideally, we should like to measure the mass, M*, and concentration, C*, of tracer in the same (tracee) units but this is not usually
practical. The curie is the accepted unit of radioactive intensity, the
curie being 3.7×10^{10} disintegrations per second. A more common unit of
radioactive intensity in biological work is the *dpm*, or the number of disintegrations per minute. Thus the units of mass, M*, and concentration, C*,
of tracer are commonly the dpm and dpm per millilitre.

Specific activity, a, may properly be defined as the ratio of the concentration of tracer to that of tracer plus tracee; ie.

$$a \equiv \frac{C*}{C + C*} \qquad\qquad (3.1)$$

In this book we shall adopt the alternative definition

$$a \equiv \frac{C*}{C} \quad . \qquad\qquad (3.2)$$

Since usually C* << C, the two definitions are often identical. Thus ideally,
specific activity should be a dimensionless number since it is the ratio of
two concentrations. However, due to the different units in which tracee and
tracer concentrations are measured, the units of specific activity are
commonly *dpm per mg*. Thus, for example, if one millilitre of blood plasma
is sampled, the specific activity is obtained by dividing the number of disintegrations per minute occurring in the sample by the mass of tracee within
it. The above discussion of ideal and non-ideal units is summarized in
Table 3.1. Bearing in mind that the practical unit of specific activity is
the dpm per mg, we may now overlook this inconvenience temporarily. It is
simpler, for purposes of elaborating the theory, to regard specific activity
as a dimensionless quantity with C and C* sharing common units.

Although little discussion of the techniques of measurement will be given
here, it goes without saying that the measurements of radioactivity must be
made with scrupulous care, and that the information derived from mathematical

TABLE 3.1

	Ideal Units	Some Practical Units
C (tracee concentration)	moles per litre	mg per ml. or µg per ml. or moles per l.
C* (tracer concentration)	moles per litre	dpm per ml.
a (specific activity)	none	dpm per mg. or per µg or per mole

Comparison of the ideal units of measurement with the practical units which
are more commonly used. The *dpm*, although it is still in common usage,
will probably be phased out in favour of the *microcurie*.
1 microcurie = 2.22 x 10^6 dpm.

analysis of data can only be as good as the data itself. The application of
the theory, as it will be developed here, will generally require the measurement of the level of radioactivity within a small sample of liquid. The
method currently employed for such measurement is to count the scintillations
in a liquid scintillation spectrometer. Then, by comparison with standards
of known activity and by application of quench corrections, the actual number
of disintegrations per minute within the sample may be calculated. No
emphasis will be placed, in this book, on measurements of activity within an
organ or gland made by placing a counter externally in the region of the
organ or gland; e.g. placing a counter adjacent to the skin overlying the
kidney or thyroid gland regions to measure the activity of radioisotope
sequestered by kidney or thyroid. Such techniques are useful in nuclear
medicine, but no discussion of the technique or interpretation of external
counting will be attempted here.

SECTION 4

PHYSICAL VS. BIOLOGICAL DECAY
OF RADIOISOTOPES

When a radioisotopically labelled tracer is injected suddenly into an
organism and its concentration C* at some sampling site is measured at intervals thereafter, C* is usually found, eventually, to fall monotonically with
time. This progressive decrease in C*, or decay of activity, is due to the
concurrence of two processes: tracer is removed from the system by the
physiological process which tends to "turn over" body constituents (to be
discussed in more detail below), and the activity of the tracer is reduced
by the progressive physical decay of the *labelling* radioisotope into a stable
form. The purely biological decay process is the one of interest to us here
and the nuclear decay process is a contaminant. Nuclear disintegration is
governed by the first order differential equation

$$\frac{dP*}{dt} = -\lambda P*, \quad \lambda > 0 \tag{4.1}$$

where P* is the mass of radioactive atoms present and λ the decay constant.
That is

$$P*(t) = P*_o \, e^{-\lambda t} \tag{4.2}$$

If $C_M^*(t)$ is the function giving the measured tracer concentration at time t,
then the tracer concentration corrected for nuclear decay is

$$C*(t) = C_M^*(t) \, e^{\lambda t} \tag{4.3}$$

The *half life* or *half time* is the time taken for the activity to decay to
one-half of its original level. ^{131}I, ^{32}P and $^{99}T_e$ are radioactive atoms
which decay with half lives of the order of magnitude of the duration of
biological experiments and allowance for the nuclear decay may have to be
made. Conversely ^{14}C and ^3H have half lives of many years and their nuclear
decay is ignored in the analysis of most biological experiments. In Table
4.1 are listed a number of radioactive nuclides commonly used in biology and
medicine, together with their half lives. The figures were taken from a
book edited by Wagner, and the reader is referred to this book for a more
complete table.

TABLE 4.1

Element	Mass Number	Half Time
Calcium	45	164 days
Carbon	14	5570 years
Chromium	51	27.8 days
Hydrogen	3 (tritium)	12.26 years
Iodine	125	60 days
	131	8.1 days
Iron	55	2.94 years
Phosphorus	32	14.3 days
Potassium	42	12.5 hours
Technetium	99	6 hours
Xenon	133	5.27 days

Some commonly used radionuclides and their half times

Problem

4.1   The element Fictitium has a stable isotope, Fi, and an unstable isotope Fi*, which has a half-life of H hours (decaying into soft lambda-rays). The salt Fictitium phosphate, when injected intravenously, is detoxified rapidly in the liver by a first order process and excreted in the bile as Fictitium farceate. The biological half-life of $FiPO_4$ is h hours.

Find the equation relating decay constant and half-life. What is the differential equation governing the disappearance of $Fi*PO_4$ from plasma?

SECTION 5

SOME SCALAR AND VECTOR POINT FUNCTIONS

For our purposes we may regard *point functions* as functions of the 3 spatial

dimensions x, y and z and of time, t, which are single-valued for every set
of (x, y, z, t). We shall usually regard these functions and their
derivatives as continuous. With the origin of coordinates fixed, the
position of any point within the organism may be designated by the position
vector, $\vec{r}$, where $\vec{r}$ has the coordinates (x, y, z) (see Fig. 5.1). The arguments of the point functions are then $(\vec{r}, t)$.

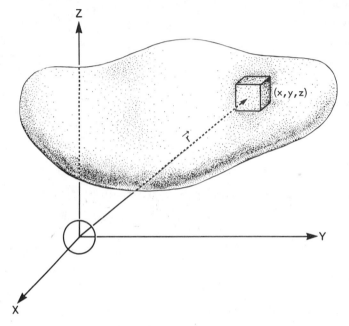

Fig. 5.1. The position vector $\vec{r}(x,y,z)$ designates a general point within an
organism (depicted of arbitrary shape). A small element of volume
is shown at position $\vec{r}$.

The following scalar point functions will be used frequently:

$C(\vec{r}, t)$ designates the concentration of tracee at position $\vec{r}$ at time t;

$C^*(\vec{r}, t)$ designates the concentration of tracer;

$a(\vec{r}, t)$ designates the specific activity;

$R_{d,v}(\vec{r}, t)$ designates the rate of disappearance of tracee per unit volume;

$R^*_{d,v}(\vec{r}, t)$ designates the rate of disappearance of tracer per unit volume.

The functions $R_{d,v}(\vec{r}, t)$ and $R^*_{d,v}(\vec{r}, t)$ may be appreciated by visualizing each
small element of volume (Fig. 5.1) within the organism as degrading or de-
stroying the substance S at a rate varying with position within the organism

and with time. The dimensions of both these functions are
mass·volume^{-1} · time^{-1}. $R_{d,v}(\vec{r},t)$ commonly has the units mg. ml^{-1} min^{-1},
and $R^*_{d,v}(\vec{r},t)$ the units dpm. ml^{-1} min^{-1}. Similarly we can define the
functions $R_{a,v}(\vec{r},t)$ as the rate of appearance of tracee per unit volume, and
$R^*_{a,v}(\vec{r},t)$ as the rate of appearance of tracer per unit volume. The functions
$R_{a,v}$, $R^*_{a,v}$, $R_{d,v}$, $R^*_{d,v}$ are defined in such a way that they are always greater
than zero. Because it is rather cumbersome to write the arguments of these
functions, they will frequently be omitted. However all functions should be
regarded as varying in space and in time unless they are specifically
designated as restricted in some way.

If we integrate $R_{d,v}$ and $R^*_{d,v}$ etc. over the entire volume V, of the organism,
we obtain the total rate of disappearance of tracee and tracer respectively.
That is

$$\int_V R_{d,v}\ (\vec{r},t)\ dV = R_d(t) \tag{5.1}$$

$$\int_V R^*_{d,v}\ (\vec{r},t)\ dV = R^*_d(t) \tag{5.2}$$

$$\int_V R_{a,v}\ (\vec{r},t)\ dV = R_a(t) \tag{5.3}$$

$$\int_V R^*_{a,v}\ (\vec{r},t)\ dV = R^*_a(t) \tag{5.4}$$

where the single integration sign $\int_V$ represents the triple integral $\int_x \int_y \int_z$
taken over the enclosed volume (see Appendix C). By integrating $R_{d,v}$ over a
volume, we remove the volume-dependence of the function and are left with
R_d, which is only a function of t.

There are also a number of vector point functions which will be found useful.
Define $\vec{J}(\vec{r},t)$ as the flux of tracee at position $\vec{r}$ at time t; and
        $\vec{J}*(\vec{r},t)$ as the flux of tracer.
These fluxes have the dimensions of mass·area^{-1} · time^{-1} and commonly have
the units of mg. cm^{-2} min^{-1} and dpm. cm^{-2} min^{-1}. These functions should
more properly be written $\vec{J}(j_1[\vec{r},t], j_2[\vec{r},t], j_3[\vec{r},t])$ etc. where j_1, j_2, and
j_3 are the 3 components of the flux vector $\vec{J}$, but such pedantry is hardly
warranted and hence $\vec{J}(\vec{r},t)$. The flux $\vec{J}(\vec{r},t)$ indicates that tracee is flowing
in the direction designated by cosines $j_1/|\vec{J}|$, $j_2/|\vec{J}|$, $j_3/|\vec{J}|$ at the rate

$|\vec{J}|$ where $|\vec{J}| = \sqrt{j_1{}^2 + j_2{}^2 + j_3{}^2}$. This may be seen in Fig. C.1 where the cosines are $\dfrac{a_1}{|\vec{A}|}$ etc. Ony may visualize a small, square window of unit area oriented in space in such a way that $\vec{J}$ lies along the normal to the window drawn at its centre (Fig. 5.2). The flux at $\vec{r}$ in the direction of some unit vector $\vec{\lambda}$ is then $\vec{\lambda}\{\vec{\lambda} \cdot \vec{J}(\vec{r},t)\}$, as we shall demonstrate below.

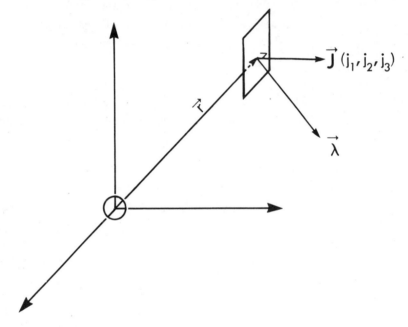

Fig. 5.2. A small, square window of unit area is drawn at the point $\vec{r}$ orthogonal to the flux vector $\vec{J}(j_1, j_2, j_3)$. $\vec{J}$ then gives the amount of tracee which would flow through the window per unit time, if $\vec{J}$ were constant over the window. $\vec{\lambda}$ is a unit vector in some arbitrary direction (see Fig. 5.3).

In order to apply the physics of continua in the study of biological transport, it is sometimes necessary temporarily to suppress our view of the particulate or molecular structure of matter. We may view matter as continuously distributed through space. To each *material point* at position $\vec{r}$ at time t, we may assign a velocity $\vec{v}(\vec{r},t)$ (length/time) such that $\vec{v}(\vec{r},t)$ is a continuous vector point function. We may consider the material point to be large with respect to a molecule, so that each material point contains many molecules. The motion of the material point with the velocity $\vec{v}$ is then not the motion of any single molecule but rather the net motion of a "point" made up of many molecules (Landau and Lifshitz, 1959). We shall regard the substance S as continuously distributed throughout the volume of the organism and $\vec{v}(\vec{r},t)$ will designate its motion.

The vector point functions $\vec{J}$, $\vec{J}*$, $\vec{v}$ and $\vec{v}*$ are simply related by the equations

$$\vec{J} = C\vec{v} \tag{5.5}$$

and

$$\vec{J}* = C*\vec{v}* \tag{5.6}$$

In order to find the mass of material which passes per unit time through the small window of unit area orthogonal to $\vec{\lambda}$ (not the same window shown in Fig. 5.2) let us construct the parallelepiped shown in Fig. 5.3 by translating the window through space a distance $-\vec{v}$ (Desloge, 1966). In one unit of time, all the particles contained within the parallelepiped will pass through the upper window. The volume of the parallelepiped is equal to the area of the base times the perpendicular height = $1\vec{\lambda} \cdot \vec{v}$. If the parallelepiped is small, the concentration throughout is C; hence the total mass within is $C\vec{\lambda} \cdot \vec{v} = \vec{\lambda} \cdot \vec{J}$. Thus the total mass passing through the window of unit area is $\vec{\lambda} \cdot \vec{J}$. This quantity has been called the *scalar resolute* of $\vec{J}$ in the direction of $\vec{\lambda}$.

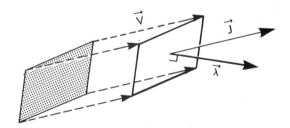

Fig. 5.3. A small window of unit area is now drawn orthogonal to the unit vector $\vec{\lambda}$. The window is translated in the direction $-\vec{v}$ to form a parallelepiped of volume $\vec{\lambda} \cdot \vec{v}$.

Speaking Practically

If you integrate the rate of disappearance per unit volume over the entire volume of distribution of the substance, you obtain the total rate of disappearance.

The flux or mass flow rate of a substance, $\vec{J}$, is in the direction of a velocity, $\vec{v}$. Specifically, $\vec{J} = C\vec{v}$. The component of flux in the direction of some arbitrary unit vector $\vec{\lambda}$ is $\vec{\lambda} \cdot \vec{J}$.

References

Brownell, G.L., Berman, M. and Robertson, J.S., Nomenclature for tracer kinetics, *Int. J. Appl. Radiat.* 19, 249-262 (1968).

Desloge, E.A., *Elementary Statistical Physics*, Holt, Rinehart and Winston Inc., New York, 1966.

Landau, L.D. and Lifshitz, E.M., *Fluid Mechanics*, Pergamon Press, London, 1959.

Wagner Jr., H.N., (ed.), *Principles of Nuclear Medicine*, W.B. Saunders Co., Philadelphia, p. 862, 1968.

CHAPTER II

THE RELATIONSHIP BETWEEN TRACER AND TRACEE

SECTION 6
INDISTINGUISHABILITY OF TRACER AND TRACEE

In section 3 it was explained that S and S*, because of the similar nature of the atoms comprising them, could be expected to be "indistinguishable" to the biological system. But how do we introduce this property of indistinguishability into the equations of transport? Some people have an inherent tendency to think (incorrectly) that (Any function related to tracer) = (Specific activity) · (The corresponding function of tracee). But the above relationship is much too general. Let us divide tracer-tracee indistinguishability into two parts:

chemical indistinguishability and

physical indistinguishability

Indistinguishability of Tracer and Tracee
with Respect to Chemical Properties

Two assumptions are made:

(i) Vectorial Assumption

$R_{a,v}$ and $R_{d,v}$ (defined in section 5) may be treated as if they were collinear vectors pointing in opposite directions. For purposes of introducing these rates into differential equations, we shall express them in alternative form:

$$\frac{\partial C}{\partial t}\bigg|_A \equiv R_{a,v} \qquad\qquad (6.1)$$

$$-\frac{\partial C}{\partial t}\bigg|_D \equiv R_{d,v} \qquad\qquad (6.2)$$

15

Two similar definitions are made for the tracer functions $R^*_{a,v}$ and $R^*_{d,v}$. Then, taking unit vector $\vec{e}$ along the appearance-disappearance, or "existence" axis,

$$\frac{\partial C}{\partial t}\bigg|_N \vec{e} = \frac{\partial C}{\partial t}\bigg|_A \vec{e} + \frac{\partial C}{\partial t}\bigg|_D \vec{e} \qquad (6.3)$$

where $\dfrac{\partial C}{\partial t}\bigg|_N$ designates the net rate of change of C with respect to time due to metabolic activity.

That is

$$\frac{\partial C}{\partial t}\bigg|_N = R_{a,v} - R_{d,v} \qquad (6.4)$$

(ii) Assumption of Chemical Indistinguishability

Specific activity will not change *due to* the process of chemical degradation of tracee and tracer. Expressed mathematically

$$\frac{\partial a}{\partial t}\bigg|_D \equiv \frac{\partial}{\partial t}\bigg|_D \left(\frac{C^*}{C}\right) = 0 \qquad (6.5)$$

If we expand Eq. (6.5) it assumes a more usual form:

$$\frac{\partial}{\partial t}\bigg|_D \left(\frac{C^*}{C}\right) \frac{C\dfrac{\partial C^*}{\partial t}\bigg|_D - C^*\dfrac{\partial C}{\partial t}\bigg|_D}{C^2} = 0$$

$$\frac{\partial C^*}{\partial t}\bigg|_D = \frac{C^*}{C}\frac{\partial C}{\partial t}\bigg|_D \qquad (6.6)$$

or

$$R^*_{d,v} = aR_{d,v} \qquad (6.7)$$

using (6.2). Eq. (6.7) is fundamental in tracer kinetics and may be seen to rest upon Eq. (6.5) which assumes that at each material point the ratio of concentrations or densities of tracer to tracee will not change as a consequence of chemical reaction – a tracer molecule is as likely as a tracee molecule to enter a chemical reaction and the number of each species to actually react during an element of time is proportional to the total number of molecules of the species present. No assumption regarding the order of the chemical reaction has been made. No assumption regarding the relative magnitudes of C and C* has been made (but see problem 6.1).

The assumption that specific activity does not change due to the process of chemical degradation may perhaps be viewed more clearly by considering a mechanical analogy. Suppose there is a barrel containing many balls of ab-solutely uniform physical dimensions - the same size, mass, texture, etc. A fraction, a, of the balls are black and the remainder are white. The black and white balls are well-mixed within the barrel. The assumption is made that a blindfolded person drawing balls *randomly* from the barrel cannot, while there are still many balls left, "appreciably" change the fraction, a, of black balls remaining in the barrel. Analogously, a chemical reaction in-volving S and S* cannot change the ratio of C* to C since it cannot perceive the difference between molecules of S and S*. The chemical reactions of which we are speaking here are assumed to be irreversible.

Indistinguishability of Tracer and Tracee
with Respect to Physical Properties

It is assumed that isotope effects are negligible. The molecules of the two substances S and S* are, therefore, assumed to manifest no differences in behaviour based upon differences in their mass. Why then was it necessary to write in Eq. (5.5) and (5.6) the velocity $\vec{v}$ for tracee and $\vec{v}*$ for tracer rather than writing the same velocity for both substances? In fact, as we shall see, it is often quite proper to state that

$$\vec{v} = \vec{v}* \qquad (6.8)$$

In biological systems, the two most important forces producing transport of non-ionized particles are probably those of *convection* and of *diffusion*. We shall, therefore, in a subsequent section develop the equations of *convective diffusion* for tracer and tracee. "Active transport" and the mechanisms postulated to account for it will not be discussed in this book. S and S* are carried by the fluids in which they are dissolved or mixed. This *carrier fluid* is either a gas or a liquid, the former being compressible and the latter incompressible.

(i) Convection
In this discussion we shall assume that the particles of tracer and tracee are completely *entrained* by the carrier fluid (Levich, 1962); that is, at every point in space, the motion of the carrier fluid is transferred to the tracer and tracee which are present. It is assumed that no "slipping" occurs

between the fluid and the substances S and S*. If S and S* represent
labelled and unlabelled hemoglobin molecules and the hemoglobin is contained
within red blood cells, then there will indeed be some slipping of the
carrier fluid - blood plasma in this case - over the red blood cell, and the
no-slipping assumption would not be well taken. But for any of the thousands
of small molecules dissolved in the plasma the no-slipping approximation is
probably reasonable. If we represent the velocity of a material point *of
fluid* by the vector $\vec{u}(\vec{r},t)$, then as a consequence of complete entrainment
of the two "identical" molecules of S and S* we can write for the equations
of convection

$$\vec{J}_{conv} = C\vec{u} \tag{6.9}$$

$$\vec{J}^*_{conv} = C^*\vec{u} \tag{6.10}$$

where $\vec{J}_{conv}$ and $\vec{J}^*_{conv}$ are the fluxes of the two substance due to the process
of convection. $\vec{J}_{conv}$ and $\vec{J}^*_{conv}$ are each only the first of the two fluxes
which must be added vectorially to give the total or resultant fluxes $\vec{J}$ and
$\vec{J}^*$ defined by Eqs. (5.5) and (5.6) (see Fig. 6.1).

(ii) Diffusion

The law of diffusion, as formulated by Fourier for heat flow and by Fick for
material flow, states that

$$\vec{J}_{diff} = -D(C) \text{ grad } C \tag{6.11}$$

$$= -D(C)\nabla C$$

D(C) is the coefficient of diffusion and is, in general, a function of concentration C. For tracer we can write

$$\vec{J}^*_{diff} = -D^*\nabla C^* \tag{6.12}$$

without specifying, for the moment, the arguments of D*. D may be referred
to as the coefficient of "bulk" diffusion, while D* is the coefficient of
"self"-diffusion governing the diffusion of labelled substance within a
matrix of unlabelled material. Can we write

$$D = D^* \tag{6.13}$$

as a consequence of the indistinguishability of the molecules of S and S*?
Is the relationship between the two coefficients this simple? The answer is:
yes and no. Yes we can and no it isn't. To explore the relationship between
D and D* we must reach into physical chemistry, and the following discussion
rests heavily upon the very lucid discussions given by J. Crank (1956) and
L.S. Darken (1948).

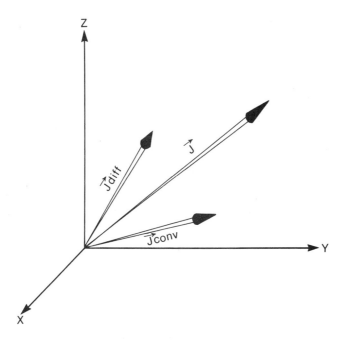

Fig. 6.1. A non-ionized particle in a biological system is acted upon by
        convective and diffusive forces. The resultant flux is depicted
        as the vector sum of convective and diffusive fluxes.

A difference between the magnitudes of D and D* may arise because the force
which is responsible for the diffusive transport of a substance is related to
the gradient in the chemical potential of this substance. Hence the
diffusion coefficient is a function of the chemical potential. The chemical
potential, in turn, is defined in terms of the activity coefficient of the
substance. But the activity coefficient is a function of the concentration,
varying appreciably with concentration for the unlabelled substance, while
remaining nearly constant (at unity) over the concentration range of the
labelled substance which is present only in trace amounts. Thus D will vary
with the concentration of S while D* will not vary much with the concentration of S*. This is the gist of the mathematical development which follows.

Let us deal with the case where diffusion is confined to one dimension and turn first to the diffusion of unlabelled molecules. Following Crank we represent the force per gram mole of S which tends to impart diffusive motion to S by F, where

$$F = -\frac{\partial \mu}{\partial x} \tag{6.14}$$

μ is the chemical potential of S. The flux of S in the direction of x is given by

$$j_x = \frac{FC}{R} \tag{6.15}$$

where R is the resistance offered to the motion. But

$$\frac{C}{R} F = \frac{-C}{R} \frac{\partial \mu}{\partial x} = \frac{-C}{R} \frac{\partial \mu}{\partial C} \cdot \frac{\partial C}{\partial x}$$

using (6.14). Therefore

$$j_x = \frac{-C}{R} \frac{\partial \mu}{\partial C} \cdot \frac{\partial C}{\partial x} \tag{6.16}$$

From Fick's law, Eq. (6.11)

$$j_x = -D \frac{\partial C}{\partial x} \tag{6.17}$$

From (6.16) and (6.17) we obtain

$$D = \frac{C}{R} \frac{\partial \mu}{\partial C} \tag{6.18}$$

The substance S* is present in very, very much lower concentrations than that of S. Thus the behaviour of the labelled material may be approximated by that of an ideal gas in solution *(ideal solution)*. From the definition of activity,

$$d\mu* = RTd(\ln a*)$$

where $a*$ is the activity of the gas molecules, R is the gas constant and T the absolute temperature. Thus

$$\frac{\partial \mu*}{\partial x} = RT \frac{\partial(\ln a*)}{\partial x} \tag{6.19}$$

If $\gamma*$ is the activity coefficient and N* is the mole fraction of the tracer gas molecules, then

$$a* = \gamma* \ N* \tag{6.20}$$

Since we are dealing with an "ideal solution"

$$\gamma* = 1 \tag{6.21}$$

so that

$$a* = N*$$

and

$$\frac{\partial \mu^*}{\partial x} = RT \frac{\partial(\ln N^*)}{\partial x}$$

$$= RT \frac{1}{N^*} \frac{\partial N^*}{\partial x} \qquad (6.22)$$

Since the solution is of low concentration, we write

$$\frac{1}{N^*} \frac{\partial N^*}{\partial x} = \frac{1}{C^*} \frac{\partial C^*}{\partial x}$$

and hence (6.22) becomes

$$\frac{\partial \mu^*}{\partial x} = \frac{RT}{C^*} \frac{\partial C^*}{\partial x} \qquad (6.23)$$

We now write for S* the analogues of the various equations describing the diffusion of S. In place of Eqs. (6.14), (6.15) and (6.17) we have

$$F^* = \frac{-\partial \mu^*}{\partial x} \qquad (6.24)$$

$$j_X^* = \frac{F^* C^*}{R} \qquad (6.25)$$

$$j_X^* = -D^* \frac{\partial C^*}{\partial x} \qquad (6.26)$$

where R^* has been set equal to R because of the identical physical characteristics of the molecules of S and S*.
Combining (6.23) and (6.24),

$$F^* = \frac{-RT}{C^*} \frac{\partial C^*}{\partial x} \qquad (6.27)$$

so that from (6.25) and (6.27)

$$j_X^* = \frac{-RT}{R} \frac{\partial C^*}{\partial x} \qquad (6.28)$$

and from (6.26) and (6.28)

$$D^* = \frac{RT}{R} \qquad (6.29)$$

Eliminating R between Eqs. (6.18) and (6.29) leaves

$$D = D^* \left(\frac{C}{RT} \frac{\partial \mu}{\partial C} \right) \qquad (6.30)$$

which is Crank's Eq. (11.43).

Darken derived two equations

$$D^* = kTB^* \left(1 + N^* \frac{d(\ln \gamma^*)}{dN^*} \right) \qquad (6.31)$$

$$D = kTB \left(1 + N \frac{d(\ln \gamma)}{dN} \right) \qquad (6.32)$$

where $N*$ and N are the respective mole (or atom) fractions of $S*$ and S, $B*$
and B the mobilities, and k is Boltzmann's constant. When only small
quantities of $S*$ are present $\frac{d(\ln \gamma*)}{dN*} \to 0$. Since S and $S*$ are physically
identical,

$$B = B*$$

Therefore from (6.31) and (6.32)

$$D = D* \left[1 + N \frac{d\ln\gamma}{dN}\right] \tag{6.33}$$

Eqs. (6.30) and (6.33) both describe the relationship between D and $D*$, the
diffusion coefficients for S and $S*$.

We can now show that the Eqs. (6.30) and (6.33) are identical. Since

$$\alpha = \gamma N$$

as in (6.20)

$$\ln\alpha = \ln\gamma + \ln N$$

$$\frac{d\ln\gamma}{dN} = \frac{d\ln\alpha}{dN} - \frac{1}{N} \tag{6.34}$$

The relationship between chemical potential and activity is, again,

$$d\mu = RT \, d \ln\alpha$$

so that

$$\frac{d \ln\alpha}{dN} = \frac{1}{RT} \frac{d\mu}{dN} \tag{6.35}$$

Eliminating $\frac{d\ln\alpha}{dN}$ between (6.34) and (6.35),

$$\frac{d\ln\gamma}{dN} = \frac{1}{RT} \frac{d\mu}{dN} - \frac{1}{N} \tag{6.36}$$

If we now replace $\frac{d\ln\gamma}{dN}$ in (6.33) by the expression given in (6.36), we
obtain,

$$D = D* \left[1 + \frac{N}{RT} \frac{d\mu}{dN} - 1\right];$$

that is

$$D = D* \left[\frac{N}{RT} \frac{d\mu}{dN}\right] \tag{6.37}$$

If the tracee solution is dilute, then

$$D = D* \left[\frac{C}{RT} \frac{d\mu}{dC}\right]$$

which is identical with (6.30). Thus Eqs. (6.30) and (6.33) are identical
and both describe the relationship between D and $D*$.

As the concentration, C, or the mole fraction, N, approach zero, it may be
seen from (6.33) that

$$D = D^* = \text{a constant} \qquad (6.38)$$

$$\text{very nearly.}$$

Hence as long as we confine our discussion to very dilute solutions, we may
adopt Eq. (6.38).

Eqs. (6.31) and (6.32) were not derived here because their derivation
requires a digression into physical chemistry which would take us rather too
far afield. The interested reader is referred either to the original paper
or to Van Holde's very readable text. Eq. (6.32) is effectively given by
Van Holde's Eq. (4.16) which is fully derived in the preceding sections of
his text.

Curran, Taylor and Solomon (1967) have approached the same problem using the
Onsager equations. Their results, summarized in Table I of their paper (%Δ),
show again that the two diffusion coefficients D and D* are of nearly the
same magnitude for the (lower) "concentrations of biological interest".

Chemical identity has been interpreted to mean that specific activity, or the
ratio of tracer to tracee densities at a point, cannot be altered by the process of irreversible chemical reaction regardless of how high this ratio may
be. Physical identity of tracee and tracer has been embodied in the above
discussion by equating the resistances R and R^* and the mobilities B and B*.
If tracer is present in very low concentrations so that the tracer solution
approaches an ideal solution in composition, then D and D* are related by
Eqs. (6.30) and (6.33). If tracee is also present in fairly low concentrations, then we may take D and D* to be equal and constant. The question
naturally arises, How low is a "low concentration"? Table 7.1 is an attempt
to answer this question.

Speaking Practically

The ratio of the rate of disappearance per unit volume of tracer
        to the rate of disappearance per unit volume of tracee
is equal to the ratio of the concentration of tracer
        to the concentration of tracee (Eq. 6.7)).
(The ratio of the total rates of disappearance is *not* necessarily equal to
a ratio of the concentrations.)

The convective flux is equal to the product of concentration and convective velocity.

The diffusive flux is equal to minus the product of the concentration gradient and the coefficient of diffusion.

The magnitude of the coefficient of diffusion is, generally, dependent upon the magnitude of the concentration.

The coefficient of diffusion which governs the diffusion of tracer may be taken as equal to the coefficient which governs the diffusion of tracee provided that the concentration of tracee is low enough.

Problem

6.1   Formulate Eq. (6.5) for the case where specific activity is defined as $C^*/(C + C^*)$, and show that Eq. (6.7) emerges for this case as well, but with the definition of $R_{d,v}$ now changed to mean $\frac{\partial}{\partial t} (C + C^*)$.
$$D$$

SECTION 7
CHEMICAL REACTION KINETICS

Suppose we deal with the chemical reaction
$$\ell A \rightleftarrows mB + nE$$
where a substance A reacts to form two products B and E. ℓ, m and n are stoichiometric coefficients. The rate at which the reaction proceeds from left to right in moles per unit volume per unit time may be given by

$$r = k[A]^\lambda - k'[B]^\mu [E]^\nu \qquad (7.1)$$

where square brackets designate a concentration, k and k' are velocity constants and the indices λ, μ and ν assume the values of the stoichiometric coefficients ℓ, m and n respectively, but they need not in general be so

restricted. (In fact, even (7.1) is not perfectly general.) An example of
a simple chemical reaction is cited by Denbigh (1966),

$$2HI \rightleftharpoons H_2 + I_2$$

where r has been shown experimentally to be given accurately by the equation

$$r = k[HI]^2 - k'[H_2][I_2] \tag{7.2}$$

In biology we are frequently concerned with biochemical reactions which proceed very nearly unidirectionally, in the manner

$$\ell A \rightarrow mB + nE + \ldots$$

Here we would expect that (if no further additions to the solution of the
substance A are made)

$$r = \frac{-d[A]}{dt} = k[A]^\lambda \tag{7.3}$$

where λ is known as the *order* of the chemical reaction. When $\lambda = 1$, the
reaction is of the first order, which is a particularly simple case. The
disappearance of free fatty acids from plasma is an example of a first
order process. The differential equation describing a first order reaction
is

$$\frac{dC}{dt} = -kC , \quad k > 0 \tag{7.4}$$

whose solution is

$$C(t) = C(0)e^{-kt} \tag{7.5}$$

or

$$\ln C = \ln C_0 - kt \tag{7.6}$$

Thus a reaction may be tested to see if it is of the first order by plotting
the logarithm of concentration against time. If the reaction is of the
first order, then such a graph should be linear. It must be understood that
after t = 0 no more of the reagent S is to be added to the solution if such
a test is carried out. Eq. (7.4) has been written, for simplicity, with an
ordinary derivative implying a well-mixed solution with no concentration
gradients. But in greater generality, the first order reaction should be
described by the equation

$$\frac{\partial C}{\partial t}_D = -kC \tag{7.7}$$

using the nomenclature of section 6. When written in this form, there can be
no doubt that the process refers only to disappearance (no source or appearance terms present) and that the concentration is at liberty to vary with

position within the solution. First order reactions will assume a special position in our subsequent discussion of tracer kinetics.

In biochemistry we often deal with enzyme-catalyzed reactions whose kinetics are more complex than those described by the equations given above. These enzymic reactions are presumed to occur in several steps, and involve the formation of intermediary enzyme-substrate complexes. The total rate at which such reactions occur is often well-described by the Michaelis-Menten equation (e.g. White et al., 1964),

$$r = \frac{V_{max}\ C}{k_m + C} \tag{7.8}$$

where V_{max} and k_m are positive constants, and C is the substrate concentration. That is, the substance S whose concentration is C reacts to form a new substance at the rate defined by (7.8). Clearly r is not expressible as a finite polynomial in C. For $C < k_m$ we might expand r in a binomial series:

$$r = \frac{V_{max}\ C}{k_m} \left[\frac{1}{1 + \frac{C}{k_m}} \right] = \frac{V_{max}\ C}{k_m} \left(1 + \frac{C}{k_m} \right)^{-1}$$

$$r = \frac{V_{max}\ C}{k_m} \left[1 - \frac{C}{k_m} + \left(\frac{C}{k_m} \right)^2 - \cdots \right] \tag{7.9}$$

Since all powers of C are represented, no specific order can be assigned to the reaction. For $C > k_m$

$$r = \frac{V_{max}}{1 + \frac{k_m}{C}} = V_{max} \left(1 + \frac{k_m}{C} \right)^{-1}$$

$$r = V_{max} \left[1 - \frac{k_m}{C} + \left(\frac{k_m}{C} \right)^2 - \cdots \right] \tag{7.10}$$

which looks even less like an elementary reaction since it contains powers of reciprocals of C.

When the mechanism of a reaction is not known, one may try to fit his experimental data to an nth order equation such as

$$\frac{dC}{dt} = -kC^n \tag{7.11}$$

where k and n are unknown and n is not necessarily an integer. Integrating

(7.11),

$$c^{1-n} - c_0^{1-n} = (n-1)kt, \quad n \neq 1 \tag{7.12}$$

from which k and n may be estimated (Appendix B). These matters are discussed in Levenspiel's text.

Let us return to the irreversible reaction
$$\ell A \rightarrow mB + nE + \ldots$$
Introducing $R_{d,v}$, the rate of disappearance of A per unit volume, we can write

$$r = R_{d,v} = \frac{-d[A]}{dt} = k[A]^\lambda \tag{7.13}$$

or more generally

$$r = R_{d,v} = \frac{-\partial[A]}{\partial t}\Bigg|_D = k[A]^\lambda \tag{7.14}$$

Recalling that the principle of chemical indistinguishability (6.7) was derived by analogy with the process of withdrawing balls randomly from an urn *but not returning them*, we can apply (6.7) to the irreversible reaction described by (7.14). That is

$$R^*_{d,v} = \frac{[A^*]}{[A]} R_{d,v} \tag{7.15}$$

If in general we deal with the irreversible reaction described by
$$R_{d,v} = f(C) \tag{7.16}$$
then

$$R^*_{d,v} = a \cdot f(C) \tag{7.17}$$

$$= \frac{f(C)}{C} \cdot C^*$$

If we then apply the constraint that $C(\vec{r},t) = C(\vec{r})$ —i.e. that C is constant in time — then

$$R^*_{d,v} = \text{constant} \cdot C^*$$
or

$$\frac{\partial C^*}{\partial t}\Bigg|_D = -\text{constant} \cdot C^* \tag{7.18}$$

Thus the disappearance of tracer at every point in solution is governed by first order reaction kinetics provided that tracee reacts by an essentially irreversible process and that its concentration is held constant in time. Such a system must be *open* with respect to tracee; that is tracee must be

added continuously to the system from the outside. Tracer must be present
only in small quantities.

We have spoken about a *reversible* reaction such as
$$A \rightleftharpoons B + E + \ldots \quad ,$$
and a reaction which is essentially *irreversible* such as
$$A \rightarrow B + E + \ldots \quad .$$
Let us now introduce a third concept: the *cyclic* reaction, which plays a
prominent role in metabolic studies. A cyclic reaction is one such as
$$A \rightarrow B \rightleftharpoons E \ldots \rightarrow A$$
Here we say that A is *recycled*. If A is labelled, giving A*, and C* is the
concentration of A*, then (7.18) is still valid. But great efforts are
often made, in the analysis of experiments, to separate the recycled A* from
that which was present at the beginning of the experiment (*correcting for
recycling of label*). The biochemical techniques used for correction depend
upon the particular substance used. When we are certain that we deal with
the concentration of non-recycled A* only, then and only then can (7.18)
be simplified to

$$\frac{\partial C^*}{\partial t} = - \text{ constant} \cdot C^* \qquad\qquad (7.18a)$$

Throughout the remainder of this book we shall assume, unless otherwise
stated, that C^* refers to non-recycled tracer only; ie. the correction for
recycling has been made.

In chemical equilibria the disappearance of tracer is not necessarily a first
order or *linear* process, as demonstrated by Fleck (1972). Fleck showed that
in chemical equilibria "first-order kinetics will be observed for mechanisms
in which labelled molecules enter the reaction with a molecularity of unity
in elementary reaction steps". To see the gist of his argument let us
examine his example where two labelled molecules appear on the same side of
the chemical equation:

$$2NO + Br_2 \underset{k'}{\overset{k}{\rightleftharpoons}} 2NOBr$$

Suppose that isotopic bromine is added to such an equilibrium. The tracer
will then participate in two types of reaction:

$$2NO + Br* - Br* \underset{k'}{\overset{k}{\rightleftharpoons}} 2NOBr* \qquad\qquad (7.19)$$

and

$$2NO + Br* - Br \underset{k'}{\overset{k}{\rightleftarrows}} NOBr + NOBr* \qquad (7.20)$$

To the extent that the first of these reactions (7.19) may occur, the rate of disappearance of labelled bromine (Br* — Br* or Br* — Br) will not necessarily be a first order process despite the constancy of tracee. This may be seen easily by applying elementary kinetic theory (as in (7.2)) to (7.19).

$$\frac{d[Br* - Br*]}{dt} = -k[NO]^2[Br* - Br*] + k'[NOBr*]^2 \qquad (7.21)$$

Two more differential equations are necessary to define completely the behaviour of the three labelled substances Br* — Br*, Br* — Br and NOBr*, but the appearance of a quadratic term on the right - hand side of (7.21) indicates that the kinetics will *not necessarily* be linear. If [NOBr*] is low enough so that the reaction of two labelled molecules of NOBr is improbable, then the differential equations will reduce to the first order. While the reaction of two labelled molecules may always be a rare occurence in biological systems, one should bear in mind the theoretical constraints upon one's equations.

The reader who is totally confused in the matter of just how low a concentration must be in order that a given mathematical relationship be valid should be reassured that he is not alone. Table 7.1 is an attempt to collate some of the material presented in sections 6 and 7.

Speaking Practically

The concentration of a tracer will only decline by a first order process

$$\frac{dC*}{dt} = -kC*$$

when the reaction in which the tracer takes part is non-cyclic, and at least partly irreversible (tracer is not resynthesized from its immediate products), and when the tracee concentration, C, remains constant. This is the only case in which a "monoexponential" decay in specific activity can be expected.

TABLE 7.1

	Magnitude of C	Magnitude of C*	Equation
1	No restriction	————	$D = kTB\left[1 + N\frac{d(\ln\gamma)}{dN}\right]$ 6.32
2	————	No restriction	$D^* = kTB\left[1 + N^*\frac{d(\ln\gamma^*)}{dN}\right]$ 6.31
3	No restriction	Low enough that no interaction between molecules of S* occurs; $\frac{d(\ln\gamma^*)}{dN^*} = 0$	$D = D^*\left[1 + N\frac{d(\ln\gamma)}{dN}\right]$ 6.33
4	Low enough that $N\frac{d\mu}{dN} = C\frac{d\mu}{dC}$	Restricted as in 3	$D = D^*\left(\frac{C}{RT}\frac{\partial\mu}{\partial C}\right)$ 6.30
5	Low enough that $\frac{d\ln\gamma}{dN} \approx 0$	Restricted as in 3	$D = D^*$ 6.38
6	No restriction	No restriction	$\frac{\partial C^*}{\partial t}\Big|_D = a\frac{\partial(C + C^*)}{\partial t}\Big|_D$ 6.6 Subscript "D" implies irreversibility; a = C*/(C + C*)
7	Constant	Negligibly small in comparison with C	$\frac{\partial C^*}{\partial t}\Big|_D = -$ constant $\cdot$ C* 7.18
8	Constant	Small enough that a molecule of P_1 - C* will not react with a molecule of P_2 - C*, where P_1 - C* and P_2 - C* are intermediary labelled compounds (may be same compound) No recycling of label.	$\frac{\partial C^*}{\partial t} = -$ constant $\cdot$ C*

The orders of magnitude of concentrations of tracee and tracer for which the various equations can be expected to be valid. An asterisk indicates a function of tracer. The symbols C, D, μ, γ, N, k, T and B denote the

concentration, diffusion coefficient, chemical potential, activity
coefficient, mole fraction, Boltzmann constant, absolute temperature and
molecular mobility respectively.

Problem

7.1   From the very illuminating work of G.M. Fleck, we expect that if
labelled bromine (Br* - Br* or Br_2**, and Br* - Br or Br_2*) enters the
reaction

$$2NO + Br_2 \underset{k'}{\overset{k}{\rightleftarrows}} 2NOBr$$

the resulting kinetics can be described by 5 differential equations: Eq.
(7.21) and 4 others. These 5 equations are given below, with some of the
numerical coefficients replaced by a_i.

$$\frac{d[Br_2**]}{dt} = -k[NO]^2[Br_2**] + k'[NOBr*]^2 \qquad (7.21)$$

$$\frac{d[NOBr*]}{dt} = a_1k[NO]^2[Br_2**] - a_2k'[NOBr*]^2$$

$$+ a_3k[Br_2*][NO]^2 - a_4k'[NOBr*][NOBr] \qquad (7.22)$$

$$\frac{d[Br_2*]}{dt} = -a_5k[NO]^2[Br_2*] + a_6k'[NOBr*][NOBr]$$

$$\frac{d[Br_2]}{dt} = -a_7k[NO]^2[Br_2] + a_8k'[NOBr]^2$$

$$\frac{d[NOBr]}{dt} = a_9k[NO]^2[Br_2] - a_{10}k'[NOBr]^2$$

$$+ a_{11}k[NO]^2[Br_2*] - a_{12}k'[NOBr*][NOBr]$$

Your problem is first to evaluate the a_i's. You can then test the validity
of your answer in the following way.

Define $[Br_2^{total}] \equiv [Br_2**] + [Br_2*] + [Br_2]$

$\qquad [NOBr^{total}] \equiv [NOBr*] + [NOBr]$

Show using your differential equations that the derivatives $\frac{d[Br_2^{total}]}{dt}$ and
$\frac{d[NOBr^{total}]}{dt}$ have values consistent with the overall reaction

$$2NO + Br_2^{total} \overset{k}{\underset{k'}{\rightleftarrows}} 2NOBr^{total}$$

I am most grateful to G.M. Fleck who, in a personal communication, helped me formulate this extension of his paper.

SECTION 8

A SIMPLE MODEL OF NTH ORDER DISAPPEARANCE

Suppose that we deal with a well-mixed system so that ordinary time derivatives may be used in describing it. Suppose moreover that the system is *closed* so that after zero time no further material is added to it. Thus we can drop the subscript "D" and designate the disappearance operator $\frac{d}{dt}_D$ by the complete, ordinary, differential operator $\frac{d}{dt}$. Both S and S* are present and no restriction is placed, for purposes of this example, on the magnitude of specific activity. We shall not rule out the possibility that specific activity exceed a value of one-half or even approach its upper bound of unity. The reagent is observed experimentally to disappear from the system in accordance with n^{th} order reaction kinetics, where n is an integer, so that if we define a total concentration

$$C_T = C + C* \qquad (8.1)$$

then

$$\frac{dC_T}{dt} = -kC_T^n \qquad (8.2)$$

(cf. Eq. (7.3)).

A simple physical model of an nth order chemical reaction process where n is integral views the reaction as occuring when n molecules of the reagent collide. Not every such collision will result in chemical reaction but the assumption is that a constant fraction of collisions will result in reaction.

In fact, it is uncommon to find chemical reactions of integral orders higher than 2, and in biology reactions of this type do not play a prominent rôle anyway. We are considering this model for its heuristic rather than for its practical importance.

Substituting for C_T from (8.1) into (8.2),

$$\frac{d(C + C^*)}{dt} = -k(C + C^*)^n \tag{8.3}$$

The right-hand side of (8.3) can be expanded in a binomial series:

$$\frac{d(C + C^*)}{dt} = -k \sum_{r=0}^{n} \binom{n}{r} C^{n-r} C^{*r} \tag{8.4}$$

$$= -k[C^n + nC^{n-1} C^*$$

$$+ \frac{n(n-1)}{2!} C^{n-2} C^{*2} + \dots + C^{*n}]$$

The significance of the terms in this expansion will soon become evident.

Since we are modelling the reaction by the random removal of n-tuplets of molecules from a well-mixed solution of tracee and tracer, we might ask the question: What is the probability of removing an n-tuplet consisting of exactly r labelled and n-r unlabelled molecules? The probabilities of drawing a tracee or tracer molecule in a single trial are proportional to their respective concentrations or densities, C and C^*. Therefore (referring to any introductory text on probability, e.g. Freund), the probability of drawing exactly r tracer molecules in a group of n molecules withdrawn at random is proportional to the $(r + 1)$th term in the binomial expansion of $(C + C^*)^n$. That is the probability is proportional to $\binom{n}{r} C^{n-r} C^{*r}$. If then n-tuplets are being withdrawn randomly (reacting chemically), at what rate is the concentration of tracer diminishing? One is tempted to write (incorrectly)

$$- \frac{dC^*}{dt} \propto \sum_{r=0}^{n} \binom{n}{r} C^{n-r} C^{*r}$$

This equation is not correct because it fails to take account of the fact that tracer concentration falls at a slower rate when an n-tuplet of 1 tracer molecule and n-1 tracee molecules is removed than when an n-tuplet of 2 tracer molecules and n-2 tracee molecules is removed etc. Therefore each term in the series must be weighted by the fraction of tracer molcules in the n-tuplet which it represents. That is

$$-\frac{dC^*}{dt} = \gamma \sum_{r=0}^{n} \frac{r}{n} \binom{n}{r} C^{n-r} C^{*r} \tag{8.5}$$

where γ is a proportionality constant.

The reaction is observed to be described by (8.3) or (8.4). A simple physical model leads to (8.5). The proportionality constant, γ, may be evaluated by setting $C=0$, and $C_T = C^*$ so that (8.3) becomes

$$\frac{dC^*}{dt} = -kC^{*n} \tag{8.6}$$

and (8.5) becomes

$$\frac{dC^*}{dt} = -\gamma C^{*n} \tag{8.7}$$

Since these equations must be identical if the model is to describe the experimental process, therefore

$$\gamma = k \tag{8.8}$$

and (8.5) becomes

$$-\frac{dC^*}{dt} = k \sum_{r=1}^{n} \frac{r}{n} \binom{n}{r} C^{n-r} C^{*r} \tag{8.9}$$

The summation extends from $r = 1$ to $r = n$ since the term in $r = 0$ was always equal to zero. The right-hand side of (8.9) can be simplified,

$$-\frac{dC^*}{dt} = k \sum_{r=1}^{n} \frac{(n-1)!}{(n-r)! \, (r-1)!} C^{n-r} C^{*r}$$

$$= k \sum_{r=1}^{n} \binom{n-1}{r-1} C^{(n-1)-(r-1)} C^{*(r-1)} C^*$$

Let $m = r-1$. Then dividing both sides by C^*,

$$-\frac{1}{C^*} \frac{dC^*}{dt} = k \sum_{m=0}^{m=n-1} \binom{n-1}{m} C^{(n-1)-m} C^{*m}$$

$$= k(C + C^*)^{n-1}$$

i.e.

$$\frac{1}{C^*} \frac{dC^*}{dt} = -k \, C_T^{n-1} \tag{8.10}$$

Since we are treating tracer and tracee in a symmetrical fashion, we may write the equation giving the rate of change of tracee concentration by substituting C for C^* in (8.10):

$$\frac{1}{C} \frac{dC}{dt} = -kC_T^{n-1} \tag{8.11}$$

Hence we may combine (8.10) and (8.11),

$$\frac{1}{C^*}\frac{dC^*}{dt} = \frac{1}{C}\frac{dC}{dt} \tag{8.12}$$

or

$$\frac{dC^*}{dt} = \frac{C^*}{C}\frac{dC}{dt} = a\frac{dC}{dt} \tag{8.13}$$

which is again

$$R^*_{d,v} = aR_{d,v} \quad , \tag{6.7}$$

the principle of chemical indistinguishability of tracer and tracee.

Returning to the case when $C^* \ll C$ so that

$$C_T = C \tag{8.14}$$

very nearly, we see from (8.10) and (8.14)

$$\frac{dC^*}{dt} = -k\ C^{n-1}\ C^* \tag{8.15}$$

This equation has been discussed by Bergner (1961) and Norwich and Hetenyi (1971). It is not necessary to go through a binomial expansion to arrive at (8.15) *if we invoke Eq. (6.7)*[†] since

$$\frac{dC}{dt} = -k\ C^n \tag{8.16}$$

from (8.2) with the constraint (8.14), and since

$$R_{d,v} = \frac{dC}{dt} \tag{8.17}$$

under the experimental circumstances described, thus, using (6.7)

$$\frac{dC^*}{dt} = -\frac{C^*}{C}\ k\ C^n$$

$$= -k\ C^{n-1}\ C^* \tag{8.15}$$

as before.

To state the significance of (8.15) we must return to the open system where C is held constant in time. Then (8.15) may be written more generally

---

[†]The crux of the issue is that we must use *either* Eq. (6.7) *or* the weighted binomial expansion Eq. (8.9). The same assumptions are implicit in both.

$$\frac{\partial C*}{\partial t}\Bigg|_D = - (k\ C^{n-1})\ C* \tag{8.16}$$

That is, if tracee vanishes by an nth order irreversible process, then
tracer will react by a first order process with a rate constant equal to
$k\ C^{n-1}$ (cf Eq. (7.18)).

This example is helpful in crystallizing the association between the balls-
in-an-urn model (or simple collision model) of a chemical reaction and the
equation

$$R*_{d,v} = aR_{d,v} \tag{6.7}$$

which is ubiquitous in tracer theory. We have already seen in section 6 how,
beginning with the assumption that specific activity cannot change due to
the process of chemical reaction, (6.7) is deduced. We can now see clearly
that by drawing black and white balls in groups of n out of an urn containing many such balls we reach the same conclusion. Using the urn (or
simple collision) model we must, of course, be able to show that specific
activity does not change, and this follows from (8.12)

$$C\frac{dC*}{dt}\Bigg|_D - C*\frac{dC}{dt}\Bigg|_D = 0 \tag{8.12}$$

Dividing both sides by C^2, $C \neq 0$,

$$\frac{C\frac{dC*}{dt}\Big|_D - C*\frac{dC}{dt}\Big|_D}{C^2} = 0$$

or

$$\frac{da}{dt}\Bigg|_D = 0$$

Speaking Practically

Tracer theory, as it exists today, is based upon the premise that the
specific activity of a molecular species does not change *due to* the process
of chemical degradation of this species. In our terminology,

$$\frac{da}{dt}\Bigg|_D = 0$$

There are a number of ways of re-formulating this premise, but they do not
alter its conjectural nature.

References

Bergner, P.-E. E., Tracer dynamics: I. a tentative approach and definition of fundamental concepts, *J. Theor. Biol.* 2, 120-140 (1961). Note especially Eq. (26).

Crank, J., *The Mathematics of Diffusion*, Oxford University Press, London, 1956.

Curran, P.F., Taylor, A.E. and Solomon, A.K., Tracer diffusion and uni-directional fluxes, *Biophys. J.* 7, 879-901 (1967).

Darken, L.S., Diffusion mobility and their interrelation through free energy in binary metallic systems, *Trans. Am. Inst. Mining Metall. Eng.* 175, 184-194 (1948).

Denbigh, K.G., *Chemical Reactor Theory, An Introduction*, Cambridge University Press, London, 1966.

Fleck, G.M., On the generality of first-order rates in isotopic tracer kinetics, *J. Theoret. Biol.* 34, 509-514 (1972).

Freund, J.E., *Mathematical Statistics*, Prentice-Hall, Englewood Cliffs, N.J., p. 67, 1962.

Levenspiel, O., *Chemical Reaction Engineering*, Wiley, New York, 1962.

Levich, V.G., *Physicochemical Hydrodynamics*, Prentice-Hall, Englewood Cliffs, N.J., 1962.

Norwich, K.H. and Hetenyi Jr., G.J., Basic studies on metabolic steady state. Linearity of a metabolic tracer system, *Biophysik.* 7, 169-180 (1971).

Van Holde, K.E., *Physical Biochemistry*, Prentice-Hall, Englewood Cliffs, N.J., 1971.

White, A., Handler, P. and Smith, E.L., *Principles of Biochemistry*, 4th edition, Blakiston Division, McGraw-Hill, New York, 1964.

CHAPTER III
THE BIOKINETICS OF
DISTRIBUTED SYSTEMS

SECTION 9
THE DIVERGENCE THEOREM (GAUSS' THEOREM)

Gauss' theorem is demonstrated either for the simple region bounded by a
closed surface or for the region bounded by an outer, closed surface and any
number of inner, closed surfaces in the manner illustrated in Fig. 9.1. The
latter type of region, with many inner bounding surfaces, provides a convenient representation of the extracellular fluid; the outer bounding surfaces
may literally represent skin, which prevents leakage of extracellular fluid
to the exterior of the organism, and the inner bounding surfaces represent
the cellular membranes which separate the intracellular from the extracellular
fluid. It is awe-inspiring that the theorem we are about to demonstrate
appears to be valid for such a complex region which may contain countless
numbers of cellular units. The reader who is already familiar with the demonstration of this well-known theorem may pass on to the next section.

Consider a vector point function $\vec{G}$ which, together with its derivative, is
continuous and finite. The unit outward normal at any point in the surface
is represented by $\vec{n}$. The component $\vec{G}$ in the direction of $\vec{n}$ is given by the
scalar product $\vec{G} \cdot \vec{n}$. Representing an element of surface area by dS, the
normal surface integral of $\vec{G}$ over any boundary is given by

$$\int_S \vec{G} \cdot \vec{n} \, dS \qquad (9.1)$$

The significance of this surface integral is easy to appreciate if we
consider the corresponding summation. The normal scalar resolute (see
section 5), $\vec{G}_p \cdot \vec{n}_p$, is evaluated at some point on the surface and multiplied
by an element of surface area ΔS_p to give $\vec{G}_p \cdot \vec{n}_p \, \Delta S_p$. The product is then
summated for all the elements, ΔS_p, which comprise the total surface area
and the sum is given by (Appendix C)

39

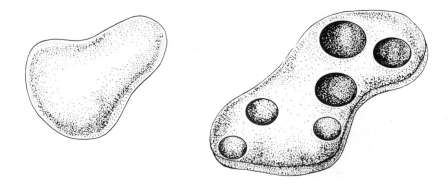

Fig. 9.1. (a) A simple region enclosed by an outer surface.
        (b) A region enclosed by an outer closed surface and six inner
            closed surfaces. This region (depicted here in cross sec-
            tion) may represent, for example, the extracellular fluid,
            the inner boundary surfaces representing the cells.

$$\sum_p \vec{G}_p \cdot \vec{n}\ \Delta S_p \tag{9.2}$$

which in the limit may be represented in the integral form (9.1). Clearly
also

$$\sum_p \Delta S_p = \text{total surface area} \tag{9.3}$$

The divergence theorem states that

$$\int_S \vec{G} \cdot \vec{n}\ ds = \int_V \text{div}\ \vec{G}\ dV$$

$$= \int_V \nabla \cdot \vec{G}\ dV \tag{9.4}$$

where, as usual, the integral $\int_V$ represents the volume integral (triple

integral) over the bounded region.

To prove the theorem, let us represent the three components of the vector $\vec{G}$ by α, β and γ where each of the components is again finite and continuous. Thus

$$\nabla \cdot \vec{G} = \frac{\partial \alpha}{\partial x} + \frac{\partial \beta}{\partial y} + \frac{\partial \gamma}{\partial z} \qquad (9.5)$$

by definition. Let us now break up the volume integral $\int_V \nabla \cdot \vec{G}\ dV$ into its

three parts and consider the third part,

$$\int_V \frac{\partial \gamma}{\partial z}\ dV = \int_V \frac{\partial \gamma}{\partial z}\ dx\ dy\ dz \qquad (9.6)$$

This integral may be evaluated by summing the contributions of a large number of rectangular prisms of sectional area dx dy which are arrayed paralled to the z-axis. Consider one such rectangular prism which passes right across the bounded region from outer surface to outer surface, cutting any number of interior surfaces (Fig. 9.2). The prism cuts elements of surface area dS_1, dS_2 ... dS_m on each of m boundary surfaces. A glance at Fig. 9.2 will show that any such prism must cut an even number of boundary surfaces (6 in Fig 9.2). Carrying out the integration in (9.6) partially with respect to z for the portion of the prism contained between the outer boundary surface and the first inner surface, we obtain

$$\int_{xy} (\gamma_2 - \gamma_1)\ dx\ dy \qquad (9.7)$$

where $\int_{xy}$ denotes the double integral with respect to x and y and γ_2, γ_1 the

value of γ on the second and first surface respectively. Integrating over the regionbetween the third and fourth surfaces (both inner) we obtain

$$\int_{xy} (\gamma_4 - \gamma_3)\ dx\ dy \qquad (9.8)$$

etc. Combining all the contributions to the integral (9.6) from the one rectangular prism,

$$\int_V \frac{\partial \gamma}{\partial z}\ dV = \int_{xy} (-\gamma_1 + \gamma_2 - \gamma_3 + \gamma_4 \ ... \ \gamma_m)dxdy \qquad (9.9)$$

The area element dx dy can be removed from the right-hand side of (9.9) by expressing it in terms of the elements of surface area cut by the prism. This element dx dy is the projection of an element of surface area onto the plane orthogonal to the axis of the prism (Fig. 9.3); thus if $\vec{k}$ is a unit vector parallel to the z-axis,

$$dxdy = -\vec{k} \cdot \vec{n}_1 \, dS_1 = \vec{k} \cdot \vec{n}_2 \, dS_2$$

$$= -\vec{k} \cdot \vec{n}_3 \, dS_3 = \ldots \tag{9.10}$$

Eq. (9.9) then becomes

$$\int_V \frac{\partial \gamma}{\partial z} \, dV = \int_S \vec{k} \cdot (\gamma_1 \vec{n}_1 \, dS_1$$

$$+ \gamma_2 \vec{n}_2 \, dS_2 + \ldots + \gamma_m \vec{n}_m \, dS_m) \tag{9.11}$$

If we now include the contributions from all the rectangular prisms which can be drawn within the bounded region paralled to the z-axis, we have

$$\int_V \frac{\partial \gamma}{\partial z} \, dV = \int_S \gamma \vec{k} \cdot \vec{n} \, dS \tag{9.12}$$

where S denotes the total surface area.

In similar fashion it may be proved that

$$\int_V \frac{\partial \alpha}{\partial x} \, dV = \int_S \alpha \, \vec{i} \cdot \vec{n} \, dS \tag{9.13}$$

$$\int_V \frac{\partial \beta}{\partial y} \, dV = \int_S \beta \, \vec{j} \cdot \vec{n} \, dS \tag{9.14}$$

where $\vec{i}$ and $\vec{j}$ are unit normal vectors parallel to the x- and y-axes respectively. From (9.12), (9.13) and (9.14) then, we have the required result

$$\int_V \left(\frac{\partial \alpha}{\partial x} + \frac{\partial \beta}{\partial y} + \frac{\partial \gamma}{\partial z}\right) dV = \int_S (\alpha \, \vec{i} + \beta \, \vec{j} + \gamma \, \vec{k}) \cdot \vec{n} \, dS$$

or

$$\int_V \nabla \cdot \vec{G} \, dV = \int_S \vec{G} \cdot \vec{n} \, dS \tag{9.4}$$

Speaking Practically

The importance of this theorem will be to permit us to transpose our discussion of tracer and tracee fluxes from the interior of a region (volume integral) to the surface of a region (surface integral).

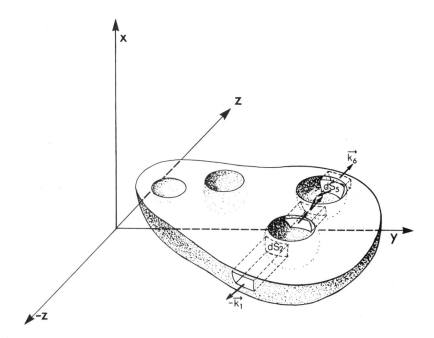

Fig. 9.2. A rectangular prism passing right across the bounded region
           parallel to the z-axis, intersecting a number of inner surfaces.
           The prism cuts elements of surface area dS_1, dS_2, ... dS_6 on each
           of the boundary surfaces. The unit basis vectors $\vec{k}_1$, ... $\vec{k}_6$ are
           shown parallel to the z-axis. Some of the unit outward normals
           to the surface elements are drawn in but not labelled.

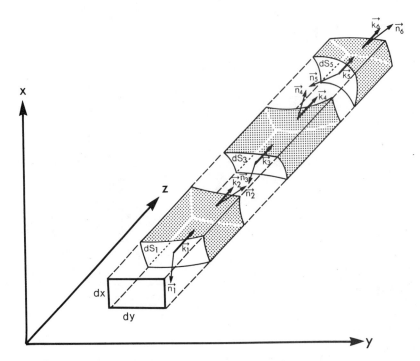

Fig. 9.3. An enlargement of the same prism shown in Fig. 9.2. The unit outward normal vectors, n_p, are all drawn in explicitly. Each of
the elements of surface area dS_p, p = 1,m project onto the area
element dx dy which lies in the x-y plane in accordance with the
relation

$$dxdy = (-1)^p \, \vec{k} \cdot \vec{n}_p \, dS_p$$

SECTION 10
THE EQUATION OF CONTINUITY

The equation of continuity expresses the principle of conservation of mass which will be familiar to most readers. We shall derive this equation in the form (see, for example, Menzel 1961)

$$\frac{\partial C}{\partial t} = -\nabla \cdot (C\vec{v}) + R_{a,v} - R_{d,v}$$

$$= -\text{div} (C\vec{v}) + R_{a,v} - R_{d,v} \tag{10.1}$$

Please note: In this section we shall represent the substance S as a compressible fluid, spread out continuously throughout its total region of distribution. The velocity of this "fluid" at any point (the velocity of a material point) is given by $\vec{v}(\vec{r},t)$. This velocity is distinct from that of any incompressible carrier fluid which may be present and in which the substance S is dissolved or suspended.

Because Eq. (10.1) will be so important in later sections, two derivations will be given. The first derivation is the simpler but less general; the second is more complex but demonstrates the applicability of (10.1) in biological systems of changing shape.

(i) Consider the material contained within an arbitrary closed surface *fixed in space*; ie. in fixed relation to the coordinate axes. This enclosed space is a conceptual enclosure; it is drawn in our minds; material may flow freely across its boundaries. The enclosure contains a mass of material given by the volume integral

$$\int_V C(\vec{r},t) \, dV,$$

where V is the volume of the enclosed space. The rate of change of mass within the enclosure is given by

$$\frac{d}{dt} \int_V C dV = \int_V \frac{\partial C}{\partial t} \, dV \tag{10.2}$$

The partial derivative is brought under the integral sign because of the constant boundaries of the enclosure. Now the rate of change of mass within the region is equal to the net source strength within less the rate at which mass traverses the boundaries. That is, rate of change of mass

$$= \int_V (R_{a,v} - R_{d,v})dV - \int_S C\vec{n} \cdot \vec{v} \, dS \qquad (10.3)$$

where $\vec{n}$ is the unit outward normal to the surface S of the enclosed space. The normal scalar resolute, $\vec{n} \cdot \vec{v}$, gives the component of velocity in the direction of the outward normal; the product $C\vec{n} \cdot \vec{v}$ (mass $\cdot$ area^{-1} $\cdot$ time^{-1}) is the flux of material across (unit area of) the boundary normal to the surface (see Fig. 5.3 and 10.1). The surface integral in (10.3) may now be transformed using the divergence theorem (section 9):

$$-\int_S C\vec{n} \cdot \vec{v} \, dS = -\int_V \nabla \cdot (C\vec{v})dV \qquad (10.4)$$

Replacing the surface by the volume integral in (10.3) and combining the two expressions giving the rate of change of mass, (10.2) and (10.3):

$$\int_V \frac{\partial C}{\partial t} \, dV = \int_V (R_{a,v} - R_{d,v}) \, dV$$

$$- \int_V \nabla \cdot (C\vec{v}) \, dV \qquad (10.5)$$

But since the closed surface was arbitrarily chosen (within the distribution space of a substance, S), the integrands must be equal at all points. That is

$$\frac{\partial C}{\partial t} = R_{a,v} - R_{d,v} - \nabla \cdot (C\vec{v}) \qquad (10.6)$$

When there are no sources or sinks present, this equation assumes its more familiar form,

$$\frac{\partial C}{\partial t} + \nabla \cdot (C\vec{v}) = 0 \qquad (10.7)$$

(ii) The second derivation of the equation of continuity is due to Euler (as perhaps the preceding one was also). Let us designate the components of the velocity vector $\vec{v}$ by u, v and w, where each component is, as usual, a function of x,y,z and t. We shall assume that u,v, and w are finite and continuous and their space derivates $\frac{\partial u}{\partial x}$, $\frac{\partial u}{\partial y}$, $\frac{\partial u}{\partial z}$ etc. are always finite. Motion in accordance with these restrictions is, in the terminology of H. Lamb, 'continuous motion'. In continuous motion, infinitesimal separations of material points always remain infinitesimal; the order of magnitude of the separation never increases. A closed surface is now taken to *move with the fluid* (in contrast to the earlier representation where a closed surface was taken to be fixed in space) and separates for all time the fluid inside the surface from that outside.

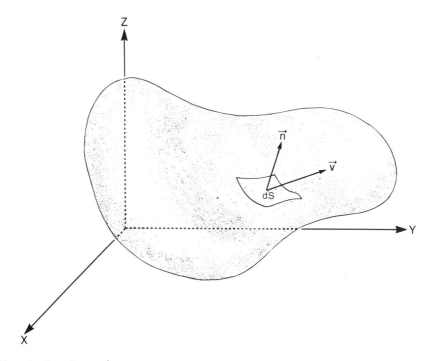

Fig. 10.1.   A small conceptual volume is shown in fixed relation to the
             coordinate axes. The unit outward normal $\vec{n}$ is drawn through the
             surface element dS. The velocity $\vec{v}$ is that of a material point
             at the element dS.

How does any function, G(x,y,z,t) change with small changes in each of its
arguments? We can represent the change as a complete differential,

$$dG = \frac{\partial G}{\partial x}\, dx + \frac{\partial G}{\partial y}\, dy + \frac{\partial G}{\partial z}\, dz + \frac{\partial G}{\partial t}\, dt \qquad (10.8)$$

where dx, dy, and dz are changes in position and dt a change in time. Hence

$$\frac{dG}{dt} = \frac{\partial G}{\partial x} \cdot \frac{dx}{dt} + \frac{\partial G}{\partial y} \cdot \frac{dy}{dt} + \frac{\partial G}{\partial z} \cdot \frac{dz}{dt} + \frac{\partial G}{\partial t} \qquad (10.9)$$

where $\frac{dG}{dt}$ is the *total* rate of change of G with time, consequent upon a
change both in time and in position. Now, if changes in position are due to
motion of a "fluid" of continuously distributed material, then $\frac{dx}{dt}$, the
change in position measured parallel to the x-axis which occurs with a

change in time, is equal to the x-component of the velocity vector $\vec{v}$. That
is

$$\frac{dx}{dt} = u \tag{10.10}$$

etc., so that (10.9) becomes

$$\frac{dG}{dt} = u\frac{\partial G}{\partial x} + v\frac{\partial G}{\partial y} + w\frac{\partial G}{\partial z} + \frac{\partial G}{\partial t} \tag{10.11}$$

or

$$\frac{dG}{dt} = \frac{\partial G}{\partial t} + \vec{v} \cdot \nabla G \tag{10.12}$$

$\frac{dG}{dt}$ is sometimes referred to as the *substantial derivative* of G and gives the
change per unit time in some function G which would be recorded by an
observer moving with a particle of "fluid".

Suppose now that we define G as the mass of the substance S contained at time
t within a small rectangular parallelepiped of dimensions δx, δy and δz.
Then

$$G(x,y,z,t) = C(x,y,z,t) \cdot \delta x \cdot \delta y \cdot \delta z \tag{10.13}$$

where C is, as usual, the concentration. The rectangular parallelepiped is
then a closed surface which separates the material within it from that outside it. Although its shape and volume may change, it must enclose for all
time the mass G(x,y,z,t). In this derivation of the equation of continuity
we shall assume no creation or destruction of the substances S; that is

$$R_{a,v} = R_{d,v} = 0 \tag{10.14}$$

To express mathematically the perpetual enclosure of the same mass we set
the substantial derivative equal to zero:

$$\frac{dG}{dt} = 0$$

or

$$\frac{d}{dt} (C \cdot \delta x \cdot \delta y \cdot \delta z) = 0 \tag{10.15}$$

That is

$$\frac{1}{\delta x \cdot \delta y \cdot \delta z} \frac{d(\delta x \cdot \delta y \cdot \delta z)}{dt} + \frac{1}{C}\frac{dC}{dt} = 0 \tag{10.16}$$

Let us now evaluate the first term in (10.16). In a short time δt, the
rectangular parallelepiped will, in general, become distorted and form to a
first approximation an oblique parallelepiped whose sides are $\delta x'$, $\delta y'$ and

$\delta z'$ (Fig. 10.2). Although the obliqueness of the parallelepiped has been
exaggerated in Fig. 10.2, we shall assume that its angles do not deviate
much from right angles. Suppose that the vertices of one side of length δy
of the original parallelepiped are P and Q. Then the side $\delta y'$ differs in
magnitude and orientation from δy as a consequence of differences in
velocities of the two material points P and Q. If the coordinates of P are
(x,y,z) and its velocity is $\vec{v}(u,v,w)$, then the velocity of point Q *relative*
to P has the components $\frac{\partial u}{\partial y} \delta y$, $\frac{\partial v}{\partial y} \delta y$, $\frac{\partial w}{\partial y} \delta y$. The lengths of the sides of
the oblique parallelepiped may now be expressed in terms of those of the
rectangular one. Let us consider first $\delta y'$. The projection of $\delta y'$ on the
y-axis has the length

$$\delta y + \left[\frac{\partial v}{\partial y} \delta y\right] \cdot \delta t = \left[1 + \frac{\partial v}{\partial y} \delta t\right] \delta y$$

as a first order approximation and the projections of $\delta y'$ on the x- and z-
axes are $\frac{\partial u}{\partial y} \delta y \cdot \delta t$ and $\frac{\partial w}{\partial y} \delta y \cdot \delta t$ respectively. Thus the length of $\delta y'$ is
given by

$$\delta y' = \delta y \left[\left(\frac{\partial u}{\partial y}\right)^2 \delta t^2 + \left(1 + \frac{\partial v}{\partial y} \delta t\right)^2 + \left(\frac{\partial w}{\partial y}\right)^2 \delta t^2\right]^{\frac{1}{2}}$$

$$= \delta y \left[1 + 2\frac{\partial v}{\partial y} \delta t + \delta t^2 \left(\frac{\partial u}{\partial y}\right)^2 + \delta t^2 \left(\frac{\partial v}{\partial y}\right)^2 + \delta t^2 \left(\frac{\partial w}{\partial y}\right)\right]^{\frac{1}{2}}$$

$$(10.17)$$

Terms containing δt^2 are of the second order of smallness and may be dropped
in the first approximation, so that (10.17) becomes

$$\delta y' = \delta y \left(1 + 2\frac{\partial v}{\partial y} \delta t\right)^{\frac{1}{2}} \tag{10.18}$$

Expanding the right-hand side in a binomial series and retaining only first
order terms

$$\delta y' = \delta y \left(1 + \frac{\partial v}{\partial y} \delta t\right) \tag{10.19}$$

which gives $\delta y'$ as a function of δy, v and δt. Proceeding in similar
fashion we can obtain expressions for $\delta x'$ and $\delta z'$:

$$\delta x' = \delta x \left(1 + \frac{\partial u}{\partial x} \delta t\right) \tag{10.20}$$

$$\delta z' = \delta z \left(1 + \frac{\partial w}{\partial z} \delta t\right) \tag{10.21}$$

Since the angles of the oblique parallelepiped are nearly right angles, its
volume is given very nearly by the product

$$\delta x' \cdot \delta y' \cdot \delta z'$$

$$= \delta x \cdot \delta y \cdot \delta z \left[1 + \frac{\partial u}{\partial x} \delta t\right]\left[1 + \frac{\partial v}{\partial y} \delta t\right]\left[1 + \frac{\partial w}{\partial z} \delta t\right]$$

which, to the first order in δt becomes

$$\delta x' \cdot \delta y' \cdot \delta z'$$

$$= \delta x \cdot \delta y \cdot \delta z \left[1 + \left(\frac{\partial u}{\partial x} + \frac{\partial v}{\partial y} + \frac{\partial w}{\partial z}\right) \delta t\right] \qquad (10.22)$$

Hence

$$\frac{1}{\delta x \cdot \delta y \cdot \delta z} \cdot \frac{\delta x' \cdot \delta y' \cdot \delta z' - \delta x \cdot \delta y \cdot \delta z}{\delta t}$$

$$= \frac{\partial u}{\partial x} + \frac{\partial v}{\partial y} + \frac{\partial w}{\partial z}$$

or

$$\frac{1}{\delta x \cdot \delta y \cdot \delta z} \cdot \frac{d(\delta x \cdot \delta y \cdot \delta z)}{dt} = \nabla \cdot \vec{v} \qquad (10.23)$$

Substituting (10.23) into (10.16) we obtain

$$\frac{1}{C}\frac{dC}{dt} + \nabla \cdot \vec{v} = 0 \qquad (10.24)$$

Expanding the substantial derivative $\frac{dC}{dt}$ and multiplying through by C,

$$\frac{\partial C}{\partial t} + \vec{v} \cdot \nabla C + C \nabla \cdot \vec{v} = 0 \qquad (10.25)$$

Using the vector identity (Appendix C, Eq. C.19)

$$\nabla \cdot (C\vec{v}) = \vec{v} \cdot \nabla C + C \nabla \cdot \vec{v} \qquad (10.26)$$

(10.25) becomes

$$\frac{\partial C}{\partial t} + \nabla \cdot (C\vec{v}) = 0 \qquad (10.27)$$

which is, again, the equation of continuity for a region with no sources of
material. This completes the second derivation of this equation.

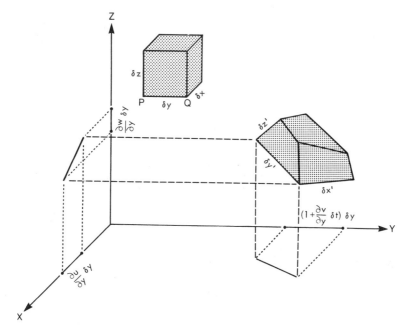

Fig. 10.2. In time interval δt, the rectangular parallelepiped of sides
δx, δy and δz is transformed into the oblique parallelepiped of
sides $\delta x'$, $\delta y'$ and $\delta z'$. The projections of $\delta y'$ on each of the
coordinate axes is shown. The point P has coordinates (x,y,z),
and Q has coordinates (x,y + δy,z) etc.

The first derivation was simpler and is the one commonly found in textbooks.
But it required the concept of a volume fixed in space and through which the
continuously distributed material passed. It is difficult to apply this
concept of a fixed volume to a portion of the region close to a distensible
boundary. A fixed element of volume may be inside the region one moment and
outside it the next. However, applying the second derivation, which follows
the history of a given mass of fluid, no such problem arises near a boundary.
An element of mass in the shape of a rectangular parallelepiped which is
close to a moving boundary will be distorted into an oblique parallelepiped,
but will continue to remain inside the boundary. It will later be found
useful to apply the equation of continuity to portions of the vascular

system - where the boundaries are not rigid.

Speaking Practically

In this section we demonstrated the theorem expressed by Eq. (10.1): the rate of change of concentration at any point is equal to the sum of changes induced by physical and chemical means. Physically, concentration is changed by the transport of material (e.g. by convection and diffusion), represented by the term $div(C\vec{v})$. Chemically, concentration is changed in response to the difference between the rates of production and degradation of material, represented by the terms $R_{a,v} - R_{d,v}$. This equation of continuity will, in the next section, be combined with the laws of convection and diffusion to give an overall equation of convection-diffusion-reaction. This latter equation will then be applied to the analysis of experimental data.

SECTION 11

THE EQUATION OF CONVECTIVE DIFFUSION

One of the first attempts to treat the transport of tracee and tracer in a continuously distributed system was made by Sheppard and Householder in 1951. It seemed reasonable that just as a single substance will be impelled to move by a gradient in its own concentration such that

$$\vec{J} = - D\nabla C \tag{11.1}$$

in accordance with Fick's first law (discussed in section 6), so too in a system consisting of tracee and tracer, the tracer would be impelled to move in response to a gradient in specific activity. That is

$$\vec{J}^* = -I\nabla a \tag{11.2}$$

where I was termed the *interfusion coefficient*. The interested reader will find this coefficient discussed more fully in the appendix to the paper of Sheppard and Householder (1951) as well as in Sheppard's excellent text (1962).

It transpires that the interfusion coefficient is not quite as simple a
construct as the diffusion coefficient, and is strongly dependent on the
magnitude of the concentration of tracee. As we have seen in section 6,
tracer in dilute solutions of tracee will diffuse in accordance with the
equation

$$\vec{J}* = -D\nabla C*$$ (11.3)

Let us consider the simplest case, where tracee is uniformly distributed
such that C, the tracee concentration, is everywhere constant and

$$\nabla C = 0$$ (11.4)

Then

$$\nabla \left(\frac{C*}{C}\right) \equiv \text{grad} \left(\frac{C*}{C}\right) \equiv \frac{\partial}{\partial x} \left(\frac{C*}{C}\right) \vec{i} + \frac{\partial}{\partial y} \left(\frac{C*}{C}\right) \vec{j} + \frac{\partial}{\partial z} \left(\frac{C*}{C}\right) \vec{k}$$

$$= \frac{C\frac{\partial C*}{\partial x} - C*\frac{\partial C}{\partial x}}{C^2} \vec{i} + \text{similar terms in y and z}$$

$$= \frac{1}{C} \frac{\partial C*}{\partial x} \vec{i} + \frac{1}{C} \frac{\partial C*}{\partial y} \vec{j} + \frac{1}{C} \frac{\partial C*}{\partial z} \vec{k}$$

That is

$$\nabla a \equiv \nabla \left(\frac{C*}{C}\right) = \frac{\nabla C*}{C}$$ (11.5)

Hence Eq. (11.3) may be written

$$\vec{J}* = -DC\frac{\nabla C*}{C} = -DC\nabla a$$ (11.6)

Comparing (11.2) with (11.6) it is seen that

$$I = DC$$ (11.7)

That is, the interfusion coefficient is equal to the product of the tracee
concentration and the diffusion coefficient. This relationship was derived
by Hearon (1968). The nucleus of the *interfusion* concept of Sheppard and
Householder is very attractive, and we shall see in a later section how much
progress we can make toward deriving a differential equation in which the
indpendent variable is time and the sole dependent variable is specific
activity.

Let us now pass on to an alternative representation of the transport of
tracee and tracer. We shall first set up two independent equations, one
governing tracee transport and the second governing tracer transport. These
equations allow only for transport by the mechanisms of convection and of
diffusion and therfore no allowance is made for transport of pure solvent

against a concentration gradient. Examples of such *active transport* are very common physiologically but are not embraced by this present mathematical structure. Having set up these two independent equations for tracee and tracer, we shall invoke the various equations expressing the chemical and physical indistinguishability of tracee and tracer, and in this way, couple the two independent equations.

The Diffusion Equation

Let us first consider the case where transport occurs entirely by the process of diffusion. The derivation of the equation of diffusion is simple with our present background. Figuratively speaking,

the equation of diffusion

= Fick's first law + the equation of continuity   (11.8)

That is (from (6.11))

$$\vec{J} = -D\nabla C \qquad (11.1)$$

and

$$\frac{\partial C}{\partial t} = -\nabla \cdot (C\vec{v}) \qquad (10.7)$$

Since

$$\vec{J} = C\vec{v} \qquad (5.5)$$

thus (10.7) becomes

$$\frac{\partial C}{\partial t} = -\nabla \cdot \vec{J} \qquad (11.9)$$

Substituting for $\vec{J}$ from (11.1) into (11.9), we obtain

$$\frac{\partial C}{\partial t} = -\nabla \cdot (-D\nabla C)$$

or

$$\frac{\partial C}{\partial t} = \nabla \cdot (D\nabla C) \qquad (11.10)$$

Eq. (11.10) is the equation of diffusion. When D is constant, (11.10) simplifies to the form

$$\frac{\partial C}{\partial t} = D\nabla \cdot \nabla C$$

or

$$\frac{\partial C}{\partial t} = D\nabla^2 C \qquad (11.11)$$

The Equation of Convective Diffusion

Suppose we no longer confine ourselves to the case of pure diffusion and allow for the more general process of concurrent convection and diffusion (convective diffusion). We can now state figuratively

the equation of convective diffusion =

Fick's first law

+ the rule for convection

(11.12)

+ the equation of continuity (with source included)

In place of (11.1) we now have (combining (6.9) with (6.11))

$$\vec{J} = -D\nabla C + C\vec{u} = C\vec{v}$$
(11.13)

where $\vec{u}$ is the velocity of the carrier fluid. That is, the total flux $C\vec{v}$ is equal to the vector sum of the diffusive and convective fluxes. Substituting for $\vec{J}$ from (11.13) into (11.9),

$$\frac{\partial C}{\partial t} = -\nabla \cdot (-D\nabla C + C\vec{u})$$

$$\frac{\partial C}{\partial t} = D\nabla^2 C - \nabla \cdot (C\vec{u})$$
(11.14)

for constant D. The divergence of the product of a scalar with a vector may be expanded using the identity (10.26), (C.19, Appendix C)

$$\nabla \cdot (C\vec{u}) = \vec{u} \cdot \nabla C + C\nabla \cdot \vec{u}$$
(11.15)

Digression

If the carrier fluid is incompressible, we may apply the constraint

$$\nabla \cdot \vec{u} = 0$$
(11.16)

The divergence of the velocity vector is always equal to zero for an incompressible fluid, and this is easily seen by returning to the Eulerian model and considering a moving mass of incompressible fluid consisting always of the same particles. Eq. (10.23) states

$$\frac{1}{\delta x \cdot \delta y \cdot \delta z} \cdot \frac{d(\delta x \cdot \delta y \cdot \delta z)}{dt} = \nabla \cdot \vec{u}$$
(11.17)

where $\vec{u}$, the velocity of an incompressible fluid, has been substituted for $\vec{v}$, the velocity of a compressible "gas" of distributed material. Since the fluid cannot be compressed,

$$\frac{d(\delta x \cdot \delta y \cdot \delta z)}{dt} = 0$$
(11.18)

and we obtain

$$\nabla \cdot \vec{u} = 0$$

from (11.17).

End of Digression

Returning now to (11.15) and substituting (11.16)

$$\nabla \cdot (C\vec{u}) = \vec{u} \cdot \nabla C \qquad (11.19)$$

Substituting (11.19) into (11.14)

$$\frac{\partial C}{\partial t} = D\nabla^2 C - \vec{u} \cdot \nabla C \qquad (11.20)$$

which is the usual form of the equation of convective diffusion.

Finally, if in place of (10.7) we use the more general form of the equation of continuity, (10.6),

$$\frac{\partial C}{\partial t} = -\nabla \cdot (C\vec{u}) + R_{a,v} - R_{d,v} \qquad (10.6)$$

then we simply get two extra terms in the equation of convective diffusion:

$$\frac{\partial C}{\partial t} = D\nabla^2 C - \vec{u} \cdot \nabla C + R_{a,v} - R_{d,v} \qquad (11.21)$$

This equation might properly be termed the *equation of metabolic convective diffusion* or the *equation of convection-diffusion-reaction*. Attention was drawn to the significance of this equation in biology by N. Rashevsky more than thirty-five years ago, and it is discussed in his book *Mathematical Biophysics* (1960).

Corresponding to (11.21), which has been formulated for an unlabelled substance or tracee, we might set up the equation of convective diffusion for tracer:

$$\frac{\partial C^*}{\partial t} = D^*\nabla^2 C^* - \vec{u} \cdot \nabla C^* + R^*_{a,v} - R^*_{d,v} \qquad (11.22)$$

Using (6.13)

$$D^* = D$$

and (6.7)

$$R^*_{d,v} = aR_{d,v}$$

(11.22) becomes

$$\frac{\partial C^*}{\partial t} = D\nabla^2 C^* - \vec{u} \cdot \nabla C^* + R^*_{a,v} - aR_{d,v} \qquad (11.23)$$

Eqs. (11.21) and (11.23) describe the transport of non-ionized tracee and tracer in a biological system. They were set down in this form by Norwich (1971). Neither "active" transport nor transport through small pores has been *explicitly* allowed for. We have assumed the presence of dilute solutions of tracee and exceeding low concentrations of tracer.

Speaking Practically

In this section we have compounded the principle of mass balance (conservation of mass), the laws governing convection and diffusion, and the principle of physical and chemical indistinguishability of tracer and tracee, to obtain the two equations of convection-diffusion-reaction, (11.21) and (11.23). While we might go on to express energy and momentum balance, the resulting equations would probably not be applicable in the intact organism. Therefore, we have really reviewed much of the *applicable* physics and chemistry of the subject, in the current state of its development. We shall, however, return again in Chapter IV, to consider what is physics and what figment in the development of the theory of compartments.

SECTION 12

LINEAR SYSTEMS

A *system* may be thought of as something which has an *input* and an *output*. For example, the input may be what the investigator does, and the output the response that he observes to what he does. The output is assumed to be causally related to the input, but no mechanism relating input to output is postulated. The term "system" when defined in this way is perfectly general.

It may be a control system where the input is a temperature and the output
a voltage; it may be an economic system where the input is the cost of raw
material and the output the price of a product; it may be a biological
system where the input is the intensity of a visual stimulus and the output
a cortical potential. A *linear system* is a particular type of system which
may be defined by certain specific criteria. Let us denote by X the input
to the system and by Y the output from the system. The arrow → will denote
"gives rise to". A system may be said to be linear if it obeys the rules of
superposition and to *homogeneity*. If X_1 and X_2 are arbitrary inputs such
that

$$X_1 \rightarrow Y_1 \qquad\qquad\qquad (12.1)$$

and

$$X_2 \rightarrow Y_2 \qquad\qquad\qquad (12.2)$$

then the system is "superposable" if

$$X_1 + X_2 \rightarrow Y_1 + Y_2 \qquad\qquad\qquad (12.3)$$

— that is if a superposition of inputs gives rise to a superposition of
outputs. If

$$\gamma X_1 \rightarrow \gamma Y_1 \qquad\qquad\qquad (12.4)$$

for arbitrary γ, then the system is homogeneous. The above two criteria
may be combined as follows.

If X_1 and X_2 are arbitrary inputs such that

$$X_1 \rightarrow Y_1 \qquad\qquad\qquad (12.1)$$

and

$$X_2 \rightarrow Y_2 \qquad\qquad\qquad (12.2)$$

then the system is linear if, for all real α and β,

$$\alpha X_1 + \beta X_2 \rightarrow \alpha Y_1 + \beta Y_2 \qquad\qquad\qquad (12.5)$$

In practice, no physical system could be found to satisfy relation (12.5)
for *all* α and β. There is always a range over which a system is linear and
outside of which it is not. In biokinetics, we shall often be constrained
by the necessity that our inputs and outputs be positive quantities, thus
restricting the magnitudes of α and β if they are negative numbers. If α and
β are not functions of time, the system is *stationary* or *time invariant*.

As an example of a stationary linear system, consider a resistor of R ohms
resistance. If a D.C. voltage of V is applied across it, a current of I
amperes will flow through it in accordance with Ohm's law,

$$V = IR \qquad\qquad\qquad (12.6)$$

Let us regard V as the input and I as output and assume that the resistance R remains constant. Then suppose that

$$V_1 \to I_1 \qquad (12.7)$$

and

$$V_2 \to I_2 \qquad (12.8)$$

That is

$$V_1 = I_1 R$$

and

$$V_2 = I_2 R$$

So that

$$\alpha V_1 = \alpha I_1 R \qquad (12.9)$$

and

$$\beta V_2 = \beta I_2 R \qquad (12.10)$$

for all α and β.

Adding (12.9) to (12.10)

$$\alpha V_1 + \beta V_2 = (\alpha I_1 + \beta I_2)R \qquad (12.11)$$

That is

$$\alpha V_1 + \beta V_2 \to \alpha I_1 + \beta I_2$$

and the system is linear. However, if we pass sufficiently high currents through the resistor, it will heat and the resistance will change, thus possibly producing a deviation from linearity.

As we shall see in later sections, a linear differential equation may be associated with a linear system.

The Dirac and Heaviside Distributions

The symbol $\delta(x)$ represents the so-called *Dirac delta function* used by P.A.M. Dirac in his work on quantum theory. However, the delta function is not really a function but a *distribution*, and the interested reader is referred to a text on operational mathematics for more information on this subject. For our purposes, we shall regard the delta function simply as described by Eqs (12.12) and (12.13):

$$\delta(x - x') = 0, \qquad x \neq x' \qquad (12.12)$$

$$\int_{-\infty}^{\infty} f(x')\, \delta(x - x')\, dx' = f(x) \qquad (12.13)$$

Eq. (12.13) may be taken as a definition of the delta function. Hence it follows that

$$\delta(x) = \delta(-x) \tag{12.14}$$

and

$$\int_{-\infty}^{\infty} \delta(x) \ dx = 1 \tag{12.15}$$

(The dimensions of $\delta(x)$ are those of x^{-1} so that $\int_{-\infty}^{\infty} \delta(x) \ dx = 1 = a$ dimensionless number.) We cannot really draw a graph of $\delta(x)$ versus x. Such a graph must show a "curve" which has the value zero for all values of x except for x = 0; at x = 0, it suddenly breaks upwards to infinity. We can represent $\delta(x)$ schematically as in Fig. (12.1). A number of real functions have been suggested to represent the delta function; for example

$$\lim_{\lambda \to 0} \ \frac{1}{\pi} \frac{\lambda}{\lambda^2 + x^2} \tag{12.16}$$

has some of the required attributes
and

$$\frac{\lambda}{\pi} \int_{-\infty}^{\infty} \frac{1}{\lambda^2 + x^2} \ dx = 1$$

The delta function is used to denote an *impulse*, or sudden, brief burst of activity. Therefore we shall use this function primarily as a means of designating the bolus or slug or sudden injection of tracer or tracee. The expression $M^*\delta(t)$ (**dpm/time**) denotes the injection of a mass M^* of tracer very suddenly into the system and

$$\int_{-\infty}^{\infty} M^* \ \delta(t) \ dt = M^* \tag{12.17}$$

showing that the total mass of tracer injected over all time is M^*. M^* might be called the *strength* of the delta function, $\delta(t)$.

We can define a delta function in three dimensions, $\delta(x) \cdot \delta(y) \cdot \delta(z)$ such that

$$\iiint_V \delta(x) \cdot \delta(y) \cdot \delta(z) \ dx \ dy \ dz = 1 \tag{12.18}$$

or identically

$$\int_V \delta(\vec{r}) \ d\vec{r} = 1 \tag{12.19}$$

where (12.19) is just a simpler way of writing (12.18). Similarly,

$$\int_V f(\vec{r}) \ \delta(\vec{r} - \vec{r}_0) \ d\vec{r} = f(\vec{r}_0) \tag{12.20}$$

We shall use $R_a \delta(\vec{r} - \vec{r}_0)$ (mass $\cdot$ volume^{-1} $\cdot$ time^{-1}) to designate a point source of strength R_a at position $\vec{r} = \vec{r}_0$.

The symbol $H(x)$ represents the Heaviside distribution, which is defined by the equations

$$H(x - x') = 0 \qquad x' > x$$
$$= 1 \text{ otherwise} \tag{12.21}$$

Thus the Heaviside distribution represents the unit step function as illustrated in Fig. (12.2). The relationship between the Dirac and Heaviside distributions is

$$\delta(x - x') = \frac{dH(x - x')}{dx} \tag{12.22}$$

which may be appreciated intuitively.

The Laplace transformations of the Dirac and Heaviside distributions are of importance. Designating the Laplace transformation by L

$$L\delta(t) = \int_0^\infty e^{-st} \delta(t) \, dt = 1 \tag{12.23}$$

$$LH(t) = \int_0^\infty e^{-st} H(t) \, dt = \frac{1}{s} \tag{12.24}$$

More information on linear systems analysis may be found in the texts by B.M. Brown and D.K. Cheng. The use of operators such as the Laplace operator is treated well in Churchill's text.

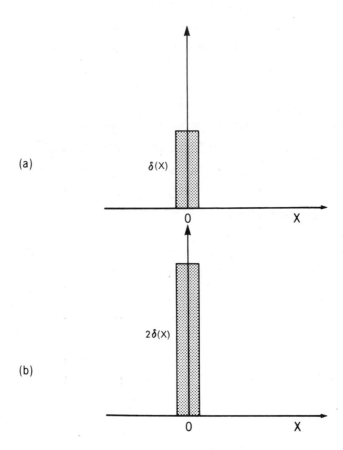

Figs. 12.1(a) and (b). In (a) the function $\delta(x)$ has been represented as a
long, narrow rectangle of area 1. In (b) the function $2\delta(x)$ has
been represented as a rectangle of the same width as in (a) but of
area 2.

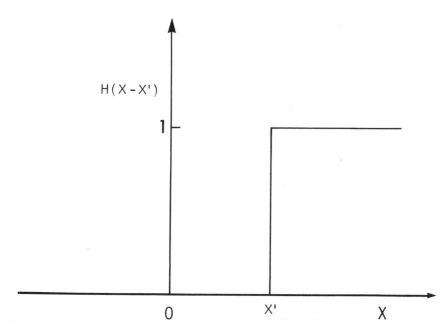

Fig. 12.2. The Heaviside distribution: H(x - x') is equal to zero for
            negative values of its argument and is equal to unity for
            positive values.

The Convolution Integral

The great advantage in dealing with a linear system is that once the *unit
impulse response function* has been determined (i.e. the output corresponding
to unit input), one can then readily determine the output corresponding to
any given input. We might designate the determination of outputs from inputs
as the problem of *output discovery* which, as we shall later see, is simpler
than the converse problem of input discovery.

If an impulse of magnitude $P\delta(t)$ is chosen as the input to a linear system,
and the corresponding output is $Q(t)$, then $1.\delta(t)$ is the unit input and
$\frac{Q(t)}{P}$ is termed the *unit impulse response function* of the system. The unit
impulse response function is a characteristic of the system and we shall

designate it by $h(t)$. This function can be used to map any input, $X(t)$, onto a corresponding output $Y(t)$. The equation connecting the three variables is

$$Y(t) = \int_0^t h(\tau) \; X(t - \tau) \; d\tau \qquad\qquad (12.25)$$

or equivalently

$$Y(t) = \int_0^t h(t - \tau) \; X(\tau) \; d\tau \qquad\qquad (12.26)$$

It is assumed that the input prior to zero time was equal to zero. The above integral is known as the *convolution integral* and a demonstration of the convolution theorem (probably not rigorous enough to be called a "proof") now follows. Again we shall use an arrow → to designate "gives rise to", where an input appears on the left side and an output on the right side of the arrow.

$$\delta(t) \rightarrow h(t) \qquad\qquad (12.27)$$

by definition of the unit impulse response function. Since the impulse input can be applied at any time τ,

$$\delta(t - \tau) \rightarrow h(t - \tau) \qquad . \qquad\qquad (12.28)$$

From the principle of homogeneity,

$$X\delta(t - \tau) \rightarrow Xh(t - \tau) \qquad ; \qquad\qquad (12.29)$$

i.e. an impulse of strength X applied at time, τ, gives rise to a response of strength X beginning at the time, τ^+. From the principle of superposition

$$\int_0^t X(\tau) \; \delta(t - \tau)d\tau \rightarrow \int_0^t X(\tau) \; h(t - \tau)d\tau \qquad ; \qquad (12.30)$$

i.e. a summated sequence of impulses, each applied briefly, gives rise to a summated sequence of responses. The input integral can be simplified using the definition of the delta function (12.13)[‡],

$$\int_0^t X(\tau) \; \delta(t - \tau)d\tau = X(t) \qquad\qquad (12.31)$$

[†]The function $h(t)$ may be throught of as always multiplied by a Heaviside function. That is, $h(t)$ is really $H(t) \cdot h(t)$; $h(t)$ is equal to zero for $t < 0$.

[‡]One can think of $X(\tau)$ as being $H(\tau) \cdot H(t-\tau) \cdot X(\tau)$, so that $\int_{-\infty}^{\infty} X(\tau) \; ...$

$= \int_0^t X(\tau) \; ...$

which just means that any input described by the continuous function X(t)
can by analysed into an infinite number of impulses of varying strengths
applied sequentially at time intervals which are infinitesimally brief.
Thus

$$X(t) \longrightarrow \int_0^t X(\tau)h(t - \tau)d\tau = Y(t) \qquad (12.26)$$

and we have demonstrated (12.26).

An Alternative (Rather Simple) Derivation of the Convolution Integral

To obtain a clearer understanding of the convolution integral, let us take a
simple example. Suppose that the unit impulse response function has the form
of a simple decay curve as shown in Fig. (12.3). Suppose now that the input
to this system X(t) consists of 6 sequential impulses of various strengths
as illustrated in Fig. (12.4). The output Y(t), will then be composed of 6
decay curves of different amplitudes as shown in Fig. (12.5). From the
principle of homogeneity we know that the amplitudes of the 6 decay curves
are proportional to the strengths of the corresponding input impulses. The
observed output will be a superposition of these 6 curves as shown in Fig.
(12.6). Let us calculate the observed output at t = 5 directly from the
principles of homogeneity and superposition and without using the convolution
integral. Recalling that a unit impulse input gives rise to the output h(t),
we see that an impulse input of strength X gives rise to an output Xh(t)
(homogeneity). Therefore the output at t = 5, Y(5), consists of the sum

$X(5)h(0)$
 (the output produced by the impulse input applied at t = 5 evaluated at t=0)[†]
$+X(4)h(1)$
 (the output produced by the impulse input applied at t = 4 evaluated at t=1)
$+X(3)h(2)$
 (the output produced by the impulse input applied at t = 3 evaluated at t=2)
$+X(2)h(3)$
 (the output produced by the impulse input applied at t = 2 evaluated at t=3)
$+X(1)h(4)$
 (the output produced by the impulse input applied at t = 1 evaluated at t=4)
$+X(0)h(5)$
 (the output produced by the impulse input applied at t = 0 evaluated at t=5)

[†] i.e. evaluated at t = 0 *relative to this impulse response curve,* etc.

i.e.

$$Y(5) = \sum_{\tau=0}^{5} X(\tau)h(5 - \tau) \qquad (12.32)$$

or in general

$$Y(t) = \sum_{\tau=0}^{t} X(\tau)h(t - \tau) \qquad (12.33)$$

Eq. (12.33) defines the *convolution summation*. To obtain this summation we have designated the strength of an impulse given at time τ by $X(\tau)$, which was simply represented by a diagram. But recalling that the strength of a delta function is given by an area, in designating a continuous input we change the representation of the input to $X(\tau)d\tau$ and write (12.33) in the form

$$Y(t) = \int_{0}^{t} X(\tau)h(t - \tau)d\tau \qquad (12.34)$$

which is the convolution integral (12.26) as before. The relationship between the Roman and italic x's is

$$X(\tau)d\tau = X(\tau) \qquad (12.35)$$

The italic X defines an area; the Roman X defines an amplitude. Thus we have a second derivation of the convolution integral.

It should be stated here that if

$$Y(t) = \int_{0}^{t} X(\tau)h(t - \tau)d\tau \qquad (12.34)$$

then by taking Laplace transformations of both sides,

$$LY(t) = LX(t) \cdot Lh(t) ; \qquad (12.36)$$

that is, the Laplace transform of the unit impulse response function is equal to the ratio of the Laplace transform of the output to that of the input. The Laplace transform of the unit impulse response function, $Lh(t)$, is known as the *transfer function* of the system.

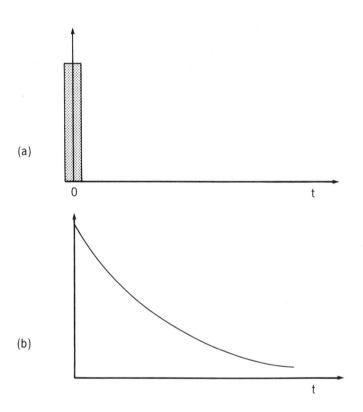

Fig. 12.3. An example of a particular unit impulse response function
            (a) The unit impulse input.
            (b) The output corresponding to the input (a) is the unit
                impulse response function, which here has the form of
                a simple decay curve.

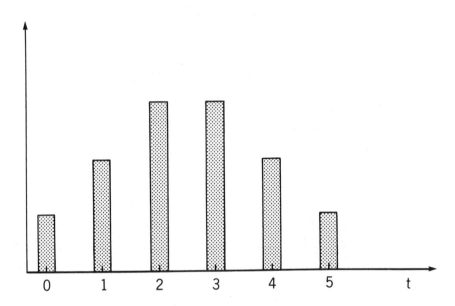

Fig. 12.4.   Example of an input consisting of sequential impulses applied at
             intervals of one time unit. The impulses have different
             strengths.
             What is being demonstrated here is that *any* smooth input
             function can be broken up into a series of narrow rectangles.
             Each narrow rectangle can then be treated as an independent
             impulse input whose strength is proportional to its height.

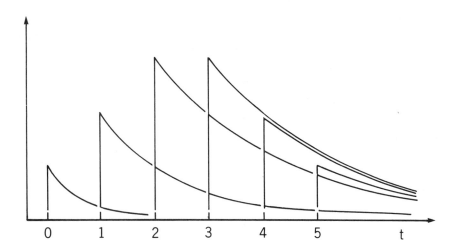

Fig. 12.5. The output corresponding to the input shown in Fig. (12.4) in a
            system whose unit impulse response function is of the type shown
            in Fig. (12.3). The output consists of 6 simple decay curves
            whose amplitudes are proportional to the impulses which gave
            rise to them. The observed output will be a superposition of
            these 6 curves.

            What is being demonstrated here is that the output corresponding
            to *any* smooth input can be treated *as if* it were composed of
            (i.e. the sum of) a very large number of impulse response curves
            of different heights. We do not observe these component curves.
            We observe their sum.

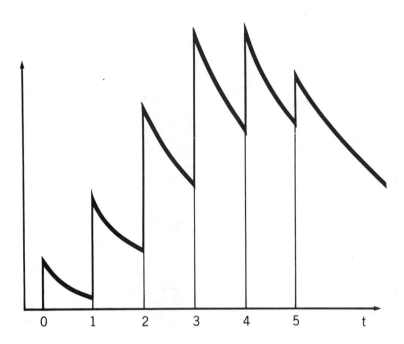

Fig. 12.6.   Superposition of the 6 impulse response curves of varying weight
             depicted in Fig. (12.5). This is the output which would be
             observed in such a system.

             In general, we observe a smooth output curve, because it results
             from a smooth input curve. That is, if the input curve is
             broken down into an infinite number of rectangles of infini-
             tesimal width, then the output will consist of the superposition
             of an infinite number of impulse response curves of very low
             amplitude. Because there are so many curves, they blend into
             a smooth output curve.

The convolution integral plays an important role in statistics, which might
be mentioned here in passing.
Let $f_1(p_1)$ be the continuous probability density function governing the
    random variable p_1, and
let $f_2(p_2)$ be the continuous probability density function governing the
    random variable p_2.
That is $f_1(p_1)dp_1$ is the probability that the variable p_1 takes on a value
between p_1 and $p_1 + dp_1$ etc.
What, then, is the probability density function governing the variable
$p = p_1 + p_2$?
Let us represent this function by $f(p)$ and suppose that p_1, $p_2 > 0$. Then

$$f(p) = \int_0^\infty f_1(p_1) \, f_2(p - p_1)dp_1 \qquad (12.37)$$

which is a convolution integral. Then $f(p)$ gives the probability that the
variable p takes on a value between $(p_1 + p_2)$ and $(p_1 + p_2) + d(p_1 + p_2)$.

Finally it must be noted that in the various demonstrations given in this
section, the delta distribution and the differential dt have been treated in
a rather cavalier fashion, relying heavily on intuition. This procedure is
perhaps justified only because the various theorems have been proved
rigorously elsewhere.

Speaking Practically

We have discovered a way of predicting stimulus-response relationships. If
we are given the nature of the stimulus (input), we can predict the nature
of the response (output). It doesn't matter how complex the system is; it
doesn't matter how many elements (gears, circuits, reagents) there are.
We can predict outputs from inputs. With a little bit of practice and a
little luck we shall even be able to reverse the process and predict inputs
from outputs. BUT we must first know that the system is
     linear,
     unchanging with time (i.e. the *structure* of the system is unchanging),
and we must be able to determine the response of the system to a sudden,
brief burst of activity, this response being known as the
     unit impulse response function.
The latter function is sometimes transformed and appears as the

transfer function,
but we may regard these two functions as having the same physical content.
If we use the unit impulse response function, we obtain outputs from inputs
   by integrating (convolution).
If we use the transfer function, we obtain outputs from inputs
   by multiplying.

Problems

12.1 Prove Eq. (12.14) and Eq. (12.15) using the definition of the delta function.
Prove Eq. (12.23) and Eq. (12.24).

12.2 Show that Eqs. (12.25) and (12.26) are equivalent.

SECTION 13

TRACERS AND LINEAR SYSTEMS

In section 12 it was shown how, in a system known to be linear, one can determine the output corresponding to any given input when a fundamental characteristic of the system has once been determined. This fundamental characteristic is the *unit impulse response function* and its Laplace transform is the *transfer function* of the system. We shall now explore the linear system associated with the kinetics of tracer.

The foundations for modern biokinetics were set down by Rudolph Schönheimer in his classical book "The Dynamic State of Body Constituents", (1944). Gerald A. Wrenshall later defined the *dynamic steady state* of some body constituent or metabolite — the state where the rates of appearance and disappearance are equal and constant, in some region. We shall define the steady state of S even more strictly to be the state where

$$R_a = R_d = \text{a constant} \tag{13.1}$$

$R_{a,v}$ and $R_{d,v}$ are constant for all $\vec{r}$

and where

$$\frac{\partial C(\vec{r},t)}{\partial t} = 0 \tag{13.2}$$

and

$$\frac{\partial^2 C(\vec{r},t)}{\partial t^2} = 0 \tag{13.3}$$

for all $\vec{r}$. While all of the above conditions are required for our mathematical definition of the steady state, the experimentalist, upon whom it is incumbent to create such a steady state in the laboratory, will generally be satisfied that he has obtained it if

$$R_a = R_d = \text{a constant}$$

for a "long" period of time. The long period is believed to be sufficient to assure the vanishing of the first and second partial derivatives of concentration with respect to time, and it is unlikely that R_a would be constant if $R_{a,v}$ were changing etc. The reader should note that no restriction has been placed upon the space derivatives of concentration, and in general

$$\nabla C \neq 0$$

That is, while concentration, C, may vary from place to place within the system, it retains a fixed value at any given point for all time.

The above equations have been set up to describe the steady state for tracee or unlabelled substance in general, but it is clear that the steady state of tracer can be described in a similar fashion. The steady state of tracer has been termed by some writers "the state of tracer equilibrium", but there seems to be no particular reason why a new term should be invoked. We shall later explore the properties of the system where both tracee and tracer are in steady state.

Let us now examine the equation of convective diffusion for tracer when tracee is in steady state:

$$\frac{\partial C^*}{\partial t} = D\nabla^2 C^* - \vec{u} \cdot \nabla C^* + R^*_{a,v} - \frac{C^*}{C} R_{d,v} \tag{11.23}$$

If tracee is in steady state, then $R_{d,v}/C$ is constant in time. Let us also regard $\vec{u}$ as constant in time. Rearranging (11.23) we have

$$R^*_{a,v} = \frac{\partial C^*}{\partial t} - D\nabla^2 C^* + \vec{u} \cdot \nabla C^* + \frac{R_{d,v}}{C} \cdot C^* \qquad (11.23)$$

Suppose we now define the system whose input is $R^*_{a,v}(\vec{r}_0,t)$ and whose output is $C^*(\vec{r}_1,t)$; that is we define the system's input as the rate of appearance of tracer per unit volume at position $\vec{r}_0$ and its output as the concentration at position $\vec{r}_1$. Having fixed the sites of input and output we may now regard input and output purely as functions of time. Since (11.23) is of the first degree in the input and output variables and the operators are linear differential operators, we might suspect that the system so defined is linear. Since D, $\vec{u}$ and $R_{d,v}/C$ are constants we might suspect also that the system is stationary. Again using the $\rightarrow$ arrow to connect input and output, suppose that[†]

$$R^*_{a,v,1}(t) \rightarrow C_1^*(t) \qquad (13.4)$$

and

$$R^*_{a,v,2}(t) \rightarrow C_2^*(t) \qquad (13.5)$$

Substituting (13.4) into (11.23) and multiplying through by an arbitrary positive scalar α

$$\alpha R^*_{a,v,1} = \frac{\partial(\alpha C_1^*)}{\partial t} - D\nabla^2(\alpha C_1^*) + \vec{u} \cdot \nabla(\alpha C_1^*) + \frac{R_{d,v}}{C} \cdot (\alpha C_1^*) \qquad (13.6)$$

Substituting (13.5) into (11.23) and multiplying through by an arbitrary positive scalar β

$$\beta R^*_{a,v,2} = \frac{\partial(\beta C_2^*)}{\partial t} - D\nabla^2(\beta C_2^*) + \vec{u} \cdot \nabla(\beta C_2^*) + \frac{R_{d,v}}{C} \cdot (\beta C_2^*) \qquad (13.7)$$

[†]That is, on one occasion the input $R^*_{a,v,1}(t)$ gave rise to the output $C_1^*(t)$, and on another occasion, after the system had emptied itself of tracer, the input $R^*_{a,v,2}(t)$ gave rise to the output $C_2^*(t)$

Adding (13.6) to (13.7)

$$\alpha R^{\star}_{a,v,1} + \beta R^{\star}_{a,v,2} = \frac{\partial}{\partial t}(\alpha C_1^{\star} + \beta C_2^{\star}) - D\nabla^2(\alpha C_1^{\star} + \beta C_2^{\star})$$

$$+ \vec{u} \cdot \nabla(\alpha C_1^{\star} + \beta C_2^{\star}) + \frac{R_{d,v}}{C}(\alpha C_1^{\star} + \beta C_2^{\star}) \qquad (13.8)$$

Thus

$$\alpha R^{\star}_{a,v,1} + \beta R^{\star}_{a,v,2} \rightarrow \alpha C_1^{\star} + \beta C_2^{\star} \qquad (13.9)$$

for positive α and β, and hence the system is linear. (The constraint that both α and β be positive can probably be relaxed.)

Summarizing, then, if we have tracee steady state and if we define a system whose input is the rate of infusion of tracer at $\vec{r}_0$ and whose output is the concentration of tracer at $\vec{r}_1$, then this system on theoretical principles is expected to be linear.

We can now introduce some experimental evidence in support of the theory which has been developed. In a paper published in 1971, Norwich and Hetenyi presented experimental results to show that in dogs, when unlabelled glucose (tracee) remained in steady state, the tracer system whose input is the rate of intravenous infusion of labelled glucose and whose output is the plasma labelled glucose concentration, is in fact a linear one. Dogs were fasted 14 to 16 hours prior to the experiment, and then, under local anaesthesia, cannulae were passed into saphenous and sometimes cephalic veins. The cannulae were used for infusion of tracer and for withdrawal of blood samples. The dogs were trained to stand quietly in a Pavlov harness throughout the sampling period.

Tracer infusions of various types (see below) were given through one cannula, and the blood samples withdrawn through the other were analysed for C and C*. The blood samples were replaced by an equal volume of heparinized saline. The tracer used was either ^{14}C-1-glucose or ^{3}H-2-glucose. Various tests of linearity were carried out, three of which were reported.

(a) Testing relations (12.1), (12.2) and (12.5) for one pair of α and β values.

$$X_1 \rightarrow Y_1 \qquad (12.1)$$

$$X_2 \rightarrow Y_2 \qquad (12.2)$$

$$\alpha X_1 + \beta X_2 \to \alpha Y_1 + \beta Y_2 \qquad\qquad (12.5)$$

X_1 was chosen as an impulse input of tracer injected intravenously through one of the cannulae. On a second occasion, after all tracer from the first experiment had vanished, X_2 was chosen as a step input infused intravenously through the same cannula. These two inputs are sketched in Fig. 13.1. On a third occasion, a combined injection-infusion (impulse-step function) was given (12.5) with α chosen as 1.52 and β as 0.54, and the comparison between the observed and theoretical plasma concentration output $\alpha Y_1 + \beta Y_2$ is shown in Fig. 13.2. The very close matching of experiment with theory may be seen. This experiment was carried out 8 times on 3 dogs with different values of α and β, and the results were always similar to that shown in Fig. 13.2.

(b) Testing the output predicted by the convolution integral (12.25) or (12.26) for one particular input function $X(t)$. Using the same experimental design as cited above, a mass M_o^* of tracer was injected intravenously and the plasma ^{14}C-1-glucose concentrations were determined periodically for 120 minutes thereafter. The curve of C^* against time was of the decay type and was fitted very closely by a double exponential function:

$$C^*(t) = Ae^{-b_1 t} + Be^{-b_2 t} \qquad b_1, b_2 > 0 \qquad (13.10)$$

The method by which experimental data may be fitted by a least squares technique to a double exponential function is discussed in Appendix B. For the moment let us accept that numerical values may be assigned to A, B, b_1 and b_2 in such a way that the mathematical function (13.10) closely approximates the experimental points over the two hour interval. Two weeks later, the same dog was cannulated and replaced in the Pavlov harness and a constant infusion of ^{14}C-1-glucose was administered at a rate R_a^* dpm/min. The step response expected, using the convolution integral (12.25), is

$$C^*(t) = \int_0^t \frac{R_a^*}{M_o^*}\left(Ae^{-b_1\tau} + Be^{-b_2\tau}\right)d\tau \qquad (13.11)$$

since the *unit* impulse response function is

$$h(t) = \frac{1}{M_o^*}\left(Ae^{-b_1 t} + Be^{-b_2 t}\right) \qquad (13.12)$$

Carrying out the integration,

$$C^*(t) = \frac{R_a^*}{M_o^*}\left[\left(\frac{A}{b_1} + \frac{B}{b_2}\right) - \left(\frac{A}{b_1}e^{-b_1 t} + \frac{B}{b_2}e^{-b_2 t}\right)\right] \qquad (13.13)$$

In the limit as t → ∞, C*(t) → a constant value, as might be expected:

$$\lim_{t \to \infty} C^*(t) = \frac{R_a^*}{M_0^*} \left(\frac{A}{b_1} + \frac{B}{b_2} \right) \tag{13.14}$$

Over the two hour period during which blood samples were collected, C*(t) did not really near its asymptote but rose steeply monotonically. The very close matching of the measured plasma ^{14}C-1-glucose concentration with the concentrations predicted by the convolution integral (13.13) is shown in Fig. 13.3. There is very little left to be desired. Four experiments of this nature were carried out.

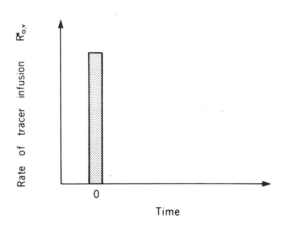

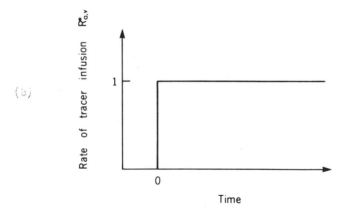

Fig. 13.1. (a) "Represents" the impulse input, X_1

           (b) Depicts the step input, X_2

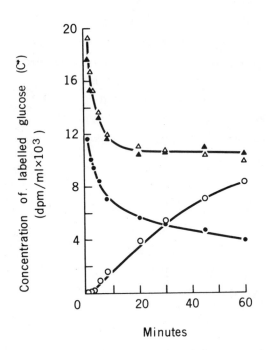

Fig. 13.2. ● Impulse response Y_1: experimental
            o Step response Y_2: experimental
            ▲ Response to $1.52X_1 + 0.54X_2$: experimental
            △ $1.52Y_1 + 0.54Y_2$: theoretical

The linear behaviour of the tracer system is evidenced by the close matching of the two curves ▲ and △.

From Biophysik 7, 169-180, 1971 by K.H. Norwich and G. Hetenyi, Jr. Reproduced by permission of Springer-Verlag.

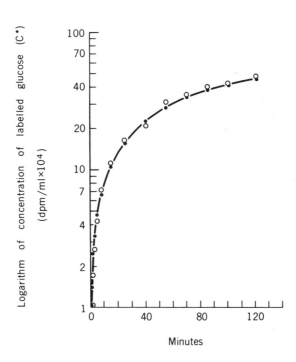

Fig. 13.3. The experimental step response (response to a steady tracer infusion) is shown by the open circles and the response predicted
from the impulse response function and convolution integral is
shown in dark circles. The two curves match nearly perfectly.
From Biophysik 7, 169-180, 1971 by K.H. Norwich and G. Hetenyi, Jr.
Reproduced by permission of Springer-Verlag.

In these experiments the impulse response function was determined 2 weeks
prior to the step response. It was necessary to assume that the dog was
"identical" in both cases — i.e. of the same body weight, state of hy-
dration etc. In order to eliminate worries about the time-invariance of the
dog's state, a slightly different approach was used in subsequent experiments.

(c) Testing the output predicted by the convolution integral for a second
input function. On this occasion a unit impulse of M_0^* dpm of ^{3}H-2-glucose

was administered and *concurrently* a time-varying tracer infusion of ^{14}C-1-glucose was begun. In this way the impulse response function was measured at the same time as the response to the test function, eliminating the peril of non-stationarity in the dog's state. This is an example of a double tracer experiment and the samples are analysed by a process of "double-label counting" in the liquid scintillation spectrometer. The ^{14}C-1-glucose curves were corrected for recycling of label in this, as in all the experiments reported (see section 7). For example, some ^{14}C from glucose will be transformed into blood lactate, and thence to liver glycogen, which appears later as newly-produced ^{14}C-glucose. This recycled tracer was discounted.

Once again the unit impulse response function was of the form (13.10), a double exponential. The input was selected as a function of the type

$$X(t) = E^*(1+\cos pt), \quad E^* \text{ constant}, \qquad (13.15)$$

a raised cosine function. (We clearly could not apply a function of the type *cos pt* since the cosine function may assume negative values and infusion rates may not[†].) The output concentration predicted by the convolution integral is given by

$$C^*(t) = \int_0^t \frac{E^*}{M_0^*} \left[Ae^{-b_1\tau} + Be^{-b_2\tau} \right] \left[\cos p(t - \tau) + 1 \right] d\tau \qquad (13.16)$$

When the required integration is carried out, we have

$$C^*(t) = \frac{E^*}{M_0^*} \left[\left(\frac{A}{b_1} + \frac{B}{b_2} \right) - \left(\frac{A}{b_1} e^{-b_1 t} + \frac{B}{b_2} e^{-b_2 t} \right) \right.$$

$$+ A \left(\frac{b_1 \cos pt + p \sin pt}{b_1^2 + p^2} - \frac{b_1 e^{-b_1 t}}{b_1^2 + p^2} \right)$$

$$\left. + B \left(\frac{b_2 \cos pt + p \sin pt}{b_2^2 + p^2} - \frac{b_2 e^{-b_2 t}}{b_2^2 + p^2} \right) \right] \qquad (13.17)$$

As may be seen in Fig. 13.4, the matching of the measured with theoretical output curves is very close between 0 and 90 minutes. During the last 30 minutes the theoretical curve exceeded the measured curve by increasing amounts and one can only speculate about the reasons for the disparity. It

---

[†]But, see section 37 for ways in which "negative" tracer may be infused.

may be due to the imperfect correction of the ^{14}C-1-glucose concentration
curve for the effects of recycling of tracer; it may be due to slight differences in the kinetics of the two tracers used. Or, of course, it may
indicate an unexpected nonlinearity in the tracer system.

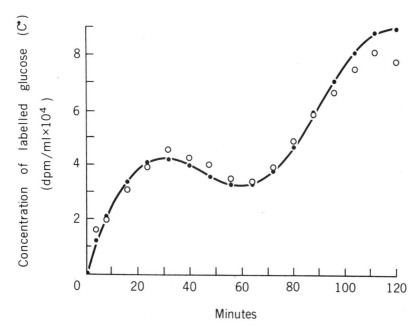

Fig. 13.4. A test of linearity

       o Measured concentrations of labelled glucose in plasma during
         a sinusoidal infusion of ^{3}H-2-glucose

       • Curve expected theoretically from the impulse response func-
         tion to simultaneously injected ^{14}C-1-glucose

From Biophysik 7, 169-180, 1971 by K.H. Norwich and G. Hetenyi, Jr.
Reproduced by permission of Springer-Verlag.

It is instructive to explore the theoretical input-output relationship as t
becomes large. (We shall avoid referring to these as the "steady state" input and output so that no ambiguity arises regarding the meaning of the term
"steady state".) Examining (13.17) for large t we have

$$\lim_{t\to\infty} C*(t) = \frac{E*}{M_0*} \left[\left(\frac{A}{b_1} + \frac{B}{b_2} \right) + \left(\frac{Ab_1}{b_1^2 + p^2} + \frac{Bb_2}{b_2^2 + p^2} \right) \cos pt \right.$$

$$\left. + \left(\frac{Ap}{b_1^2 + p^2} + \frac{Bp}{b_2^2 + p^2} \right) \sin pt \right] \qquad (13.18)$$

since the exponential terms vanish. Writing

$$F = \frac{Ab_1}{b_1^2 + p^2} + \frac{Bb_2}{b_2^2 + p^2} \qquad (13.19)$$

and

$$G = \frac{Ap}{b_1^2 + p^2} + \frac{Bp}{b_2^2 + p^2} \qquad (13.20)$$

(13.18) may assume the form

$$\lim_{t\to\infty} C*(t) = \frac{E*}{M_0*} \left[\left(\frac{A}{b_1} + \frac{B}{b_2} \right) + \left(F^2 + G^2 \right)^{\frac{1}{2}} \left(\frac{F}{(F^2 + G^2)^{\frac{1}{2}}} \cos pt + \frac{G}{(F^2+G^2)^{\frac{1}{2}}} \sin pt \right) \right]$$

$$(13.21)$$

Now

$$\cos (pt - \varepsilon) = \cos \varepsilon \cos pt + \sin \varepsilon \sin pt \qquad (13.22)$$

and comparing (13.21) with (13.22) we see that we may set

$$\frac{F}{(F^2 + G^2)^{\frac{1}{2}}} = \cos \varepsilon \qquad (13.23)$$

and

$$\frac{G}{(F^2 + G^2)^{\frac{1}{2}}} = \sin \varepsilon \qquad (13.24)$$

leaving

$$\cos^2 \varepsilon + \sin^2 \varepsilon = \frac{F^2}{F^2 + G^2} + \frac{G^2}{F^2 + G^2} = 1$$

as required. Combining Eqs (13.22), (13.23) and (13.24) we see that (13.21) may be written

$$\lim_{t\to\infty} C*(t) = \frac{E*}{M_0*} \left[\left(\frac{A}{b_1} + \frac{B}{b_2} \right) + (F^2 + G^2)^{\frac{1}{2}} \cos (pt - \varepsilon) \right] \qquad (13.25)$$

Finally defining

$$\gamma = \frac{(F^2 + G^2)^{\frac{1}{2}}}{\frac{A}{b_1} + \frac{B}{b_2}} \qquad (13.26)$$

(13.25) will assume the form

$$\lim_{t\to\infty} C*(t) = \frac{E*}{M_0*} \left(\frac{A}{b_1} + \frac{B}{b_2} \right) (1 + \gamma \cos \{pt - \varepsilon\}) \qquad (13.27)$$

Recalling that the input, X(t) has the form

$$X(t) = E*(1 + \cos pt) \qquad (13.15)$$

we see that apart from scaling factors, the input and output for large t differ with respect to the quantity ε: where the input has cos pt, the output has cos (pt - ε). The input leads the output by the *phase shift* ε as shown schematically in Fig. 13.5.

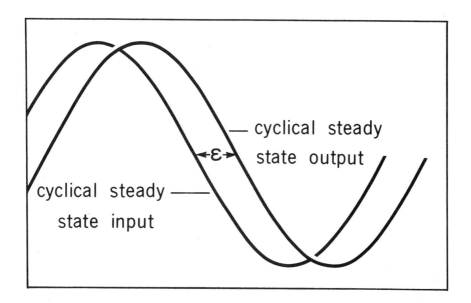

Fig. 13.5.   When the input to a tracer system is an infusion of tracer at a
             rate proportional to an elevated cosine wave, then the output
             tracer concentration for large values of time is a wave of the
             same form but delayed by a phase shift, ε (Schematic repre-
             sentation).

From Biophysik 7, 169-180, 1971, by K.H. Norwich and G. Hetenyi, Jr.
Reproduced by permission of Springer-Verlag.

We can see from this example that when a repetitive input is applied to a tracer system, the output to be expected is of "similar" form but delayed in time.

It is of interest to verify Eq. (12.36), the relationship between the transfer function of the system and the Laplace transforms of the input and output

$$LC*(t) = LX(t) \cdot Lh(t) \tag{13.28}$$

Let us take example (b) above, where the input, $X(t)$, is a constant tracer infusion, R_a^*, and the unit impulse response function has the form given in (13.12). The Laplace transform of any function $f(t)$, is given by

$$Lf(t) = \int_0^\infty e^{-st} f(t)dt \tag{13.29}$$

Thus,

$$LX(t) = LR_a^* = \int_0^\infty e^{-st} R_a^* \, dt$$

$$= \frac{R_a^*}{s} \tag{13.30}$$

and

$$Lh(t) = \int_0^\infty \frac{e^{-st}}{M_0^*} \left[Ae^{-b_1 t} + Be^{-b_2 t} \right] dt$$

$$= \frac{1}{M_0^*} \left(\frac{A}{s + b_1} + \frac{B}{s + b_2} \right) \tag{13.31}$$

Hence

$$LR_a^* \cdot Lh(t) = \frac{R_a^*}{M_0^*} \cdot \frac{1}{s} \left(\frac{A}{s + b_1} + \frac{B}{s + b_2} \right) \tag{13.32}$$

Also from (13.13)

$$LC*(t) = L \left\{ \frac{R_a^*}{M_0^*} \left[\left(\frac{A}{b_1} + \frac{B}{b_2} \right) - \left(\frac{A}{b_1} e^{-b_1 t} + \frac{B}{b_2} e^{-b_2 t} \right) \right] \right\}$$

$$= \frac{R_a^*}{M_0^*} \int_0^\infty e^{-st} \left(\frac{A}{b_1} + \frac{B}{b_2} \right) dt - \frac{R_a^*}{M_0^*} \int_0^\infty e^{-st} \left(\frac{A}{b_1} e^{-b_1 t} + \frac{B}{b_2} e^{-b_2 t} \right) dt$$

$$= \frac{R_a^*}{M_0^*} \cdot \frac{1}{s} \left(\frac{A}{b_1} + \frac{B}{b_2} \right) - \frac{R_a^*}{M_0^*} \left(\frac{A}{b_1} \cdot \frac{1}{s + b_1} + \frac{B}{b_2} \cdot \frac{1}{s + b_2} \right)$$

$$= \frac{R_a^*}{M_0^*} \cdot \frac{1}{s} \left(\frac{A}{s + b_1} + \frac{B}{s + b_2} \right) \tag{13.33}$$

which is identical with (13.32), thus verifying (13.28).

In this section we have examined the theoretical evidence to support the linearity of tracer kinetics under certain circumstances. We have then examined the experimental validation of the theory using glucose kinetics as an example. The reader will find the linearity of indicator kinetics discussed by Bassingthwaite and Ackerman (1967), and various facets of the linearity of metabolic tracers analyzed by Reiner (1953), Robertson (1957) and Bergner (1961 and 1966).

Speaking Practically

We begin with a system which is in steady state with respect to tracee. We then select one site for the infusion of tracer and one site for the sampling of tracer. We can then show that the system whose input is the rate of infusion of tracer at the infusion site and whose output is the concentration of tracer at the sampling site is a linear one. We determine the output of the system corresponding to an input of a unit "bolus" of tracer; this output is the unit impulse response of the system. We curve-fit to conform the data of the experimental unit impulse response to some convenient and reasonable (see Problem 24.4) mathematical function, such as a double exponential. We can now predict the output (concentration-time curve) of the system corresponding to any desired input. This can be a lot of fun. We can while away rainy afternoons predicting outputs, and then testing the predictions experimentally. But we shall see that it is quite practical too, because it permits us to deduce equations for calculating rates of appearance of metabolites based upon the use of an infinite variety of tracer infusions or injections, and to demonstrate the equivalence of these equations. It also permits us to deduce a means of achieving a steady state in tracer almost instantaneously, which, as we shall see, can be very useful.

Problem

13.1 Let $F(s)$ be the Laplace transform of $f(t)$. A theorem called the *final value theorem* states that

$$\lim_{t \to \infty} f(t) = \lim_{s \to 0} sF(s).$$

This theorem can be used to give the steady state response to a tracer infusion, constant or otherwise.

a) Show that Eq. (13.14) can be obtained by applying the final value theorem to Eq. (13.32).

b) Derive Eq. (13.27) by applying the final value theorem to Eqs. (13.12) and (13.15).

SECTION 14
AN ANALOGUE WITH CLASSICAL MECHANICS

Generally speaking, this book is restricted to those aspects of the theory of tracers which can be applied in the laboratory or clinic. However, in this section we shall make exception and study some of the implications of the equations of convective diffusion — without consideration of applicability. Eq. (11.21), the equation of (metabolic) convective diffusion for tracee has been derived in the form

$$\frac{\partial C}{\partial t} = D\nabla^2 C - \vec{u} \cdot \nabla C + R_{a,v} - R_{d,v} \tag{14.1}$$

which, as a consequence of (11.19) may be written

$$\frac{\partial C}{\partial t} = D\nabla \cdot \nabla C - \nabla \cdot (C\vec{u}) + R_{a,v} - R_{d,v} \tag{14.2}$$

Suppose that the bounding surfaces of the distribution space are rigid and impermeable to carrier fluid and to tracee. Integrating over the entire distribution volume,

$$\int_V \frac{\partial C}{\partial t} dV = D\int_V \nabla \cdot \nabla C \, dV - \int_V \nabla \cdot (C\vec{u})dV + \int_V (R_{a,v} - R_{d,v})dV \tag{14.3}$$

Since the boundaries are rigid,

$$\int_V \frac{\partial C}{\partial t} dV = \frac{d}{dt}\int_V C dV = \frac{dM(t)}{dt} \tag{14.4}$$

where M(t) is the total mass of tracee in the system at any time, t. We also know from the definitions of $R_{d,v}$, R_d etc. ((5.1) and (5.3))

$$\int_V (R_{a,v} - R_{d,v})dV = R_a - R_d \tag{14.5}$$

Applying the divergence theorem (section 9) and introducing (14.4) and (14.5) Eq. (14.3) becomes

$$\frac{dM(t)}{dt} = \int_S D\nabla C \cdot \vec{n}dS - \int_S C\vec{u} \cdot \vec{n}dS + R_a(t) - R_d(t) \qquad (14.6)$$

where $\vec{n}$ is, as usual, the unit outward normal at the surface. The condition for impermeability of the surface to carrier fluid is expressed by the boundary condition

$$\vec{u} \cdot \vec{n} = 0 \qquad (14.7)$$

which simply states that fluid cannot flow outward perpendicularly to the surface *at* the surface. From Fick's law we have

$$\vec{J} = -D\nabla C \qquad (6.11)$$

so that

$$-D\nabla C \cdot \vec{n}$$

is the component of diffusive flux of tracee in an outward direction, perpendicular to the surface *at* the surface. Since the surface is impermeable to tracee we may write

$$-D\nabla C \cdot \vec{n} = 0 \qquad (14.8)$$

In a later section we shall construct more realistic boundary conditions for a metabolite. Introducing, now, (14.7) and (14.8) into (14.6) we emerge with the rather simple equation,

$$\frac{dM(t)}{dt} = R_a(t) - R_d(t) \qquad (14.9)$$

That is, the rate of change of total mass within the system is equal to the difference between the total production rate and total destruction rate — obviously true for the system we chose, which possessed impermeable boundaries. We can immediately set down an equation similar to (14.9) for tracer.

Let us now set down the equations of convection-diffusion-reaction for tracee and tracer (11.21) and (11.23):

$$\frac{\partial C}{\partial t} = D\nabla^2 C - \vec{u} \cdot \nabla C + R_{a,v} - R_{d,v} \qquad (14.10)$$

$$\frac{\partial C^\star}{\partial t} = D\nabla^2 C^\star - \vec{u} \cdot \nabla C^\star + R^\star_{a,v} - aR_{d,v} \qquad (14.11)$$

$R_{d,v}$ may be eliminated between Eqs. (14.10) and (14.11). Muliplying (14.10) by a and subtracting,

$$a\frac{\partial C}{\partial t} - \frac{\partial C^\star}{\partial t} = aD\nabla^2 C - D\nabla^2 C^\star - a\vec{u} \cdot \nabla C + \vec{u} \cdot \nabla C^\star + aR_{a,v} - R^\star_{a,v} \qquad (14.12)$$

This expression can now be simplified. Since

$$\frac{\partial}{\partial t}(aC) = a\frac{\partial C}{\partial t} + C\frac{\partial a}{\partial t} = \frac{\partial C^*}{\partial t}$$

therefore

$$a\frac{\partial C}{\partial t} - \frac{\partial C^*}{\partial t} = -C\frac{\partial a}{\partial t} \qquad (14.13)$$

If U and V are two scalar points functions, then

$$\text{grad}(UV) \equiv \nabla(UV) = U\nabla V + V\nabla U, \qquad (14.14)$$

an identity which is easily proved (C.13, Appendix C). Since

$$C^* = aC$$

therefore

$$\vec{u} \cdot \nabla C^* = \vec{u} \cdot \nabla(aC)$$

$$= a\vec{u} \cdot \nabla C + C\vec{u} \cdot \nabla a \qquad (14.15)$$

using (14.14). Finally,

$$\nabla^2 C^* = \nabla \cdot \nabla C^* = \nabla \cdot \nabla(aC) \qquad (14.16)$$

and

$$\nabla \cdot \nabla(aC) = \nabla \cdot (a\nabla C + C\nabla a)$$

$$= \nabla a \cdot \nabla C + a\nabla \cdot \nabla C + \nabla C \cdot \nabla a + C\nabla \cdot \nabla a$$

$$= 2\nabla a \cdot \nabla C + a\nabla^2 C + C\nabla^2 a \qquad (14.17)$$

using vector identity (10.26). Substituting (14.13), (14.15), (14.16) and
(14.17) into (14.12) yields

$$-C\frac{\partial a}{\partial t} = aD\nabla^2 C - 2D\nabla a \cdot \nabla C - Da\nabla^2 C - DC\nabla^2 a - a\vec{u} \cdot \nabla C + a\vec{u} \cdot \nabla C + C\vec{u} \cdot \nabla a$$

$$+ aR_{a,v} - R^*_{a,v}$$

or

$$\frac{\partial a}{\partial t} = \frac{1}{C}R^*_{a,v} - \frac{a}{C}R_{a,v} - \vec{u} \cdot \nabla a + D\left[\nabla^2 a + 2\frac{\nabla a \cdot \nabla C}{C}\right]$$

$$C \neq 0 \qquad (14.18)$$

But

$$\frac{\nabla \cdot (C^2\nabla a)}{C^2} = \frac{C^2\nabla^2 a + 2C\nabla a \cdot \nabla C}{C^2}$$

$$= \nabla^2 a + 2\frac{\nabla a \cdot \nabla C}{C} \qquad (14.19)$$

by (10.26), and substituting (14.19) into (14.18) gives the simpler form

$$\frac{\partial a}{\partial t} = \frac{1}{C}R^*_{a,v} - \frac{a}{C}R_{a,v} - \vec{u} \cdot \nabla a + \frac{D}{C^2}\nabla \cdot (C^2\nabla a) \qquad (14.20)$$

a partial differential equation involving derivatives of specific activity, and of tracee concentration. If we multiply both sides of (14.20) by C^2, we obtain the more symmetrical form

$$C^2 \frac{\partial a}{\partial t} = CR^*_{a,v} - C^*R_{a,v} - \vec{u} \cdot (C^2\nabla a) + D\nabla \cdot (C^2\nabla a) \quad (14.21)$$

Comparing (14.21) to the general equation of convection-diffusion-reaction (14.10) and recalling the derivation of the latter from Fick's law and the equation of continuity, we see that $C^2\nabla a$ plays the part of a *force*.

i.e. define

$$\vec{G} = C^2\nabla a \quad (14.22)$$

and

$$\vec{J} = -D\vec{G} \quad (14.23)$$

The terms $CR^*_{a,v} - C^*R_{a,v}$ play the part of a net source strength:

$$\sigma = CR^*_{a,v} - C^*R_{a,v} \quad (14.24)$$

We can then define a concentration analogue, C, such that

$$\frac{\partial C}{\partial t} = C^2 \frac{\partial a}{\partial t} \quad (14.25)$$

Assembling (14.21), (14.22), (14.23), (14.24) and (14.25)

$$\frac{\partial C}{\partial t} = \sigma - \vec{u} \cdot \vec{G} - \nabla \cdot \vec{J} \quad (14.26)$$

In order to complete the analogy of (14.21) with (14.10) we must assume the liberty to define

$$\vec{G} = C^2\nabla a \equiv \nabla C \equiv \text{grad } C \quad (14.27)$$

so that C is jointly defined by (14.25) and (14.27). C then plays the part of a *potential* associated with the force $\vec{G}$. However, the condition to be satisfied when any vector is equated to the gradient of some scalar is that the curl of the vector must vanish identically. That is, if

$$\vec{G} = \nabla C$$

then

$$\text{curl } \vec{G} \equiv \nabla \times \vec{G} = 0 \quad (14.28)$$

or

$$\text{curl } (C^2\nabla a) = 0 \quad (14.29)$$

There is a vector identity,

$$\text{curl } (w\vec{v}) = w \text{ curl } \vec{v} - \vec{v} \times \text{grad } w, \quad (14.30)$$

where w is any scalar point function and $\vec{v}$ any vector point function. Therefore

$$\text{curl } \vec{G} = C^2 \text{ curl grad } a - \text{grad } a \times \text{grad } C^2$$

$$= -2\nabla a \times \nabla C \qquad (14.31)$$

since

$$\text{curl grad } a = 0$$

Eqs. (14.29) and (14.31) then give for the condition that $\vec{G}$ may be equated to a gradient

$$\nabla a \times \nabla C = 0 \qquad (14.32)$$

or

$$(C\nabla C^* - C^*\nabla C) \times \nabla C = 0$$

Since the cross product of any vector with itself is equal to zero we have

$$\nabla C^* \times \nabla C = 0 \qquad (14.33)$$

or

$$|\nabla C^*| \, |\nabla C| \sin \theta = 0 \qquad (14.34)$$

where θ is the angle between the two gradient vectors ∇C^* and ∇C. Since the magnitudes of the two gradients are arbitrary and hence not necessarily equal to zero, we obtain from (14.34)

$$\sin \theta = 0 \; ; \qquad (14.35)$$

that is the vectors ∇C^* and ∇C are parallel. Hence we may write

$$\nabla C^* = \lambda \nabla C \qquad (14.36)$$

where λ is a scalar. Eq. (14.36) is, then, the condition which must be satisfied if we set

$$\vec{G} = C^2 \nabla a = \nabla C \qquad (14.27)$$

Since

$$C^2 \nabla a = C\nabla C^* - C^*\nabla C \qquad (14.37)$$

then introducing (14.36)

$$C^2 \nabla a = \nabla C(\lambda C - C^*)$$

or

$$C\nabla a = \nabla C(\lambda - a) \qquad (14.38)$$

Equations (14.36) and (14.38), then, follow directly from (14.28).

Let us now backtrack a little and ask the question: what is the condition under which C can be found to satisfy both (14.25) and (14.27)? If we regard t as the coordinate x_4 and x, y, z as x_1, x_2 and x_3 respectively, then (14.25) and (14.27) in unison ask the question: is there a C such that

$$\nabla_4 C \equiv \frac{\partial C}{\partial x_1} \vec{i} + \frac{\partial C}{\partial x_2} \vec{j} + \frac{\partial C}{\partial x_3} \vec{k} + \frac{\partial C}{\partial x_4} \vec{\tau} = C^2 \nabla_4 a \ ? \qquad (14.39)$$

In other words, can the 4-dimensional vector $C^2\nabla_4 a$ be equated to the 4-dimensional gradient of some quantity? The answer seems to be that it can if the 4-dimensional curl of the vector $C^2\nabla_4 a$ vanishes (see, for example, Weatherburn, 1938); that is if

$$\frac{\partial (C^2\nabla_4 a)_j}{\partial x_i} - \frac{\partial (C^2\nabla_4 a)_i}{\partial x_j} = 0 \qquad (14.40)$$

for all i and j. Eq. (14.40) embodies the 3-dimensional constraint expressed by (14.29), (14.38) etc. Provided that (14.40) is met, we may introduce (14.27) into (14.23) and then the latter into (14.26):

$$\vec{J} = -D\nabla C \qquad (14.41)$$

$$\frac{\partial C}{\partial t} = \sigma - \vec{u} \cdot \nabla C + D\nabla^2 C \qquad (14.42)$$

Eq. (14.42) is now in the form of the equation of convective diffusion (convection-diffusion-reaction).

The condition under which $C^2\nabla a$ may be expressed as the gradient of some scalar C, namely

$$\text{curl } (C^2\nabla a) = 0 \qquad (14.29)$$

suggests a continuation of the analogy with the mechanics of a particle moving in a potential field. If we consider any simple, closed curve within our defined region and represent a line or contour integral around this closed curve by $\oint$, then we shall have under the condition (14.29)

$$\oint C^2\nabla a \cdot \vec{dl} = 0 \qquad (14.43)$$

Eq. (14.43) follows from Stokes theorem,

$$\oint C^2\nabla a \cdot \vec{dl} = \int_S \text{curl } (C^2\nabla a) dS \qquad (14.44)$$

since the integrand on the right-hand side is always equal to zero. Just as a material particle moving along a closed curve within a conservative field under influence of a force $\vec{F}$ loses no energy, — i.e. obeys the law

$$\oint \vec{F} \cdot \vec{dl} = 0,$$

so too an imagined particle moving along a closed curve within a field analogous to a conservative field under influence of a force $C^2\nabla a$ obeys the law expressed by (14.43).

The reader should not feel besieged by all this mathematical persiflage. While it has all been in good fun, the hard-core development of the mathematical theory has been arrested at the level of Eq. (14.21). Let us now return to this equation and explore some of its properties.

(i) Case where convection is negligible and diffusive transport predominates. When $R_{a,v} = 0$ and $R_{a,v}^* = 0$ (14.21) becomes

$$\frac{\partial a}{\partial t} = \frac{D}{C^2} \nabla \cdot (C^2 \nabla a) \tag{14.45}$$

which is a type of interfusion equation. When C is constant, (14.45) immediately becomes identical with C.W. Sheppard's Eq. (4) (Chapter 8, 1962)

$$\frac{\partial a}{\partial t} = \left[\frac{1}{[C]}\right] \text{div} \left[\mathcal{I}\nabla a\right]$$

where

$$\mathcal{I} = DC \tag{11.7}$$

Equivalently (for constant C)

$$\frac{\partial C^*}{\partial t} = D\nabla^2 C^* \tag{14.46}$$

the equation of diffusion for tracer (cf (11.11)).

(ii) Case where diffusion is negligible and convective transport predominates. When $R_{a,v} = 0$ and $R_{a,v}^* = 0$ (14.21) becomes

$$\frac{\partial a}{\partial t} + \vec{u} \cdot \nabla a = 0 \tag{14.47}$$

or

$$\frac{da}{dt} = 0 \tag{14.48}$$

where $\frac{d}{dt}$ is the substantial derivative (cf (10.12)). Thus an observer travelling with a particle of carrier fluid would perceive no change in specific activity.

(iii) Case where

$$\nabla_4 a = 0; \tag{14.49}$$

that is, where both the space and the time derivatives of specific activity are equal to zero. Eq. (14.21) then becomes simply

$$CR_{a,v}^* = C^* R_{a,v} \tag{14.50}$$

Let us apply yet another constraint and suppose that both tracee and tracer have but a single source point, tracee at $\vec{r}_1$ and tracer at $\vec{r}_2$. Thus

$$R_{a,v} = Ra \ \delta(\vec{r} - \vec{r}_1) \tag{14.51}$$

and

$$R_{a,v}^* = R_a^* \, \delta(\vec{r} - \vec{r}_2) \tag{14.52}$$

[Integrating (14.51) and (14.52) over the entire volume shows that each of these equations is identically true:

$$\int_V R_{a,v} \, dV = R_a \tag{5.3}$$

by definition, and

$$\int_V R_a \, \delta(\vec{r} - \vec{r}_1) \, dV = R_a$$

from the definition of the delta function (12.20).]

Substituting (14.51) and (14.52) into (14.50) and integrating both sides over the entire volume,

$$C(\vec{r}_2) R_a^* = C^*(\vec{r}_1) R_a \tag{14.53}$$

or

$$R_a = \frac{C(\vec{r}_2)}{C^*(\vec{r}_1)} \cdot R_a^* \tag{14.53}$$

We shall see in a later section by what experimental techniques we try to achieve the condition (14.49) which leads us to (14.53). Eq. (14.53) permits us to calculate the rate of appearance of tracee by measuring the quantities in the right-hand side: the rate of infusion of tracer and the concentrations of tracee and tracer at two specified points. When the two source points are, effectively, located in the blood stream, S and S* are quite rapidly intermixed within the plasma and (14.53) becomes

$$R_a = \frac{R_a^*}{a_p} \tag{14.54}$$

where a_p is the plasma specific activity.

Summarizing this section then — we have combined the equations of convection-diffusion-reaction for tracee and tracer eliminating the term governing rate of disappearance. We have shown how the resulting equation, under conditions of a "conservative field" $C^2 \nabla a$, could itself be cast into the form of the equation of convection-diffusion-reaction, with the dependent variables no longer simple quantities. Finally we have constrained the combined equation in a number of ways thereby demonstrating its similarity to the interfusion equation as well as its applicability in the analysis of experiments.

Speaking Practically

You can skip the whole section.

However, it would be desirable to follow the derivation of Eq. (14.21) since we shall use it later in the development of Eq. (30.3), an equation of some practical importance.

Problems

14.1 Demonstrate the identity (14.30).

14.2 Show that the vanishing of the generalized curl of the vector, $\vec{\xi}$, (14.40) guarantees that $\vec{\xi}$ is a generalized gradient.

References

Bassingthwaite, J.B. and Ackerman, F.H., Mathematical linearity of circulatory transport, *J. Appl. Physiol.* 22, 879-888 (1967).

Bergner, P.-E.E., Tracer dynamics: a tentative approach and definition of fundamental concepts, *J. Theor. Biol.* 2, 120-140 (1961).

Bergner, P.-E.E., Tracer theory: a review, *Isotopes and Radiation Technology,* 245-262 (1966).

Brown, B.M., *The Mathematical Theory of Linear Systems,* 2nd edition, Chapman and Hall, London, 1965.

Cheng, D.K., *Analysis of Linear Systems,* Addison-Wesley, Reading, 1959.

Churchill, R.V., *Operational Mathematics,* 2nd edition, McGraw-Hill, 1958.

Hearon, J.Z., The differential equations for specific activities of a distributed tracer system, *Bull. Math. Biophys.* 30, 325-331 (1968).

Lamb, H., *Hydrodynamics,* 6th edition, Dover Publications, New York, 1945.

Menzel, D.H., *Mathematical Physics,* Dover Publications, New York, 1961.

Norwich, K.H., Convective diffusion of tracers, *J. Theor. Biol.* 32, 47-57 (1971).

Norwich, K.H. and Hetenyi, Jr., G., Basic studies on metabolic steady state. Linearity of a metabolic tracer system, *Biophysik.* 7, 169-180 (1971).

Rashevsky, N., *Mathematical Biophysics: Physico-mathematical Foundations of Biology,* Dover Publications, New York, 1960.

Reiner, J.M., Study of metabolic turnover rates of means of isotopic tracers. I, II, *Arch. Biochem.* 46, 53-99 (1953).

Robertson, J.W., Theory and use of tracers in determining transfer rates in biological systems, *Physiol. Rev.* 37, 133-154 (1957).

Schönheimer, R., *The Dynamic State of Body Constituents,* Reprinted by Haffner Publishing Co., New York-London, 1964.

Sheppard, C.W. and Householder, A.S., The mathematical basis of the interpretation of tracer experiments in closed steady-state systems, *J. Appl. Physics* 22, 510-521 (1951).

Sheppard, C.W., *Basic Principles of the Tracer Method,* Wiley, New York, 1962.

For other theorems related to the divergence theorem and for general reference in vector calculus, I recommend an older, but very lucid text by
C.E. Weatherburn. For a concise, one-page proof of the case with no interior
surfaces together with some simple samples, I suggest the text of G.B. Thomas.

Thomas, G.B., *Calculus and Analytic Geometry*, 4th edition, p 606, Addison
    Wesley, Reading, 1968.

Weatherburn, C.E., *Introduction to Riemannian Geometry and the Tensor
    Calculus*, Cambridge, 1938.

Weatherburn, C.E., *Advanced Vector Analysis*, Open Court, La Salle, Illinois,
    1948.

Wrenshall, G.A., Working basis for the tracer measurement of transfer rates
    of a metabolic factor in biological systems containing compartments whose
    contents do not intermix rapidly, *Can. J. Bioch. Physiol.* 33, 909-925
    (1955).

CHAPTER IV
THE THEORY OF MULTI-COMPARTMENT SYSTEMS

SECTION 15
COMPARTMENTS: AN INTRODUCTION

We have hitherto approached biokinetics purely from a base of physics and chemistry, but we must now interrupt the physicochemical development in order to discuss the theory of compartments and multicompartment systems. Compartments, as they are usually defined, are mathematical and not physical constructions. It is mandatory that any text of this era contain a description of the compartmental approach to biokinetics because the current literature abounds with examples of compartmental analysis. However, it is my opinion that the time has arrived to begin a phasing out of compartments in favour of simpler approaches more solidly grounded in physics and chemistry. At this time, I know of no results obtainable by compartmental analysis of biological systems which have been verified by experimental measurement and which cannot be obtained without the use of compartments; neither do I know of any results obtainable by a process of compartmental analysis and by no other means which have been verified by experimental measurement. These items will be developed further as we proceed.

From a historic and heuristic point of view there can be no doubt whatever about the importance of compartments. In 1866 a paper entitled *On the Laws of Connexion Between the Conditions of a Chemical Change and its Amount* by A.V. Harcourt and W. Esson appeared in the Philosophical Transactions of the Royal Society of London. Their paper was one of the first on chemical kinetics. The authors stated quite clearly that for the chemical reaction with which they dealt "The reagents and products of the reaction are all soluble, and thus the system can quickly be made and will remain *homogeneous*." The homogeneity of the solution was their licence to postulate a set of ordinary differential equations to describe the kinetics of the reaction.

"If ... the substances are not independent but are such that one of them is gradually formed from the other, we have a ... system of equations to represent the reaction. Let u, v be the residues of the substances after an interval x ... let αu be the rate of diminution of u due to its reaction with one of the other elements of the system, and βu its rate of diminution due to its reaction with another of the elements of the system, by which v is formed, and let γv be the rate of diminution of v, then

$$\frac{du}{dx} = -(\alpha + \beta)u$$

$$\frac{dv}{dx} = \beta u - \gamma v \qquad "$$

A well-known paper of A.C. Burton (1936) incorporated the concept of such irreversible chain reactions into biology. Burton analysed the kinetics of the chain reaction

$$A \xrightarrow{k_1} B \xrightarrow{k_2} C \xrightarrow{k_3} D \xrightarrow{k_4} \dots \xrightarrow{k_{m-1}} M$$

in the manner of Harcourt and Esson.

The next step in the development of the theory of compartments was taken by T. Teorell (1937). Teorell was concerned with the transport and distribution of a drug *in vivo*, - e.g. the transport of the drug from a subcutaneous depot to the blood, or from blood to tissue etc. He based his equations upon Fick's Law, which for one dimension is

$$-j_x = D\frac{\partial C}{\partial x} \qquad (15.1)$$

where the flux j_x has the usual dimensions [mass $\cdot$ area^{-1} $\cdot$ time^{-1}]. If we suppose that the flux is constant over some surface area, A, then the unidirectional rate of flow of material, ρ, will be given by the equation

$$-\rho = -j_x A = AD\frac{\partial C}{\partial x} \qquad (15.2)$$

Suppose moreover, that the two depots of drug are of uniform concentration and separated by a thin partition of thickness Δx; let the concentration in one depot be $C_1(t)$ and in the other $C_2(t)$, where $C_1 > C_2$. Then the derivative $\frac{dC}{dx}$ in the partition can be approximated quite closely using

$$\frac{dC}{dx} \doteq \frac{C_2(t) - C_1(t)}{\Delta x} \qquad (15.3)$$

If the concentration of drug inside the partition itself does not change with time and if drug diffuses through the partition, then (15.3) will hold exactly using the following arguments. From the equation of diffusion in one dimension

$$\frac{\partial C}{\partial t} = D\frac{\partial^2 C}{\partial x^2} = 0 \tag{15.4}$$

or

$$C = \lambda x + \mu, \qquad \lambda, \mu \text{ constant}$$

(see Fig. 15.1). Since

$$C(0) = C_1 \tag{15.5}$$

and

$$C(\Delta x) = C_2$$

we can solve for λ and μ obtaining

$$C = \left[\frac{C(\Delta x) - C(0)}{\Delta x} \cdot x + C(0)\right] \tag{15.6}$$

so that

$$\frac{\partial C}{\partial x} = \frac{C_2 - C_1}{\Delta x} \tag{15.7}$$

which is the form suggest by (15.3). Teorell therefore expressed the transport of drug between two depots by the equation

$$\rho = \frac{AD}{\Delta x}\left[C_1(t) - C_2(t)\right] \quad^\dagger \tag{15.8}$$

using (15.2) and (15.3). Since $\frac{AD}{\Delta x}$ is a constant, we may write it simply as κ:

$$\rho = \kappa\left[C_1(t) - C_2(t)\right] \tag{15.9}$$

Recalling that the drug depots are presumed to be homogeneous in concentration (no concentration gradients) we express the concentration as the ratio of mass of drug present in the depot (M_1 and M_2) to the volumes of the depot (V_1 and V_2 respectively). The rate of flow of material from C_1 to C_2 is equal to the rate of decrease of mass in the first depot if no exogenous sources of drug are present. Therefore (15.9) becomes

$$\frac{-dM_1}{dt} = \kappa\left(\frac{M_1(t)}{V_1} - \frac{M_2(t)}{V_2}\right) \tag{15.10}$$

which is of the form used by Teorell. He went even one step farther and

---

†We invoke here what might be called a *quasi-static* approximation. Eq. (15.7) has been derived for the steady state case; the C's are not functions of time. Eq. (15.3), however, assumes that (15.7) is valid even for the case where C changes with time.

assumed that the inactivation of drug by tissue would "follow the course of
a 'monomolecular reaction'". Hence (by analogy with (7.4)) the inactivation
of a substance whose concentration is $C_3(t)$ might be described by the
equation

$$\frac{-dM_3}{dt} = \frac{\kappa_3}{V_3} M_3 \qquad\qquad (15.10)$$

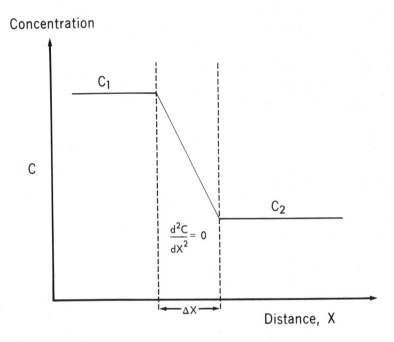

Fig. 15.1. A partition of thickness Δx separates two solutions of constant
concentrations C_1 and C_2. It is shown in the text that dC/dx, the
derivative within the partition, is given by $\frac{C_1 - C_2}{\Delta x}$. (15.7)

The equations of Teorell for the irreversible kinetics of drugs *in vivo* were
exactly of the form of the equations postulated by Burton for the irreversible kinetics of biochemical reactions. The primary difference lay in the
postulation by Teorell of depots of uniform drug concentration. The
equations set down by Teorell were always in general agreement with Fick's
law; the net flow of material was in the direction implied by a concentration
gradient.

In the 1940's, as radioisotopes became more readily available, there arose
naturally an interest in tracer kinetics. In a paper by C.W. Sheppard (1948)
the fundamentals of what we now term *multicompartmental theory* were clearly
set forth. Compartments were taken to be regions of space in which specific
activity is uniform — similar to the drug depots of Teorell where drug
concentration is uniform. A system could consist of any number of compartments, say n of them, and the substances S and S' could, in general, exchange
between any two of them. A general closed 4-compartment system is sketched
in Fig. 15.2.a. In some cases the anatomy of the system precluded exchange
of material between certain compartments. In cases where peripheral compartments exchanged material only with a central compartment (Fig. 15.2.b),
the system was termed a *mammillary system*; when the compartments were arrayed
in the form of a chain, the system was termed *catenary* (Fig. 15.2.c).
Generally, the systems studied were those in which the masses of unlabelled
material in each compartment remained constant. Unlabelled material passed
from compartment i to compartment j at the constant rate ρ_{ji} (mass $\cdot$ time^{-1}).
Tracer was assumed to be transported from compartment i to compartment j at
the rate

$$\dot{\rho}^{*}{}_{ji} = a_i \rho_{ji} \tag{15.11}$$

where a_i is the specific activity in   compartment i. We know that tracer
flux is *not*, in general, related to tracee flux by an expression as simple
as (15.11). If we confine transport to mean convective transport, then we
have

$$\vec{J}_{conv} = C\vec{u} \tag{6.9}$$

and

$$\vec{J}^{*}_{conv} = C^{*}\vec{u} \tag{6.10}$$

so that

$$\vec{J}^{*}_{conv} = a \, \vec{J}_{conv} \tag{15.12}$$

If we take flux to be constant over the partition between compartments, then
by multiplying both sides of (15.12) by the area of the partition, (15.12)
becomes identical with (15.11). However, if transport by diffusion is
allowed, then we deal with the now familiar equations

$$\vec{J}_{diff} = -D\nabla C \tag{6.11}$$

and

$$\vec{J}^{*}_{diff} = -D\nabla C^{*}$$

from which it is seen that in general

$$\vec{J}^{*}_{diff} \neq a\ \vec{J}_{diff} \tag{15.13}$$

and so (15.11) is not, in general, valid. Nevertheless, Eq. (15.11) provided a useful approximation which has permitted preliminary mathematical
analysis of the kinetics of many metabolic systems.

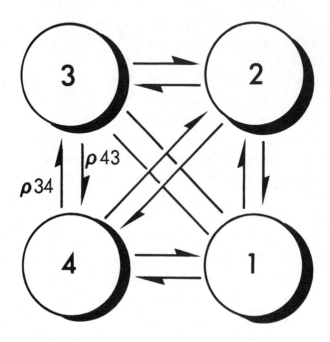

Fig. 15.2.a.   A general, closed 4-compartment system

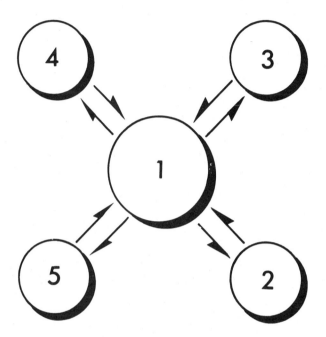

Fig. 15.2.b. A closed mammillary system.

Fig. 15.2.c. An open catenary system. The first compartment receives
material from the outside; the fourth loses material to the
outside.

Even the ostensibly benign procedure of selecting arbitrary ρ_{ji}, the steady, intercompartmental tracee fluxes, is not above indictment. The fluxes ρ_{ji} are not constrained in any way by the concentrations of the compartments i and j, and therefore no assurance is given that Fick's law will even qualitatively be obeyed. The effect of the net tracee flux, $\rho_{ji} - \rho_{ij}$, may be the movement of mass from regions of higher to regions of lower concentration or vice versa. The disadvantage of this lack of constraint upon the ρ's is that the multicompartmental model does not necessarily operate in accordance with known physical laws; the advantage is that phenomena such as active transport, transport against a concentration gradient, are admissible.

In spite of the approximations involved and in spite of the apparent neglect of some of the laws of physics, multicompartmental analysis can, under certain conditions, provide an "exact" result, - or rather one which involves few assumptions and violates no known physical laws. Though this statement may seem self-contradictory, we shall see presently why it is true.

We can now set down the differential equations governing the exchange of tracer between the compartments of an n-compartment system. If the system is *closed* (no input to the system from the outside nor loss from the system to the outside), then using (15.11)

$$\frac{dM_i^*}{dt} = \sum_{\substack{j=1 \\ j \neq i}}^{n} \rho_{ij}\, a_j \; - \; a_i \sum_{\substack{j=1 \\ j \neq i}}^{n} \rho_{ji} \qquad (15.14)$$

where $M_i^*(t)$ is the mass of tracer present in the ith compartment. We also have the condition of tracee equilibrium

$$\sum_{\substack{j=1 \\ j \neq i}}^{n} \rho_{ij} = \sum_{\substack{j=1 \\ j \neq i}}^{n} \rho_{ji} \qquad (15.15)$$

In Eq. (15.14) the terms under the summation sign on the left represent the flow of tracer *into* the ith compartment from the other n-1 compartments, and the terms under the summation sign on the right represent the flow of tracer *from* the ith compartment to the other n-1 compartments. The symbol ρ_{ii} has not been assigned a meaning and so has been excluded from the summation. Eq. (15.14) represents n equations (i = 1, n), one for each compartment in the system. In an *open* system each compartment may receive tracer from the outside, and lose tracer to the outside. We may represent the tracer input to the ith compartment by $\rho^*_{io}(t)$ and the tracer output from the ith compartment by $a_i \rho_{oi}$. Hence the n equations for a general n compartment open system are

$$\frac{dM_i^*}{dt} = \sum_{\substack{j=1 \\ j \neq i}}^{n} \rho_{ij} a_j - a_i \sum_{\substack{j=0 \\ j \neq i}}^{n} \rho_{ji} + \rho^*_{io}(t) \qquad (15.16)$$

and the tracee steady state constraint

$$\sum_{\substack{j=0 \\ j \neq i}}^{n} \rho_{ij} = \sum_{\substack{j=0 \\ j \neq i}}^{n} \rho_{ji} \qquad (15.17)$$

In order to solve the set of differential equations (15.14 or 15.15) for constant ρ_{ij}, we must further assume that the uniform specific activity which we have ascribed to each compartment is equal to the ratio of a uniform tracer concentration to a uniform constant tracee concentration. That is

$$a_i = \frac{C_i^*}{C_i} \qquad (15.18)$$

or equivalently

$$a_i = \frac{M_i^*}{M_i} = \frac{C_i^* V_i}{C_i V_i} \qquad (15.19)$$

where V_i is the volume of the ith compartment, M_i^* the mass of tracer and M_i the mass of tracee. Thus

$$\rho_{ij} a_j = \frac{\rho_{ij}}{M_j} M_j^* \qquad (15.20)$$

and we may define

$$k_{ij} = \frac{\rho_{ij}}{M_j} \qquad (15.21)$$

where k_{ij} is constant and has the dimensions (time^{-1}). Replacing the ρ's in (15.14) and (15.16) by k's using (15.21)

$$\frac{dM_i^*}{dt} = \sum_{\substack{j=1 \\ j \neq i}}^{n} k_{ij} M_j^* - \sum_{\substack{j=1 \\ j \neq i}}^{n} k_{ji} M_i^* \qquad (15.22)$$

for a closed system and

$$\frac{dM_i^*}{dt} = \sum_{\substack{j=1\\j\neq i}}^{n} k_{ij}M_j^* - \sum_{\substack{j=0\\j\neq i}}^{n} k_{ji}M_i^* + R_{ai}^*(t) \qquad (15.23)$$

for an open system, using the more familiar R_a^* notation. In the next section we shall solve (15.23) for the case of constant R_{ai}^*.

In summary then, we have followed the development of the concept of the compartment from its inception to the present. In the early formulations of Teorell, drug depots maintained uniform concentrations of a drug and the drug exchanged between depots by a mechanism in qualitative agreement with Fick's law. In recent formulations, adherence to the Fick's law has been more or less dropped; tracee will exchange between compartments at rates unrestricted by physical theories of transport. The unidirectional transport rate of tracer is taken to be equal to the corresponding unidirectional transport rate of tracee multiplied by the specific activity of the compartment of origin, regardless of the mode of transport (convection or diffusion). The *rate constants*, k_{ij}, have the dimensions of time^{-1} and sometimes may be interpreted in a purely kinematical sense as the fraction of the tracer in compartment j departing for compartment i per unit time.

In later sections we shall extend the present treatment of compartmental theory in two ways. We shall examine the effects of applying the multi-compartmental differential equations to each of the variables, mass, concentration and specific activity. And we shall extend the theory of compartments to include the nonsteady state of tracee; that is, we shall write down a set of differential equations for tracee analogous to those for tracer (15.16).

Speaking Practically

In this section we have replaced the partial differential equations, developed in the earlier part of the book, by a set of ordinary differential equations which are, ostensibly, easier to work with. This transition has been achieved by modelling the biological system as a collection of well-mixed pools in each of which the quantities c, $c*$ and a exhibit no spatial gradients. In operating with these "compartments" one does not worry too much about violating the laws of diffusion.

<u>Problems</u>

15.1. Generally, Eq. (15.13) is valid. What relationship between gradient vectors must subsist if

$$\vec{J}^*_{diff} = a\vec{J}_{diff} \qquad ?$$

15.2. Suppose we have a closed, two-compartment system in which

$$\rho_{12} = \rho_{21} = .2 \qquad V_1 = V_2 = 1.$$

$$M_1 = 2. \qquad M_2 = 1.$$

If 3.0 μCi of tracer are added to compartment 2 at t = 0, find the equilibrium distribution, $M_1^*(\infty)$ and $M_2^*(\infty)$. Observe that tracer has "climbed" against a concentration gradient.

SECTION 16

COMPARTMENTS: SOLUTION OF THE DIFFERENTIAL EQUATIONS

The set of n ordinary differential equations governing exchange of tracer between the n compartments of a general multicompartment system have been set up in section 15. Eq. (15.23) for an open system reduces to (15.22) for a closed system when $R^*_{ai} = 0$,[†] so we shall solve (15.23); Eq. (15.23) will be constrained so that the R^*_{ai} are constants.

$$\frac{dM_i^*}{dt} = \sum_{\substack{j=1 \\ j \neq i}}^{n} k_{ij}M_j^* - \sum_{\substack{j=0 \\ j \neq i}}^{n} k_{ji}M_i^* + R^*_{ai} \qquad (15.23)$$

Let us refer to the state where $\frac{dM_i^*}{dt} = 0$ for all i as the state of equilibrium and let us denote the equilibrium values of M_i^* by M_i^*. Thus at equilibrium

---

[†] To be closed we must also exclude the term $-k_{oi}M_i^*$.

$$\sum_{\substack{j=1 \\ j\neq i}}^{n} k_{ij}M_j^* - \sum_{\substack{j=0 \\ j\neq i}}^{n} k_{ji}M_i^* + R_{ai}^* = 0 \qquad (16.1)$$

Define

$$P_i = M_i^* - M_i^* \qquad (16.2)$$

so that

$$\frac{dM_i^*}{dt} = \frac{dP_i}{dt}$$

since M_i^* is constant. Transforming (15.23) using (16.2)

$$\frac{dP_i}{dt} = \sum_{\substack{j=1 \\ j\neq i}}^{n} k_{ij}P_j - \sum_{\substack{j=0 \\ j\neq i}}^{n} k_{ji}P_i$$

$$+ \sum_{\substack{j=1 \\ j\neq i}}^{n} k_{ij}M_j^* - \sum_{\substack{j=0 \\ j\neq i}}^{n} k_{ji}M_i^* + R_{ai}^* \qquad (16.3)$$

The final three expressions in (16.3) vanish due to (16.1) leaving

$$\frac{dP_i}{dt} = \sum_{\substack{j=1 \\ j\neq i}}^{n} k_{ij}P_j - \sum_{\substack{j=0 \\ j\neq i}}^{n} k_{ji}P_i \qquad (16.4)$$

The result of the transformation of variables (16.2) has been to convert
(15.23) into a similar set of equations without the absolute terms. For
simplicity, let us now define

$$k_{ii} = \sum_{\substack{j=0 \\ j\neq i}}^{n} k_{ji} \qquad (16.5)$$

so that (16.4) may be expanded in the form

$$\frac{dP_1}{dt} = -k_{11}P_1 + k_{12}P_2 + k_{13}P_3 + \ldots + k_{1n}P_n$$

$$\frac{dP_2}{dt} = k_{21}P_1 - k_{22}P_2 + k_{23}P_3 + \ldots + k_{2n}P_n$$

$$.$$
$$.$$
$$. \qquad (16.6)$$

$$\frac{dP_n}{dt} = k_{n1}P_1 + k_{n2}P_2 + k_{n3}P_3 + \ldots - k_{nn}P_n$$

where all k's are positive. Let us now take Laplace transforms of each
equation using the theorem

$$L\frac{df}{dt} = sLf - f(0) \tag{16.7}$$

and designating the Laplace transform of $P_i(t)$, $LP_i(t)$, by p_i.

$$-(k_{11}+s)p_1 + k_{12}p_2 + k_{13}p_3 + \ldots + k_{1n}p_n = -P_1(0)$$

$$k_{21}p_1 - (k_{22}+s)p_2 + k_{23}p_3 + \ldots + k_{2n}p_n = -P_2(0)$$

$$\cdot$$
$$\cdot \tag{16.8}$$
$$\cdot$$

$$k_{n1}p_1 + k_{n2}p_2 + k_{n3}p_3 + \ldots - (k_{nn}+s)\, p_n = -p_n(0)$$

Let us define

$$K = \begin{pmatrix} -k_{11} & k_{12} & k_{13} & \ldots & k_{1n} \\ k_{21} & -k_{22} & k_{23} & & k_{2n} \\ \cdot & & & & \\ \cdot & & & & \\ \cdot & & & & \\ k_{n1} & k_{n2} & k_{n3} & & -k_{nn} \end{pmatrix} \tag{16.9}$$

$$P = \begin{pmatrix} p_1 \\ p_2 \\ p_3 \\ \cdot \\ \cdot \\ \cdot \\ p_n \end{pmatrix} \tag{16.10}$$

$$I = \begin{pmatrix} 1 & 0 & 0 & \ldots & 0 \\ 0 & 1 & 0 & \ldots & 0 \\ 0 & 0 & 1 & \ldots & 0 \\ \cdot & & & & \\ \cdot & & & & \\ \cdot & & & & \\ 0 & 0 & 0 & \ldots & 1 \end{pmatrix} \tag{16.11}$$

and

$$P(0) = \begin{pmatrix} P_1(0) \\ P_2(0) \\ P_3(0) \\ . \\ . \\ . \\ P_n(0) \end{pmatrix} \qquad (16.12)$$

Now (16.8) can be written in terms of these matrices as

$$(K - sI)P = -P(0) \qquad (16.13)$$

Solving for P

$$P = -(K - sI)^{-1} P(0) \qquad (16.14)$$

as obtained by Berman and Schoenfeld (1956). Eq. (16.14) provides a formal
solution for the vector P, and thence for $M_i^*(t)$, in terms of the initial
conditions and the known values of the rates constants k.

Proceeding alternatively we may define

$$\Delta = \det (K - sI) \qquad (16.15)$$

and

$$A = (K - sI)^{-1} \Delta \qquad (16.16)$$

From a theorem of linear algebra (e.g. Heading, 1958) we know that the
adjoint matrix, A, is composed of the cofactors of the transpose of the
matrix (K - sI). Since each row and each column of (K - sI) contains only
one term in s, therefore each cofactor of (K - sI) contains s to a power no
higher than n-1. Therefore each element of A contains s to a power no
higher than n-1. From (16.14) and (16.16)

$$P = \frac{-A}{\Delta} P(0) \qquad (16.17)$$

Since Δ contains s to the power of n, therefore

$$p_i = \frac{\text{polynomial of the degree n-1 or lower}}{\text{polynomial of the degree n}} \qquad (16.18)$$

While it is possible to show mathematically that the roots of the polynomial
$\Delta(s)$ must have negative real parts and that pure imaginary roots are
impossible (see Berman and Schoenfeld) we shall not digress here to provide
proof. We shall assume, moreover, that no two roots of $\Delta(s)$ are equal.
Since P is a vector of Laplace transforms we should like to transform the
right-hand side of (16.18) into a form which can be inverted. A polynomial

fraction of that type can be decomposed into n partial fractions giving

$$P_i = \frac{\alpha_{i1}}{s + \beta_1} + \frac{\alpha_{i2}}{s + \beta_2} + \ldots + \frac{\alpha_{in}}{s + \beta_n} \qquad (16.19)$$

where the β's are the roots of the polynomial Δ. Notice that the denomi-
nators are the same for all P_i. Eq. (16.19) may now easily be inverted.

$$P_i(t) = \alpha_{i1} e^{-\beta_1 t} + \alpha_{i2} e^{-\beta_2 t} + \ldots + \alpha_{i3} e^{-\beta_3 t} \qquad (16.20)$$

It is now clear by physical reasoning that no β may have a positive real part,
for if it did, $P_i(t)$ would become indefinitely large for large t which is
physically inadmissible. The solutions for M_i* are now obtained from (16.2)
and (16.19) which completes the solution.

Example

Suppose that a closed two compartment system is displaced from equilibrium
at t = 0 by the injection of a bolus of tracer into compartment 1. The
initial condition in compartment 1 as a consequence of the injection is
$M_1*(0) = 1$, while in compartment 2, $M_2*(0) = 0$. The differential equations
are of the type (15.22) and we shall presume that the k's have known values
as indicated in (16.21) below and that k_{ii} is defined as in (16.5).

$$\frac{dM_1*}{dt} = -2M_1* + 2M_2*$$

$$(16.21)$$

$$\frac{dM_2*}{dt} = M_1* - 3M_2*$$

Solve for $M_1*(t)$ and $M_2*(t)$.

From (16.9), the matrix K is given by

$$K = \begin{pmatrix} -2 & 2 \\ 1 & -3 \end{pmatrix} \qquad (16.22)$$

and from (16.15) the determinant Δ is given by

$$\Delta(s) = \begin{vmatrix} -2 -s & 2 \\ 1 & -3 -s \end{vmatrix} \qquad (16.23)$$

The β's of (16.19) are the roots of the determinant Δ(s) and are, therefore,
obtained from

$$\begin{vmatrix} -2 -s & 2 \\ 1 & -3 -s \end{vmatrix} = 0$$

or

$$s^2 + 5s + 4 = 0 \qquad (16.24)$$

$$(s + 4)(S + 1) = 0 \qquad (16.25)$$

Thus we have from (16.25) and (16.19)

$$\beta_1 = 4 \text{ and } \beta_2 = 1 \qquad (16.26)$$

The solution (16.20) is now complete except for the coefficients α_{ij}. From the initial conditions

$$\alpha_{11} + \alpha_{12} = 1 \qquad (16.27)$$

and

$$\alpha_{21} + \alpha_{22} = 0 \qquad (16.28)$$

Eqs. (16.27) and (16.28) are two equations in the four unknown α's. We can complete the solution in two ways

(i) α's Determined the Hard Way

From (16.16)

$$\frac{A}{\Delta} = (K - sI)^{-1}$$

$$= \begin{pmatrix} -2-s & 2 \\ 1 & -3 -s \end{pmatrix}^{-1}$$

which upon taking the inverse gives

$$\frac{A}{\Delta} = \frac{-1}{s^2 + 5s + 4} \begin{pmatrix} 3 + s & 2 \\ 1 & 2 + s \end{pmatrix} \qquad (16.29)$$

Now

$$P_i(0) = M_i{}^*(0)$$

since the system is closed, so that the vector P(0) (Eq. (16.12)) is given by

$$P(0) = \begin{pmatrix} 1 \\ 0 \end{pmatrix} \qquad (16.30)$$

From (16.17) and (16.29)

$$P = \frac{1}{s^2 + 5s + 4} \begin{pmatrix} 3 + s & 2 \\ 1 & 2 + s \end{pmatrix} \begin{pmatrix} 1 \\ 0 \end{pmatrix}$$

$$= \frac{1}{s^2 + 5S + 4} \begin{pmatrix} 3 + s \\ 1 \end{pmatrix} \qquad (16.31)$$

That is

$$p_1 = \frac{(3 + s)}{s^2 + 5s + 4} \qquad (16.32)$$

and

$$p_2 = \frac{1}{s^2 + 5s + 4}$$

As required by (16.18), p_1 and p_2 each consists of the ratio of a polynomial of degree (2 - 1) or lower to a polynomial of degree 2. Breaking p_1 and p_2 into partial fractions,

$$p_1 = \frac{\frac{1}{3}}{s + 4} + \frac{\frac{2}{3}}{s + 1}$$

$$p_2 = \frac{-\frac{1}{3}}{s + 4} + \frac{\frac{1}{3}}{s + 1}$$

(16.33)

Finally, taking the inverse transforms $(P_i(t) = M_i*(t))$,

$$M_1*(t) = \frac{1}{3} e^{-4t} + \frac{2}{3} e^{-t}$$

$$M_2*(t) = -\frac{1}{3} e^{-4t} + \frac{1}{3} e^{-t}$$

(16.34)

which are the solutions for the given initial conditions.

(ii) α's Determined the Easy Way

Since the solutions are known to have the form

$$M_1*(t) = \alpha_{11} e^{-4t} + (1 - \alpha_{11}) e^{-t} \tag{16.35}$$

$$M_2*(t) = \alpha_{21} e^{-4t} - \alpha_{21} e^{-t} \tag{16.36}$$

from (16.20), (16.27) and (16.28), we may evaluate the remaining coefficients α_{11} and α_{21} by introducing (16.35) and (16.36) into *either* of the original differential equations, (16.21). Suppose we substitute into the first of the differential equations:

$$- 4\alpha_{11} e^{-4t} - (1 - \alpha_{11}) e^{-t} = -2\alpha_{11} e^{-4t}$$

$$- 2(1 - \alpha_{11}) e^{-t} \tag{16.37}$$

$$+ 2\alpha_{21} e^{-4t} - 2\alpha_{21} e^{-t}$$

Since the solutions (16.35) and (16.36) must satisfy the differential equations identically, we may equate the coefficients first of e^{-4t} and then of e^{-t} to obtain two equations in α_{11} and α_{21}:

$$-4\alpha_{11} = -2\alpha_{11} + 2\alpha_{21} \tag{16.38}$$

$$-1 + \alpha_{11} = -2 + 2\alpha_{11} - 2\alpha_{21} \tag{16.39}$$

Solving these equations gives

$$\alpha_{11} = \frac{1}{3} \tag{16.40}$$

$$\alpha_{21} = -\frac{1}{3}$$

which upon substitution into (16.35) and (16.36) gives (16.34) once again. The solutions are sketched in Fig. 16.1

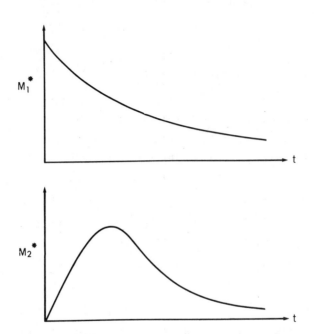

Fig. 16.1.   The solutions to the differential equations of the given example.   This is a sketch, not an exact graph.

$$M_1{}^\star = \frac{1}{3} e^{-4t} + \frac{2}{3} e^{-t}$$

$$M_2{}^\star = -\frac{1}{3} e^{-4t} + \frac{1}{3} e^{-t}$$

In this section we have discussed the methods for solution of a set of ordinary, linear differential equations with constant coefficients.   The process is straightforward and seldom causes difficulties.   Unfortunately, the solution of a set of differential equations is not the problem usually

confronting the investigator. The problem we have just solved is the one
in which the constant coefficients, k_{ij}, are given and one requires the
solution corresponding to a set of initial conditions. The problem usually
confronting the investigator is to discover the coefficients, k_{ij}, given
some of the solutions and the initial conditions. This might be termed
solving the inverse problem.

We have not discussed solutions for the case where the R^*_{ai} are functions of
time.

Speaking Practically

A set of linear differential equations such as (16.6) can be solved by a
sequence of well-defined steps which are outlined succinctly in section 18.
The only problem anticipated in their solution is finding the roots of a
polynomial equation of higher order, such as a cubic or quartic equation, and
even this is a minor inconvenience since good computer library programs for
root-finding exist. In actual fact, it is the inverse problem, to be discussed in the next section which usually confronts the investigator.

SECTION 17

COMPARTMENTS: SOLUTION OF THE
INVERSE PROBLEM

Suppose that we have conducted an experiment involving tracers and we have
reason to believe that the experimental process can be modelled by an
n-compartment system. Tracer has been added to the system at t = 0, and no
further tracer inputs have been given. We have made measurements of C_i^*,
the tracer concentration in *each* of the n compartments over the duration of
the experiment. For example, samples may have been taken at t = 0 and at
the end of every 5 minute interval until 100 minutes had passed. We have
found that each of the n concentration-time curves can be fitted by a least

squares procedure, say, to a linear combination of the same n exponential functions with real negative indices[†]. That is,

$$C_1*(t) = A_{11} e^{-\beta_1 t} + A_{12} e^{-\beta_2 t} + \ldots + A_{1n} e^{-\beta_n t}$$

$$C_2*(t) = A_{21} e^{-\beta_1 t} + A_{22} e^{-\beta_2 t} + \ldots + A_{2n} e^{-\beta_n t}$$

$$\begin{array}{c} . \\ . \\ . \end{array}$$

(17.1)

$$C_n*(t) = A_{n1} e^{-\beta_1 t} + A_{n2} e^{-\beta_2 t} + \ldots + A_{nn} e^{-\beta_n t}$$

All of the A's and β's are determined experimentally by the curve fitting process. We suppose that after an inital bolus of tracer input at $t = 0$, no further tracer is added to the system, so that the differential equations are of the form (cf. (15.22) and (15.23))

$$\frac{dM_i*}{dt} = \sum_{\substack{j=1 \\ j \neq i}}^{n} k_{ij} M_j* - \sum_{\substack{j=0 \\ j \neq i}}^{n} k_{ji} M_i* .$$

(17.2)

Introducing (16.5) into (17.2) gives

$$\frac{dM_i*}{dt} = \sum_{\substack{j=1 \\ j \neq i}}^{n} k_{ij} M_j* - k_{ii} M_i*$$

(17.3)

Our problem is to determine, if possible, the coefficients k_{ij}. I call this problem the inverse to that of solving the differential equations.

If we define

$$K_{ij} = k_{ij} \frac{V_j}{V_i} , \; i \neq 0; \; K_{ij} = k_{ij}, \; i = 0$$

(17.4)

so that

$$k_{ij} = K_{ij} \frac{V_i}{V_j} , \; i \neq 0; \; k_{ij} = K_{ij}, \; i = 0$$

then (17.3) becomes

$$\frac{dM_i*}{dt} = V_i \sum_{\substack{j=1 \\ j \neq i}}^{n} K_{ij} \frac{M_j*}{V_j} - K_{ii} \frac{M_i*}{V_i} V_i \quad \text{(see Eq. (17.20))}$$

[†]This is the function expected from Eq. (16.20).

$$\frac{dM_i{}^*}{dt} = V_i \sum_{j=1}^{n} (1 - 2\delta_{ij})K_{ij} \frac{M_j{}^*}{V_j} \tag{17.5}$$

where the V's are the constant volumes of the compartments, and δ_{ij}, the Kronecker delta is defined such that

$$\delta_{ij} = 0, \quad i \neq j \tag{17.6}$$

$$= 1, \quad i = j$$

In setting down (17.1) we have implicitly assumed that tracer is uniformly dispersed throughout *each* compartment so that (as in (15.19))

$$C_j = \frac{M_j}{V_j}$$

and hence (17.5) becomes (dividing both sides by V_i)

$$\frac{dC_i{}^*}{dt} = \sum_{j=1}^{n} (1 - 2\delta_{ij}) K_{ij} C_j{}^* \tag{17.7}$$

Eq. (17.7) is now in exactly the same form as (16.4) except that the variables are now the concentration, $C_i{}^*$, which are measurable, rather than the mass differences P_i, which are not in general measurable.

Let us now write the solution for the ith compartment from (17.1) as

$$C_i{}^* = \sum_{j=1}^{n} A_{ij} e^{-\beta_j t} \tag{17.8}$$

and taking the derivative of both sides with respect to time,

$$\frac{dC_i{}^*}{dt} = -\sum_{j=1}^{n} \beta_j A_{ij} e^{-\beta_j t} \tag{17.9}$$

By changing summation indices we can obtain the expression for $C_j{}^*$ from (17.8) and then combine (17.7), (17.8) and (17.9) to give

$$-\sum_{j=1}^{n} \beta_j A_{ij} e^{-\beta_j t}$$

$$= \sum_{j=1}^{n} \left[(1 - 2\delta_{ij})K_{ij} \sum_{m=1}^{n} A_{jm} e^{-\beta_m t} \right] \tag{17.10}$$

Expanding Eq. (17.10):

$$-\beta_1 A_{i_1} e^{-\beta_1 t} - \beta_2 A_{i_2} e^{-\beta_2 t} \ldots - \beta_n A_{in} e^{-\beta_n t} =$$

$$K_{i_1}(A_{11} e^{-\beta_1 t} + A_{12} e^{-\beta_2 t} \ldots + A_{1n} e^{-\beta_n t})$$

$$+ K_{i_2}(A_{21} e^{-\beta_1 t} + A_{22} e^{-\beta_2 t} \ldots + A_{2n} e^{-\beta_n t})$$

$$\cdot$$
$$\cdot$$
$$\cdot$$

$$- K_{ii}(A_{i_1} e^{-\beta_1 t} + A_{i_2} e^{-\beta_2 t} \ldots + A_{in} e^{-\beta_n t})$$

$$\cdot$$
$$\cdot$$
$$\cdot$$

$$+ K_{in}(A_{n1} e^{-\beta_1 t} + A_{n2} e^{-\beta_2 t} \ldots + A_{nn} e^{-\beta_n t}) \qquad (17.11)$$

Since the solutions must satisfy the differential equations identically, we may equate the coefficients of $e^{-\beta_1 t}$, $e^{-\beta_2 t}$ etc. to obtain:

$$-\beta_1 A_{i_1} = K_{i_1} A_{11} + K_{i_2} A_{21} + \ldots - K_{ii} A_{i_1} + \ldots + K_{in} A_{n1}$$

$$-\beta_2 A_{i_2} = K_{i_1} A_{12} + K_{i_2} A_{22} + \ldots - K_{ii} A_{i_2} + \ldots + K_{in} A_{n2}$$

$$\cdot$$
$$\cdot \qquad\qquad\qquad\qquad\qquad\qquad\qquad (17.12)$$
$$\cdot$$

$$-\beta_n A_{in} = K_{i_1} A_{1n} + K_{i_2} A_{2n} + \ldots - K_{ii} A_{in} + \ldots + K_{in} A_{nn}$$

If we define

$$
A = \begin{pmatrix}
A_{11} & A_{12} & \cdot\;\cdot\;\cdot & A_{1n} \\
A_{21} & A_{22} & \cdot\;\cdot\;\cdot & A_{2n} \\
\cdot & & & \\
\cdot & & & \\
\cdot & & & \\
A_{i1} & A_{i2} & \cdot\;\cdot\;\cdot & A_{in} \\
\cdot & & & \\
\cdot & & & \\
\cdot & & & \\
A_{n1} & A_{n2} & \cdot\;\cdot\;\cdot & A_{nn}
\end{pmatrix}
\tag{17.13}
$$

$$
K = \begin{pmatrix}
-K_{11} & K_{12} & \cdot\;\cdot\;\cdot & & K_{1n} \\
K_{21} & -K_{22} & \cdot\;\cdot\;\cdot & & K_{2n} \\
\cdot & & & & \\
\cdot & & & & \\
\cdot & & & & \\
K_{i1} & K_{i2} & \cdots & -K_{ii} \cdots & K_{in} \\
\cdot & & & & \\
\cdot & & & & \\
\cdot & & & & \\
K_{n1} & K_{n2} & & & -K_{nn}
\end{pmatrix}
\tag{17.14}
$$

and

$$
\beta = \begin{pmatrix}
-\beta_1 & 0 & 0 & \cdot\;\cdot\;\cdot & 0 \\
0 & -\beta_2 & \cdot & \cdot\;\cdot\;\cdot & 0 \\
0 & 0 & -\beta_3 & \cdot\;\cdot\;\cdot & \cdot \\
\cdot & & & & \\
\cdot & & & -\beta_i \cdot\;\cdot & \cdot \\
\cdot & & & & \\
0 & 0 & 0 & \cdot\;\cdot\;\cdot & -\beta_n
\end{pmatrix}
\tag{17.15}
$$

then (17.12) can be written

$$KA = A\beta \tag{17.16}$$

or

$$K = A\beta A^{-1} \tag{17.17}$$

Eq. (17.17) was derived by Berman and Schoenfeld (1956); the general approach is suggested in a paper by Hart (1955). The inverse problem is now formally solved, but the small k's can only be obtained from the large K's (using (17.4)) if the compartmental volumes, V_i, are known. Moreover, it has been assumed that experimental access to all n compartments is possible, contrary to the usual experience.

Example 1

Once again, let us take the example of a two compartment model. While an example of a higher order model would involve a greater quantity of algebra, it would not really enhance our understanding either of compartmental models or of the methods for their solution. Suppose we set up a two compartment model of the kinetics of *inulin* in the mammal. Inulin is a polysaccharide which is not found naturally in the body, but its kinetics are believed to be relatively simple. It is excreted purely by a process of glomerular filtration (with the possible exception of a small quanity which may be sequestered by the cells). We shall suppose that unlabelled inulin (tracee) has been infused intravenously at a constant rate for a "long time" so that tracee is maintained in a steady state (section 13) — although for the case of inulin, the kinetics of the tracer, which we are about to discuss, would be unaltered if there were no tracee present[†]. Suppose that an intravenous injection of a bolus of labelled inulin (tracer) is given at t = 0 and that no more tracer is added. It seems reasonable to posit that one compartment "contains" the plasma so that tracer infusions can be shown as entering this compartment. Since inulin is excreted by glomerular filteration, a process involving plasma, it also seems reasonable to show the disappearance of inulin from this same compartment. We must then show the exchange of material between the source-sink-compartment and a second compartment, about

---

[†]Inulin is excreted by a process of glomerular filtration, which we would expect to be a first order process; i.e. $R_{d,v} = \lambda C$, and hence by (6.7) $R^*_{d,v} = a \cdot R_{d,v} = \lambda C^*$. Thus, obviously, tracer is filtered, with the same rate constant, whether or not tracee is present.

which we can say very little. Presumably it encompasses some of the interstitial fluid, but anatomical localization is difficult. The total compartment configuration is, then, as shown in Fig. 17.1; the theory of compartments as developed hitherto tells us nothing about the volumes V_1 and V_2.
The exchange of material between compartments is governed by the rate constants K_{12} and K_{21} and the disappearance of material from compartment 1 is
governed by the rate constant K_{01} (see Eqs. (17.4) and (17.7)). The
differential equations may now be written down by referring to Fig. 17.1 and
Eq. (17.7):

$$\frac{dC_1^*}{dt} = -K_{11}C_1^* + K_{12}C_2^*$$

$$\frac{dC_2^*}{dt} = K_{21}C_1^* - K_{22}C_2^* \tag{17.18}$$

From (17.4)

$$K_{ii} = k_{ii} \tag{17.19}$$

and so from (16.5), (17.4) and (17.19)

$$K_{ii} = \sum_{\substack{j=1 \\ j \neq i}}^{n} K_{ji} \frac{V_j}{V_i} + K_{oi} \tag{17.20}$$

Therefore

$$K_{11} = K_{21} \frac{V_2}{V_1} + K_{01} = K_{21}' + K_{01} \tag{17.21}$$

$$K_{22} = K_{12} \frac{V_1}{V_2} + K_{02} = K_{12}' + K_{02} = K_{12}' \tag{17.22}$$

since $K_{02} = 0$.
Eq. (17.18) then becomes

$$\frac{dC_1^*}{dt} = -(K_{01} + K_{21}')C_1^* + K_{12}C_2^*$$

$$\frac{dC_2^*}{dt} = K_{21}C_1^* - K_{12}'C_2^* \tag{17.23}$$

Eq. (17.23) has been derived from (15.23), and it is, in principle, manage-
able. But no one uses it in this form. So we shall simply set $K_{12} = K_{12}'$,
and $K_{21} = K_{21}'$ and *define* the model by the resulting equations. In this form
the equations can be written down directly by inspecting Fig. 17.1.

Ideally, we should have measurements of both C_1^* and C_2^* over a time interval
sufficiently long that the measured data will define the characteristics of
the two functions, $C_1^*(t)$ and $C_2^*(t)$. But we cannot even be sure where the

second compartment is located anatomically, let alone take samples from it. And so we must be content with sequential measurements of C_1* — the concentration of tracer in plasma. Suppose that after a bolus input of tracer, the resulting impulse response in plasma is of the usual decay type, as shown in Fig. 13.2 (Y_1). We can be quite certain that the indices β_i are real, positive numbers and we can fit the experimental points to a double exponential function of the form

$$C_1*(t) = B_1 e^{-\beta_1 t} + B_2 e^{-\beta_2 t} \qquad \beta_1 > \beta_2 \qquad (17.24)$$

(see Appendix B). While we have no measurements in compartment 2, we know from the theory that $C_2*(t)$ must be a linear combination of the same two exponential functions, and from the given initial conditions, $C_2*(0) = 0$. Thus $C_2*(t)$ must have the form

$$C_2*(t) = -B_3 e^{-\beta_1 t} + B_3 e^{-\beta_2 t} \qquad B_3 > 0 \qquad (17.25)$$

We can now set up the matrices, A, K and β and define by Eqs. (17.13), (17.14) and (17.15) respectively

$$A = \begin{pmatrix} B_1 & B_2 \\ -B_3 & B_3 \end{pmatrix} \qquad (17.26)$$

$$K = \begin{pmatrix} -K_{01}-K_{21} & K_{12} \\ K_{21} & -K_{12} \end{pmatrix} \qquad (17.27)$$

$$\beta = \begin{pmatrix} -\beta_1 & 0 \\ 0 & -\beta_2 \end{pmatrix} \qquad (17.28)$$

From (17.17)

$$K = A\beta A^{-1}$$

That is

$$\begin{pmatrix} -K_{01}-K_{21} & K_{12} \\ K_{21} & -K_{12} \end{pmatrix} = \begin{pmatrix} B_1 & B_2 \\ -B_3 & B_3 \end{pmatrix} \begin{pmatrix} -\beta_1 & 0 \\ 0 & -\beta_2 \end{pmatrix} \begin{pmatrix} B_1 & B_2 \\ -B_3 & B_3 \end{pmatrix}^{-1}$$

$$= \frac{1}{B_1 B_3 + B_2 B_3} \begin{pmatrix} B_1 & B_2 \\ -B_3 & B_3 \end{pmatrix} \begin{pmatrix} -\beta_1 & 0 \\ 0 & -\beta_2 \end{pmatrix} \begin{pmatrix} B_3 & -B_2 \\ B_3 & B_1 \end{pmatrix}$$

$$= \frac{1}{B_1 B_3 + B_2 B_3} \begin{pmatrix} -\beta_1 B_1 B_3 - \beta_2 B_2 B_3 & \beta_1 B_1 B_2 - \beta_2 B_1 B_2 \\ \beta_1 B_3{}^2 - \beta_2 B_3{}^2 & -\beta_1 B_2 B_3 - \beta_2 B_1 B_3 \end{pmatrix} (17.29)$$

We now have four algebraic equations which may be solved for four unknown quantities. There are only three unknown K's but B_3, the quantity governing the amplitude of the concentrations in the inaccessible compartment 2, is also unknown. The four unknown quantities are, therefore, K_{01}, K_{21}, K_{12} and B_3; the four equations are from (17.29)

$$K_{01} + K_{21} = \frac{\beta_1 B_1 + \beta_2 B_2}{B_1 + B_2} \tag{17.30}$$

$$K_{12} = \frac{B_1 B_2 (\beta_1 - \beta_2)}{B_3 (B_1 + B_2)} \tag{17.31}$$

$$K_{21} = \frac{B_3 (\beta_1 - \beta_2)}{B_1 + B_2} \tag{17.32}$$

$$K_{12} = \frac{\beta_1 B_2 + \beta_2 B_1}{B_1 + B_2} \tag{17.33}$$

K_{12} is given explicitly by (17.33) in terms of the measured (or fitted) quantities B_1, B_2, β_1 and β_2. The remaining three unknown quantities are easily obtained from the latter four equations.

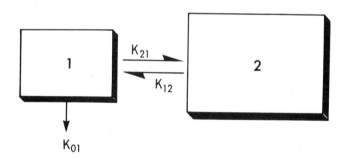

Fig. 17.1. Proposed two compartment model of inulin kinetics. The plasma
        forms a part of compartment 1 and so this compartment contains
        both source and sink. Compartment 2 is represented as sort of
        a cul-de-sac.

As in the example of section 16, the required information is more directly obtained without the use of matrices simply by substituting the expressions for the concentrations (17.24) and (17.25) directly into the differential equations (17.23) and then equating the coefficients of like exponentials.

Substituting for $C_1^*(t)$ and $C_2^*(t)$ into both differential equations gives rise to $2 \times 2 = 4$ algebraic equations in the unknown quantities K_{01}, K_{21}, K_{12} and B_3.

Evaluation of the Rate Constants

In an actual experimental (Norwich et al, 1974), a bolus of 44.4×10^6 dpm of ^{14}C-hydroxymethyl inulin was injected intravenously into a non-anaesthetized dog. Plasma tracer concentrations were measured at intervals for 120 minutes, and the results are shown in Table 17.1. When the impulse response curve in plasma was fitted to a double exponential function, the result obtained was (Fig. 17.2)

$$C^*(t) = 41,885.e^{-.278t} + 14,434.e^{-.0198t} \qquad (17.34)$$

from which we obtain (with reference to (17.24))[†]

$$
\begin{aligned}
B_1 &= 41,885 \ \text{dpm/ml} \\
B_2 &= 14,434 \ \text{dpm/ml} \\
\beta_1 &= .278 \quad \text{min}^{-1} \\
\beta_2 &= .0198 \quad \text{min}^{-1}
\end{aligned} \qquad (17.35)
$$

Introducing the numerical values of (17.35) in Eqs. (17.30) → (17.33) we obtain,

$$
\begin{aligned}
K_{21} &= 0.148 \quad \text{min}^{-1} \\
K_{12} &= 0.0860 \quad \text{min}^{-1} \\
K_{01} &= 0.0640 \quad \text{min}^{-1} \\
B_3 &= 32,200 \quad \text{dpm/ml}
\end{aligned}
$$

This completes Example 1.

[†]In some of the examples and problems we shall carry 5 digits for purposes of calculation, but the result will be reported to 2 or 3 significant figures.

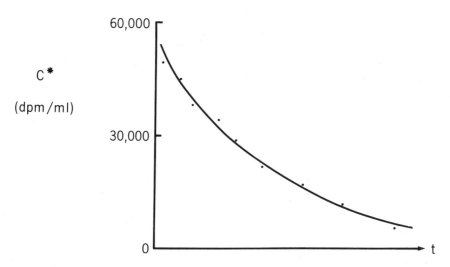

Fig. 17.2.   Response to an intravenous injection of labelled inulin
     ● Experimental Data
     Fitted Function $C*(t) = 41,885.e^{-278t} + 14,434.e^{-.0198t}$

TABLE 17.1

Time (minutes)	Plasma Tracer Concentration, C* (dpm/ml)	Fitted C* (dpm/ml)
1	47,700	45,900
3	30,800	31,800
5	22,200	23,300
8	17,400	16,900
12	13,600	12,900
20	10,300	9,900
30	8,100	8,000
45	5,700	5,900
60	4,000	4,400
90	2,500	2,400
120	1,400	1,300

Impulse response function to an intravenous injection of 44.4×10^6 dpm of ^{14}C-hydroxymethyl inulin. The actual plasma tracer concentrations C* are shown, as well as the values of C*(t) calculated from the fitted function

$$C^*(t) = 41,885.e^{-.278t} + 14,434.e^{-.0198t}$$

Example 2

Devise a solvable two compartment model for the kinetics of labelled glucose in the mammal. Such a model would be quite similar to the compartment model for inulin which we have discussed in Example 1. The plasma would form a part of compartment 1 and the interstitial fluid, presumably, would form a part of compartment 2 — although as always, anatomical definition of compartments is problematic. A major difference between the inulin and glucose models is that while inulin is excreted nearly totally by the kidneys, glucose is taken up by many tissues — red blood cells included. We therefore require four rate constants, k_{12}, k_{21}, k_{01} and k_{02}, rather than the three which sufficed for inulin (see Fig. 17.3). Moreover, we are still denied experimental access to compartment 2. With reference to Fig. 17.3 we can set down the differential equations for labelled glucose (cf. Eq. (17.23)):

$$\frac{dC_1{}^*}{dt} = -(K_{01} + K_{21})C_1{}^* + K_{12}C_2{}^*$$

$$\frac{dC_2{}^*}{dt} = K_{21}C_1{}^* - (K_{02} + K_{12})C_2{}^* \tag{17.36}$$

We can now proceed exactly as in the inulin example except that in place of
(17.33) we will have

$$K_{02} + K_{12} = \frac{\beta_1 B_2 + \beta_2 B_1}{B_1 + B_2} \tag{17.37}$$

That is, we have four algebraic equations (17.30), (17.31), (17.32) and
(17.37) in the five unknown quantities K_{01}, K_{21}, K_{02}, K_{12} and B_3. Clearly
we do not have sufficient information to solve for all five. This is the
curse of the mathematical modeller: too many parameters; too little data.
In this circumstance one usually reduces the number of parameters by one and
makes the best of a less representational model. For example, we might set

$$K_{02} = K_{01} \tag{17.38}$$

and then solve for the remaining four parameters.

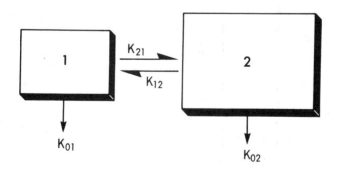

Fig. 17.3. Proposed two compartment model of glucose kinetics.

If one has measured data from more than one compartment, then it will be necessary to conform the concentration-time curves from each compartment to a linear combination of the same n exponential functions. A method for carrying out this simultaneous fitting process will be discussed in Appendix B.

Speaking Practically

Given the fact that you *want* to conform your experimental or clinical data to a multicompartmental model, this section describes the steps you take in order to identify the parameters of this model. First you must decide on the configuration of the model: How many compartments? How are they connected? How many different rate constants? You may be able to make measurements of C^* in some, but not all of these compartments. Will there be sufficient measured data to permit evaluation of all of the unknown quantities? If not, are you content with a partial evaluation of the unknown quantities, or will you reduce the complexity of the model to permit complete evaluation of the remaining parameters? Having made these decisions, this section describes the mathematical processes for carrying out the evaluations. These processes are summarized as a "recipe" in the next section. You may avoid the mathematical work nearly entirely by using the SAAM computer program of Dr. M. Berman of the National Institutes of Health in the United States.

Problems

17.1. Solve the problem of *example 1* by substituting the equations (17.24) and (17.25) directly into the differential equations (17.23). In this way evaluate K_{21}, K_{12}, K_{01} and B_3. Drop the "primes" in Eq. (17.23).

17.2(a). For the case of the glucose system described above, solve for K_{01}, K_{12} and K_{21} in terms of β_1, β_2, B_1 and B_2

The following data were obtained in an experiment conducted by Dr. G. Hetenyi Jr. in collaboration with the author. A mass of 33.34×10^6 dpm of ^{14}C-labelled glucose was injected intravenously into a dog who was being maintained at an elevated (not basal) steady state level. The steady, plasma concentration of unlabelled glucose averaged 1.10 mg/ml. The dog weighed 17.1 kg and so the *total* distribution space of glucose may be taken to be

25% of 17.1 litres, which is approximately equal to the extracellular space.
The following values of C_1*, the plasma tracer concentration were measured:

Time (from injection in minutes)	$C_1*(dpm/ml)$
1.0	13947.
2.0	12085.
3.05	10916.
5.0	9209.
8.0	8308.
12.0	6299.
20.1	5493.
30.0	4523.
45.0	3655.
60.0	3062.
75.0	2567.
90.0	2050.

17.2(b). Fit the experimental data to a double exponential function *either*
- by hand, e.g. the "peeling" method (W. Perl, 1960) *or*
- by computer using your own method or a computer library tape subroutine.

17.2(c). Evaluate K_{01}, K_{12} and K_{21}.

17.2(d). Calculate R_d using the compartmental model. Assume $C_2 = C_1$.

17.2(e). Evaluate the expression $\dfrac{M_0^* C_1}{\dfrac{B_1}{\beta_1} + \dfrac{B_2}{\beta_2}}$ = Tracer mass x tracee concen-

tration/Area under unit impulse response function.
The results of this calculation should be very nearly equal to R_d calculated
from (d). It's significance will be discussed in section 23.

SECTION 18

"RECIPES" FOR COMPARTMENTAL CALCULATIONS

The man who taught chemistry to me (and to countless others) in high school compiled a treasury of what he called "recipes". These recipes were rigid protocols for solving standard problems in elementary chemistry. Of course, if you understood chemistry you did not require a recipe, but if you were the sort for whom Boyle was what you do to water and Clerk Maxwell a candy salesman at Woolworth's, then you *thrived* on recipes. By dutifully following the instructions of the recipe you could determine the chemical formula of an unknown compound even if you were sure that "Nitrogen" was something running in the fourth race at Woodbine. Naturally, I matriculated on recipes. And so it is with reminiscences of C. "Speedy" Robinson that I offer here a few recipes for compartmental calculations.

(i) Recipe for Solving a Set of Ordinary, Linear Differential Equations with Constant Coefficients ($\dot{x}_i = \frac{dx_i}{dt}$)

Solve:

$$
\left.
\begin{aligned}
\dot{x}_1 &= \alpha_{11}x_1 + \alpha_{12}x_2 + \ldots + \alpha_{1n}x_n \\[6pt]
\dot{x}_2 &= \alpha_{21}x_1 + \alpha_{22}x_2 + \ldots + \alpha_{2n}x_n \\
&\quad \cdot \\
&\quad \cdot \\
&\quad \cdot \\
\dot{x}_n &= \alpha_{n1}x_1 + \alpha_{n2}x_2 + \ldots + \alpha_{nn}x_n
\end{aligned}
\right\} \quad \text{n equations}
$$

(18.1)

Step 1. Solve the algebraic equation in s

$$
\begin{vmatrix}
\alpha_{11} - s & \alpha_{12} & \alpha_{13} & \cdots & \alpha_{1n} \\
\alpha_{21} & \alpha_{22} - s & \alpha_{23} & \cdots & \alpha_{2n} \\
\cdot & & & & \\
\cdot & & & & \\
\alpha_{n1} & \alpha_{n2} & \alpha_{n3} & & \alpha_{nn} - s
\end{vmatrix} = 0 \qquad (18.2)
$$

to obtain n roots $s = -\beta_1, -\beta_2 \ldots -\beta_n$. (Refer to any elementary text on algebra if you are not sure about expanding determinants etc.)

Step 2. The required solutions are of the form

$$x_1 = A_{11}e^{-\beta_1 t} + A_{12}e^{-\beta_2 t} + \ldots + A_{1n}e^{-\beta_n t}$$

$$x_2 = A_{21}e^{-\beta_1 t} + A_{22}e^{-\beta_2 t} + \ldots + A_{2n}e^{-\beta_n t}$$

$$\cdot$$
$$\cdot$$
$$\cdot$$

$$x_n = A_{n1}e^{-\beta_1 t} + A_{n2}e^{-\beta_2 t} + \ldots + A_{nn}e^{-\beta_n t}$$

$\left. \right\}$ n equations

(18.3)

where the n^2 coefficients A_{ij}, $i,j = 1,n$ are to be determined. In tracer
studies, but not in general, the β's will be real, positive numbers.

Step 3. Substitute each of the n integrated equations (18.3) back into any
(n - 1) of the original differential equations (18.1) and equate the
coefficients of like exponentials (i.e. equate the coefficients of $e^{-\beta_i t}$ on
both sides of the equation etc.). Since there are n different types of
exponential functions in general and you are substituting into (n - 1)
differential equations, you will have n(n - 1) algebraic equations in the
unknown A_{ij}. But you are given n sets of initial conditions — one set for
each compartment — providing n equations of the sort

$$A_{11} + A_{12} + \ldots + A_{1n} = x_1(0) = \text{a given number}$$

Thus you have now n(n - 1) + n = n^2 equations in the n^2 coefficients A_{ij} and
the problem is solved.

(ii) Recipe for Obtaining the α's of Eq. (18.1) Given Some or All of the
A's of Eq. (18.3) and All of the β's: The Inverse Problem
Given some or all of the functions (fitted to experimental data):

$$x_1 = A_{11}e^{-\beta_1 t} + A_{12}e^{-\beta_2 t} + \ldots + A_{1n}e^{-\beta_n t}$$

$$x_2 = A_{21}e^{-\beta_1 t} + A_{22}e^{-\beta_2 t} + \ldots + A_{2n}e^{-\beta_n t}$$

$$\cdot$$
$$\cdot$$
$$\cdot$$

$$x_n = A_{n1}e^{-\beta_1 t} + A_{n2}e^{-\beta_2 t} + \ldots + A_{nn}e^{-\beta_n t}$$

$\left. \right\}$ (18.4)

where some of the n^2 coefficients A_{ij} are known. The problem is to find the
n^2 coefficients α_{ij}, $i,j = 1,n$.

<u>Step 1</u>. Substitute the functions (18.4) back into the differential
equations (18.1) and equate coefficients of like exponentials. This gives
$n \times n = n^2$ equations in the n^2 unknown α's. In the event that not all of the
n^2 α's are independent of each other (i.e. less than n^2 unknowns) then one
may not require knowledge of all n functions (18.4). In the event that there
are more unknowns than there are equations (i.e. some of the A's are not
known), then you must either reduce the complexity of the model or be content
with partial knowledge of the α's.

(iii) The SAAM Program

The extraordinary SAAM program (Simulation Analysis and Modeling) has been
written and updated by Dr. M. Berman and Ms. M.F. Weiss and is adaptable to
many digital computers. Among the manifold capabilities of the program is
the determination of the best values of the α's of (18.1) directly from the
experimental data. The investigator need only read in his experimental data
and specify the configuration of the compartmental model he wishes to test,
and the program will solve the inverse problem to provide the rate constants.
SAAM is capable of much more than this and the interested reader is referred
to the continually updated SAAM manual and the related publications of
Berman and Weiss (e.g. Berman, Shahn and Weiss, 1962). One of the great
advantages of using the SAAM program is that it relieves the investigator of
the onerous task of conforming his data to sums of exponentials in a sta-
tistically satisfactory way. One of the disadvantages is, of course, that
the investigator loses direct control of the process of analysing his own
experimental data; the SAAM program is now so complex that only one or two
persons are fully aware of its workings.

SECTION 19

A CRITICAL ANALYSIS OF THE COMPARTMENTAL APPROACH

In the preceding sections we have traced the development of the theory of compartments more or less to the present time. (Lately a number of papers have appeared dealing with a stochastic theory of compartments, e.g. Soong, 1971, 1972; Marcus, 1975, and we shall introduce that subject in section 21). It became apparent that as the differential equations governing the compartmental interchange of S^* were developed, the grounding of the transport processes in physics tended to become weakened. The constants ρ_{ij} were defined to be the rate of flow of S from the jth to the ith compartment without too much concern over the physical meaning of ρ_{ij}. For example S^* was taken to be transported from the jth to the ith compartment at the rate $a\rho_{ij}$ without regard to the physics of convection or diffusion.

Because of this undermining of the physical basis of the compartment, many authors have largely overlooked mechanisms and simply postulated that the transport of tracer between the various compartments of a multicompartmental system is governed by a set of ordinary linear differential equations with constant coefficients in the variables C_i^*. Usually no attempt is made to analyse these coefficients into simpler entities such as diffusion coefficients, velocities, areas etc. — and with good reason, for as we have seen, any such attempt must ultimately fail. *The compartment is a mathematical entity constructed as an approximation to physical reality.* In a manner of speaking, all of mathematical physics consists of mathematical constructs designed to approximate physical reality. Hence all mathematical models have their limitations, and it is the limitations of compartmental models upon which we shall now focus.

The Problem of Normalization

The determination of the coefficients of the differential equations is often the prime concern of the builder of the compartmental model. It is these coefficients (the ρ's or the k's of section 15) which provide him with the desired information about the biological system under study. It is these coefficients or rate constants which tell him the number of moles or grams

of unlabelled material which pass per unit time between any two compartments.
And yet it is often painfully difficult to induce the calculated
coefficients to speak to us in a meaningful and consistent fashion. Since
we are very seldom certain of the anatomical localization of the ith and
j^{th} compartments of a system, knowing the value of the rate constant k_{ij}
governing interchange between these compartments is much like knowing the
flow of traffic between Atlantis and Shangri-La.

Let us now contrast some of the forms in which the compartmental differential
equations have been cast. Generally speaking, the rate of change of tracer
mass in the ith compartment has been represented in one of three ways.
Following Bergner, Berman, or Rescigno and Segre (1966) we would write.

$$\frac{dM_i^*}{dt} = \sum_{\substack{j=1\\j\neq i}}^{n} \lambda_{ij}M_j^* - M_i^* \sum_{\substack{j=0\\j\neq i}}^{n} \lambda_{ji} \tag{19.1}$$

where λ_{ij} (time^{-1}) is the constant fraction of material in compartment j
passing per unit time into compartment i. Following Milsum (1966) or Bright
(1973) we might express the rate of change of tracer mass with time in the
i^{th} compartment by

$$\frac{dM_i^*}{dt} = \sum_{\substack{j=1\\j\neq i}}^{n} b_{ij}C_j^* - C_i^* \sum_{\substack{j=0\\j\neq i}}^{n} b_{ji} \tag{19.2}$$

where the b's (volume · time^{-1}) are now constant "clearances". Sheppard's
representation has already be examined. Following (15.14) while opening the
system with respect to output (i.e. extending the second summation of (15.14)
to j=0) we have

$$\frac{dM_i^*}{dt} = \sum_{\substack{j=1\\j\neq i}}^{n} \rho_{ij}a_j - a_i \sum_{\substack{j=0\\j\neq i}}^{n} \rho_{ji} \tag{19.3}$$

Transforming the a's and C's to M's using (15.19), Eqs. (19.2) and (19.3)
become respectively

$$\frac{dM_i^*}{dt} = \sum_{\substack{j=1\\j\neq i}}^{n} \frac{b_{ij}}{V_j} M_j^* - M_i^* \sum_{\substack{j=0\\j\neq i}}^{n} \frac{b_{ji}}{V_i} \tag{19.4}$$

and

$$\frac{dM_i^*}{dt} = \sum_{\substack{j=1\\j\neq i}}^{n} \frac{\rho_{ij}}{M_j} M_j^* - M_i^* \sum_{\substack{j=0\\j\neq i}}^{n} \frac{\rho_{ji}}{M_i} \qquad (19.5)$$

Eqs. (19.1), (19.4) and (19.5) are clearly identical *mathematically* (but not necessarily physically as we shall see) in the state of tracee equilibrium with

$$\lambda_{ij} = \frac{b_{ij}}{V_j} = \frac{\rho_{ij}}{M_j} \qquad (19.6)$$

In order to see more clearly the problems which arise when we try to obtain information about the nature of biological processes from the
fractional transition rates, λ,
or from the clearances, b,
or from the mass fluxes, ρ,
let us examine one of the simplest possible models — a one compartment model of glucose kinetics. The tracee will be presumed to be in steady state so that as usual, the λ's, b's, ρ's and M's are all constants. Suppose that a bolus of tracer is injected into the system intravenously at zero time and that no more tracer is added to the system. Eqs. (19.1), (19.4) and (19.5) now become respectively

$$\frac{dM^*}{dt} = -\lambda M^* \qquad (19.7)$$

$$\frac{dM^*}{dt} = -\frac{b}{V} M^* \qquad (19.8)$$

and

$$\frac{dM^*}{dt} = -\frac{\rho}{M} M^* \qquad (19.9)$$

where V is the volume of the sole compartment and hence the total distribution volume of glucose. If we divide both sides of each of Eqs. (19.7), (19.8) and (19.9) by V to convert the non-observable quantity M* into the observable quantity C*, then we obtain respectively

$$\frac{dC^*}{dt} = -\lambda C^* \qquad (19.10)$$

$$\frac{dC^*}{dt} = -\frac{b}{V} C^* \qquad (19.11)$$

and

$$\frac{dC^*}{dt} = - \frac{\rho}{M} C^* \qquad (19.12)$$

Let us confine ourselves for the moment to the analysis of a single experiment. Since the experimenter has chosen a one compartment model, he expects to find that the measured values of C^* in plasma will conform reasonably well to a function of the form.

$$C^*(t) = C^*(0)e^{-\beta t} \qquad (19.13)$$

where β takes on the value λ, b/V or ρ/M depending upon which form of the differential equation he has adopted. From (19.13)

$$\log C^* = \log C^*(0) - \beta t \qquad (19.14)$$

so that the investigator plots the logarithm of the measured values of C^* against time and expects that the plotted points will lie very nearly upon a straight line. In fact, as we shall see in a later section, the early points will probably lie on a curve above the straight line while the later points will conform quite closely to the expected linear pattern. He is not too disappointed about the deviation of the early points; after all, it is just a one compartment model and is not expected to account for all the niceties of the experimentally-measured curve. Using a least squares method he fits the best straight line to the later data points, and so obtains a value for the slope, β.

The investigator is now confronted by a problem: how can he obtain meaningful physiological information from the constant β? The selection of one of Eq. (19.10), (19.11) or (19.12) is rather arbitrary since none of these differential equations rests firmly on physicochemical principles. If he selects (19.10) as the formulation of choice, then he *may* decide that

$$\lambda = \beta \quad (time^{-1}) \qquad (19.15)$$

is a characteristic rate constant of the species of animal being studied. If in a subsequent experiment, he studied an animal of the same species but of twice the size, he might still expect to find the slope β to be of the same magnitude, since $\lambda = \beta$ is a characteristic of the species.

Suppose that the investigator selects (19.11) as the formulation of choice. He *may* then decide that the clearance b, where

$$b = V\beta \quad (volume \cdot time^{-1}), \qquad (19.16)$$

is a constant or *invariant* characteristic of the species. Expressed
alternatively, if β_1 is the slope measured in an experiment on an animal
with distribution volume V_1 and β_2 is the slope measured in a similar
experiment on an similar animal with distribution volume V_2, then this investigator will compare (and average) not β_1 and β_2 but $V_1\beta_1$ and $V_2\beta_2$.

To play out the scenario, suppose he were to select (19.12) as the formulation of choice. He may then decide that the flux ρ, where

$$\rho = M\beta \text{ (mass} \cdot \text{time}^{-1}) \qquad\qquad (19.17)$$

is a constant or invariant characteristic of the species, and may then
normalize ρ by multiplying by the *pool size*, M. That is, in two animals of
the same species, he will compare $M_1\beta_1$ with $M_2\beta_2$.

Now in this particular case, we know from many, many measurements of steady
state rate of appearance that glucose "turns over" at about 1% per minute in
the resting fasted mammal (varying a little with the species) and not varying
systematically with the size of the animal. We know, therefore, that

$$\frac{\Delta M^*}{M^*} / \Delta t = -.01 = \text{a species invariant.}$$

Hence, with reference to (19.7) we expect that λ will be, approximately, a
species invariant, and the first formulation (19.15) is probably the best.

The crux of the matter is that we were unable to make a selection between
three modes of analysing our experiment until we adduced additional information about the physiology of the system. *The undecidability arose because
the K's of Eq. (17.7) have not taken their origin from a physicochemical law.*
In general it may be said that while we may calculate, for example, the K's
of Eq. (17.7) by a process of curve fitting and inverse solution, as we have
discussed at some length, it is very difficult, if not impossible, to compare
the K's obtained in experiments performed on different animals of the same
species. It is, therefore, not generally possible to calculate a set of K's
which uniquely characterize a given matabolic system. More about this
matter will be said in the next section. It *is* possible to calculate a
unique value of R_d for glucose using this model. Since no tracer is added
after the inital injection or bolus, and since the concentration C* is, by
assumption, uniform

$$\frac{dC^*}{dt} = -R^*_{d,v} = -\frac{C^*}{C} R_{d,v} = -\frac{C^*}{CV} R_d$$

and also
$$\frac{dC^*}{dt} = -\beta C^*$$

from (19.13). Therefore
$$R_d = \beta VC \qquad\qquad (19.18)$$

to the approximation of the one compartment model. But, as we shall see later, it is not necessary to invoke the concept of the compartment to calculate steady state R_d.

Some of the problems of normalization among species of a class (Mammalia) are explored by Dedrick ("Animal scale-up", 1973).

On the Volumes of Compartments

Methods for calculating these values do not usually fall within the scope of the general theory of compartments and there is no universally accepted means for doing so. However, a commonly used technique for determining (defining?) the volume of a compartment to which one has experimental access is to inject a bolus of tracer of mass M_0^* into this compartment and measure the impulse response in this same compartment. The experimental points defining the impulse response are then fitted to a sum of exponential functions.

$$C^*(t) = \sum_{j=1}^{n} A_j e^{-\beta_j t} \qquad\qquad (19.20)$$

Thus

$$C^*(0) = \sum_{j=1}^{n} A_j \qquad\qquad (19.21)$$

The injected tracer is assumed to be instantaneously and uniformly intermixed within the compartment into which it has been injected, and at zero time all tracer is assumed to be present in this compartment. One then employs the *dilution principle* to calculate the volume, V, of this compartment:

$$V = \frac{M_0^*}{C_0^*} \qquad\qquad (19.22)$$

There is an arbitrary element involved in this method of calculating volumes. Suppose that we remain open-minded about the total number of compartments in the system, and therefore about the total number of exponential functions to which the impulse response curve is fitted. For example, if the experimental points were fitted "adequately" by a double exponential, then we should

use n = 2 in (19.20); if a double exponential is inadequate by some standard
of good fitting, then we try a triple exponential with n = 3 etc. We should
then probably find that the smaller the time of the first sample, the
greater the number of exponential functions required for an adequate curve
fitting. For example, if our first measured C* value was C*(15), we might
require only a single exponential; if the first value was C*(5), we might
require a triple exponential etc. (Fig. 19.1). The effect of each additional exponential function is to increase the value of C*(0) as calculated from
(19.21). This effect is best appreciated with reference to Fig. 19.1.
Therefore V, as calculated from (19.22), becomes progressively smaller as
we take earlier and earlier samples. What, then, is *the* volume of the compartment of access? The answer we must give is that there is no unique
volume.

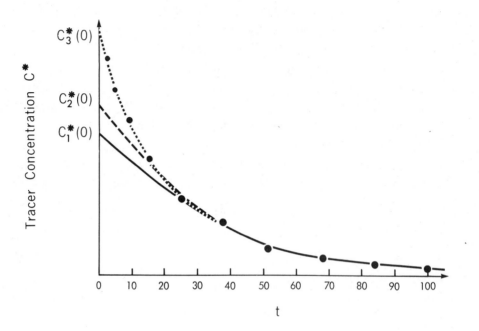

Fig. 19.1.   A bolus of tracer is injected into some compartment of access at
            zero time, and the impulse response is defined by the
            experimentally measured values of C*

            ●   ●   experimental points
            ─────── single exponential curve fit to final 6 data points
            - - - - double exponential curve fit to final 8 data points
            . . . . triple exponential curve fit to final 10 data points

            The sooner after injection the first sample for C* determination
            was taken

            (a) the greater the number of exponential functions required for
                a satisfactory curve fitting;

            (b) the greater the extrapolated C*(0) value.

On the Number of Compartments in a System and
Their Anatomical Localization

Since the number of compartments which comprise an "adequate" model of the
system seems to depend (among other factors) upon the time of the first
sample or measurement, clearly the number, n, is not unique. Different
investigators will find it convenient to use systems of different orders.
And since usually only a few of the total number of compartments are accessible experimentally, and even their value or spatial extent is not unique,
the anatomical localization of compartments can seldom be made. We can often
make the statement: "Compartment 2 is that compartment of which the plasma
forms a part", because we are infusing into the plasma and have defined all
inputs to be into compartment 2. Compartments in biological systems, as
they are usually defined, are conceptual entities, mathematical constructs
usually without real anatomical definition. For example, (section 15)
"Compartments [are] . . . regions . . . in which specific activity is
uniform". Nevertheless, when hydrodynamic models of compartments are constructed (e.g. Plentl and Gray, 1954) volumes and geometrical configurations
are, of course, unique and determinable.

Refreshingly exempt from the indictment of neglect of physics and chemistry
in the use of compartments is the work of Bischoff, Dedrick and their
colleagues (e.g. 1971, 1973). These investigators make careful measurements
of pharmaceutical substances in each of the compartments which they postulate
thus avoiding the "hypothetical compartment". They also pay strict
attention to the mode of chemical transport between compartments (e.g.
Michaelis-Menten enzyme facilitated, etc.) rather than blanketing a host of
unknowns with a simple rate constant.

Most germane to the issue of the reality of compartments are the arguments
of M.E. Wise (1971): If functions simpler than the multiexponential give
equally good fits to experimental data, how can one argue for the supremacy
of the multicompartmental model?

Input-Output Relations in Compartmental Systems

As in section 13, we shall define the input to the system as the rate of
infusion of tracer into the jth compartment, $R^*_{a\ j}(t)$, and the output as the

tracer concentration in the ith compartment, $C_i*(t)$. Let us now determine the output as a function of the input using the formulation (17.7), but open in the manner of Eq. (15.23):

$$\frac{dC_1*}{dt} = - K_{11}C_1* + K_{12}C_2* + \ldots + K_{1n}C_n*$$

$$\vdots$$

$$\frac{dC_j*}{dt} = K_{j1}C_1* + K_{j2}C_2* + \ldots - K_{jj}C_j* + \ldots$$

$$+ K_{jn}C_n* + \frac{R_{a\;j}^*(t)}{V_j}$$

$$\vdots$$

(19.23)

$$\frac{dC_i*}{dt} = K_{i1}C_1* + K_{i2}C_2* + \ldots - K_{ii}C_i* + \ldots$$

$$+ K_{in}C_n*$$

$$\vdots$$

$$\frac{dC_n*}{dt} = K_{n1}C_1* + K_{n2}C_2* + \ldots - K_{nn}C_n*$$

Rather than solving the general case directly, let us solve the special case of a three compartment system. The general result will then be seen to emerge immediately. We lose no further generality by setting $j = 1$ and $i = 2$ in (19.23). The system which we shall solve, then, is

$$\frac{dC_1*}{dt} = - K_{11}C_1* + K_{12}C_2* + K_{13}C_3* + R_{a1}^*/V_1$$

$$\frac{dC_2*}{dt} = K_{21}C_1* - K_{22}C_2* + K_{23}C_3*$$ (19.24)

$$\frac{dC_3*}{dt} = K_{31}C_1* + K_{32}C_2* - K_{33}C_3*$$

Following Ince, we now define the linear differential operators

$$F_{ii}(D) = - \frac{d}{dt} - K_{ii}$$ (19.25)

and introduce them into Eqs. (19.24):

$$F_{11}(D)C_1{}^* + K_{12}C_2{}^* + K_{13}C_3{}^* + R_{a\ 1}^*/V_1 = 0 \tag{19.26}$$

$$K_{21}C_1{}^* + F_{22}(D)C_2{}^* + K_{23}C_3{}^* = 0 \tag{19.27}$$

$$K_{31}C_1{}^* + K_{32}C_2{}^* + F_{33}(D)C_3{}^* = 0 \tag{19.28}$$

Since we wish to obtain the output, $C_2{}^*$, as a function of the input, $R_{a\ 1}^*$, we must now eliminate the variables $C_1{}^*$ and $C_3{}^*$ from the latter three equations. The reader may verify that the differential operators $F_{ii}(D)$ may be manipulated as ordinary algebraic constants. For example, the terms in $C_1{}^*$ may be eliminated between (19.26) and (19.27) by multiplying (19.26) by K_{21}, and operating upon (19.27) with $F_{11}(D)$:

$$K_{21}F_{11}(D)C_1{}^* + K_{12}K_{21}C_2{}^* + K_{13}K_{21}C_3{}^*$$
$$+ K_{21}R_{a\ 1}^*/V_1 = 0 \tag{19.29}$$

$$K_{21}F_{11}(D)C_1{}^* + F_{11}(D) \cdot F_{22}(D)C_2{}^*$$
$$+ K_{23}F_{11}(D)C_3{}^* = 0 \tag{19.30}$$

The operation on (19.27) by $F_{11}(D)$ consisted of taking first $-\frac{d}{dt}$ of both sides, then multiplying both sides by $-K_{11}$ and then adding the results. The entire procedure is represented by introducing $F_{11}(D)$ as if it were a simple constant factor, as shown in (19.30)(see problem 19.1). Subtracting (19.30) from (19.29) leaves an equation in the variables $C_2{}^*$ and $C_3{}^*$. A second equation in these two variables is found by eliminating $C_1{}^*$ from (19.27) and (19.28). These two equations in $C_2{}^*$ and $C_3{}^*$ can then be combined to leave a single equation giving $C_2{}^*$ as a function of $R_{a\ 1}^*$ as required. The whole process is similar to solving three linear algebraic equations, so we may apply Cramer's rule to obtain the solution in the form

$$\begin{vmatrix} F_{11}(D) & K_{12} & K_{13} \\ K_{21} & F_{22}(D) & K_{23} \\ K_{31} & K_{32} & F_{33}(D) \end{vmatrix} C_2{}^*(t)$$

$$= - \begin{vmatrix} K_{21} & K_{23} \\ K_{31} & F_{33}(D) \end{vmatrix} \frac{R_{a\ 1}(t)}{V_1} \tag{19.31}$$

It is now easy to generalize to the case of n compartments.

Recalling that $R_{aj}^*(t)$ is being regarded as an input, and $C_i^*(t)$ as an output, it may be seen with reference to (19.23) that the system so-defined is a linear one.

(i) Homogeneity

Multiplying all the equations (19.23) by some constant α_1 we see that since

$$R^*_{aj} \longrightarrow C^*_i$$

therefore

$$\alpha R^*_{aj} \longrightarrow \alpha C_i$$

Thus the system is homogeneous.

(ii) Superposition

Write Eq. (19.23') for a different input, $R^*_{aj}{}'$. The concentrations then become $C^*_1{}'$, $C^*_2{}'$ ... $C^*_i{}'$ ... If we now add Eqs. (19.23) + (19.23'), we obtain

$$R^*_{ai} + R^*_{ai}{}' \longrightarrow C^*_i + C^*_i{}'$$

Hence the system is homogeneous and superposable and therefore linear.

If a bolus of tracer M^*_0 is injected into the jth compartment and the resulting concentration-time curve is recorded in the ith compartment, then the function $h^*_i(t)$, expressed as a sum of exponentials, is the impulse response function of the system. The concentration-time curve $C^*_i(t)$ resulting from the general tracer infusion $R^*_{aj}(t)$ may then be obtained *either* from an equation similar to (19.31) involving the K's *or* by carrying out the convolution.

$$C^*_i(t) = \int_0^t \frac{h_i{}^*(\tau)}{M_0{}^*} \cdot R^*_{a\,j}(t - \tau)d\tau \qquad (19.32)$$

(cf. Eq. (13.11) and (13.16)).

Speaking Practically

(1) The interchange of tracer between the compartments of a multicompartmental system with tracee in a state of equilibrium is usually taken to be described by a set of n ordinary linear differential equations with constant coefficients in the variable C*. Usually no physical mechanism is postulated to account for such interchange.

(2) A result of the lack of a strong physical base for compartmental models is the inability to normalize a set of rate coefficients to allow for changes in the physical dimensions of different animals of a species.

(3) There is a large arbitrary element involved in the specification of the order, configurations and volumes of a multicompartmental system.

(4) Despite the rather weak physical base for compartmental models, if they
are used for the purpose of input or output discovery, the results obtained
should be correct because they can be shown to be in agreement with the
predictions of linear systems theory which *does* have a firm physical base.

*Input-output relationships are largely independent of the postulated internal
structure of the compartment system.*

Problem

19.1. In the second term of Eq. (19.30), C_2^* is operated on by the "product"
of the two operators, $F_{11}(D) \cdot F_{22}(D)$. Evaluate this term explicitly.

SECTION 20
DISPERSION: PHYSICS VS. COMPARTMENTS

The Saga of an Imagined Investigation

Plato conjectured that all matter was composed of four elemental substances
— earth, fire, air and water; and the Platonic theory held sway for a great
number of years. A geometrical figure was associated with each of the
elements, and all matter was regarded as composed of combinations of these
geometrical figures. But, as Bertrand Russell (1959) observed, it is less
important that Plato's elements are not the elementary particles of the
twentieth century than it is that he conceived of elementary substances
and that mathematical laws underlay their organization. To treat the theory
of compartments fairly, we must give it credit for describing biokinetic
processes correctly in a qualitative sense and input-output relations
correctly in a quantitative sense. We shall also see that it is serving as
a scaffolding by means of which the general theories of metabolic turnover
are being constructed. Compartmental models are, therefore, useful models
and may be regarded in the manner of Plato's elements as the predecessors of

more substantial representations. No scientific model is permanent; all
models are a "becoming". The well-known poem of Alfred Lord Tennyson
provides an allegory:

... all experience is an arch where thro'
   Gleams that untravell'd world whose margin fades
   For ever and for ever when I move.

from *Ulysses*

Science is the study of reproducible or verifiable measurements[†]. If
measurements of a given quantity made by two independent observers are found
to differ appreciably and repeatedly, then either (a) the methods of measurement employed are not in agreement or (b) they are not measuring the same
physical quantity.

For Example

If two observers in constant relative motion each measure the length of a
rod as the square root of the sum of the squares of its projections on three
spatial Cartesian coordinate axes at rest with respect to themselves, they
will obtain different values, because their methods of measurement are not
in agreement. To obtain the same value for the length of the rod they must
allow for their relative velocity.

For Example

If we tried to analyse a crystal of copper sulphate into its proportions of
earth, fire, air and water, we should probably never find two scientists who
agreed upon the proportions, because earth and fire particularly are not
regarded today as uniquely defined physical quantities. Our instruments
will not yield unique values for the earth or fire proportions. (We might
do surprisingly well with the air and water proportions!)

[†]This statement is not intended to be a comprehensive definition. One can
of course think of events of scientific interest recorded by a single
observer and not subject to ratification.

In the example which follows we may be able to see more clearly how the normalization problem described in the preceding section, the inability to measure a set of rate coefficients characteristic of a species or a subset of a species, reduces to a problem of the type (b) above: a lack of unique physical value in the quantities being measured.

Suppose that three investigators each decide to model a certain biological system by means of a two compartment model. In order to be able to focus entirely on the matter of physical content and avoid the knotty problems of variation due to numerical analysis of the data, let us assume that the three investigators work together on the problem until the point where all curve fitting has been completed. They then retire, each to his own study, to complete tne mathematical analysis.

The organism to be studied is the *Cylindrium*, a rare species so-named because of its peculiar cylindrical shape. Access to the fluid which occupies nearly the entire volume of the organism may be had only through the tail *canduit*, a small vessel at the tail end through which barrel-shaped respiratory cells are propelled. Fluid samples may be drawn periodically, and injections of tracer given through this canduit. The dispersion of the substance, fictitium, within the organism, is to be studied. Fictitium is known to be biologically inert within the Cylindrium. Unlabelled fictitium is uniformly distributed throughout all body fluids at the constant concentration of 50. mass units per unit volume, and its turnover rate is so small that it is nearly imperceptible. As a matter of fact, nobody is quite sure what biological function this material really serves.

Two members of the Cylindrium species are selected for experimentation. The length of the smaller specimen is 10. units and its sectional area is 1. square unit. The length and sectional radius of the larger specimen are $2^{1/3}$ times as large as the length and radius of the smaller specimen, so that the volume, V_{TOT}, of the larger is twice that of the smaller. The three investigators decide to inject an impulse of 1 mass unit of labelled fictitium (M_0^*) into the tail canduit of each Cylindrium, and subsequently draw samples for chemical assay at unit time intervals from t = 4 to t = 30 time units.

This example would not be much different from those studied in earlier
sections of the book were it not that in this instance we, you and I, are
given Privileged Information regarding the mode of transport of fictitium
in the Cylindrium. This Information is invested in us to keep with the
greatest secrecy; it is ours to enjoy but never to divulge. When labelled
fictitium is injected in the tail canduit of a Cylindrium of cross
sectional area A, it will very rapidly spread itself uniformly throughout
a "wafer" of depth h, where h has the value $1/A^{\dagger}$ (see Fig. 20.1). The
subsequent transport is governed entirely by the one-dimensional equation
of diffusion with the diffusion coefficient constant and equal to 0.5. This
Information is known to us but not to the three investigators. Not being
people to let such Information go idle, you and I decide to calculate the
concentration of tracer at the tail canduit over the duration of the
experiment. Fortunately for us, J. Crank has treated just this case in his
text "The Mathematics of Diffusion". With the x-axis drawn along the
cylindrical axis and the origin taken at the tail end, we calculate $C*(x,t)$
for x = 0 from

$$C*(0,t) = \frac{1}{2} C*(0,0) \sum_{n=-\infty}^{\infty} \left[\text{erf} \frac{h + 2n\ell}{2\sqrt{.5t}} + \text{erf} \frac{h - 2n\ell}{2\sqrt{.5t}} \right] \qquad (20.1)$$

where ℓ is the length of the cylinder. Thanks to our Privileged Information
we can evaluate h and $C*(0,0)$ ($C*(0,0) = M_0*/(hA)$). The data relating to
the start of the two experiments is summarized in Table 20.1. Fortunately
the solution (20.1) converges fairly rapidly using n = 0, ±1, ±2 ... and
$C*(0,t)$ may be calculated quite easily using a digital computer. In Table
20.2 values of $C*(0,t)$ have been listed for both Cylindria.

We are assured by our Privileged Information that these are precisely the
values that will be observed by the three investigators when they conduct
their two experiments.

---

†The depth, h, has the units of length, but the numerical value 1/A.

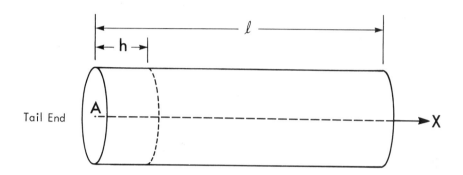

Fig. 20.1.   Diagram of a Cylindrium. The length of the organism is ℓ and
its constant cross sectional area is A. Fictitium injected at
the tail end will disperse rapidly throughout the wafer-shaped
cylinder of depth h. Subsequent transport of fictitium occurs
by a process of diffusion parallel to the axis of the cylinder.
The x-axis is drawn along the cylindrical axis with origin at
the tail end.

The three investigators, however, are totally oblivious of the "true"
mechanism of transport of fictitium and they embark upon the construction of
a closed, two compartment model of the dispersion of fictitium within the
Cylindrium. The model agreed upon by all three is shown in Fig. 20.2. The
three people work together until the point where the k's have been evaluated.
Using (15.22) they write

$$\frac{dM_1{}^*}{dt} = - k_{21}M_1{}^* + k_{12}M_2{}^* \qquad (20.2)$$

$$\frac{dM_2^*}{dt} = k_{21}M_1^* - k_{12}M_2^* \tag{20.3}$$

Specifying that compartment 1 is the sole compartment of experimental access, they then set the initial condition in compartment 2 as

$$M_2^*(0) = 0 \tag{20.4}$$

The reader can easily verify that the required solutions are of the form

$$M_1^* = E + \frac{k_{21}}{k_{12}} E\, e^{-(k_{12} + k_{21})t} \tag{20.5}$$

$$M_2^* = \frac{k_{21}}{K_{12}} E \left[1 - e^{-(k_{12} + k_{21})t} \right] \tag{20.6}$$

where the constants E, k_{12} and k_{21} are to be evaluated from the data. Dividing (20.5) by the volume V_1 of compartment 1 they obtain

$$C_1^* = \frac{E}{V_1} + \frac{k_{21}}{k_{12}} \cdot \frac{E}{V_1}\, e^{-(k_{12} + k_{21})t} \tag{20.5a}$$

The experimentally-measured values of C_1^* obtained by the investigators by sampling from the tail canduit are nearly identical with those we calculated using the diffusion equation and listed in Table 20.2. These two sets of data are fitted by a least squares procedure to the function (20.5a) providing values for E/V_1, $\frac{k_{21}}{k_{12}} \cdot \frac{E}{V_1}$ and $(k_{12} + k_{21})$ for each Cylindrium. From (20.5) and (20.6) or simply by applying the dilution principle (Eq. 19.22)

$$V_1 = \frac{M_1^*(0)}{C_1^*(0)} = \frac{M_o^*}{\dfrac{E}{V_1} + \dfrac{k_{21}}{k_{12}} \cdot \dfrac{E}{V_1}} \tag{20.7}$$

The above four quantities, as evaluated by the three investigators, have been listed in Table 20.3. Using the first three quantitites, the rate constants k_{12} and k_{21} can be calculated. But subtracting V_1 from V_{TOT}, the total volume of the Cylindrium, V_2, the volume of compartment 2 can be obtained. These quantities are also contained in Table 20.3.

The function (20.5a), which is the sum of a constant and an exponential term, matches the "experimental" data reasonably well over the experimental period $t = 4$ to $t = 30$, despite the fact that the data are given exactly by (20.1), an infinite sum of error functions. The matching of "experimental" and fitted curves may be seen in Table 20.4. There is greater deviation between the curves for lower values of t.

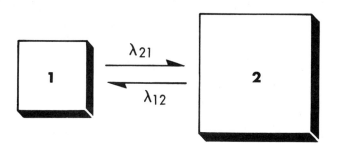

Fig. 20.2.   Closed, two compartment system postulated for the description of
             fictitium transport.

TABLE 20.1

	Smaller Cylindrium	Larger Cylindrium
(1) Length, ℓ	10.	$10 \cdot 2^{1/3}$
(2) Sectional area, A	1.	$1 \cdot 2^{2/3}$
(3) Total volume, V_{TOT}: (1)·(2)	10.	$20 \cdot$
(4) Tracer mass injected, M_o^*	1.	$1 \cdot$
(5) Initial tracer distribution depth, h: 1/(2)	1.	$2^{-2/3}$
(6) Initial tail end tracer concentration $C^*(0,0)$: (4)/{(2) · (5)}	1.	$1 \cdot$
(7) Coefficient of diffusion, D	0.5	0.5
(8) Steady tracee concentration, C	50.	50.

Data relating to the experiments on the two Cylindria

TABLE 20.2

Time	$C*(0,t)$ Smaller Cylindrium	$C*(0,t)$ Larger Cylindrium
1.	.683	.471
2.	.520	.344
3.	.436	.284
4.	.383	.247
5.	.345	.222
6.	.317	.203
7.	.295	.188
8.	.276	.176
9.	.261	.166
10.	.248	.158
11.	.237	.151
12.	.227	.144
13.	.218	.139
14.	.211	.134
15.	.204	.129
16.	.197	.125
17.	.192	.121
18.	.186	.118
19.	.181	.115
20.	.177	.112
21.	.173	.109
22.	.169	.107
23.	.165	.105
24.	.162	.102
25.	.159	.100
26.	.156	.0983
27.	.153	.0965
28.	.150	.0948
29.	.148	.0931
30.	.145	.0916
∞	.100	.0500

The concentration of tracer at the tail end of the two Cylindria at various times after injection of a bolus of tracer, calculated from the equation

(Crank, 1956)

$$C*(0,t) = \frac{1}{2} C*(0,0) \sum_{n=-\infty}^{\infty} \left\{ erf \frac{h+2n\ell}{2\sqrt{.5t}} + erf \frac{h-2n\ell}{2\sqrt{.5t}} \right\}$$

$C*(0,0)$, h and ℓ are listed in Table 20.1

TABLE 20.3

	Smaller Cylindrium	Larger Cylindrium
$\frac{E}{V_1}$	.10	.05
$\frac{k_{21}}{k_{12}} \cdot \frac{E}{V_1}$	.301	.197
$k_{12} + k_{21}$	.0661	.0555
$\dfrac{M_0 *}{\frac{E}{V_1} + \frac{k_{21}}{k_{12}} \cdot \frac{E}{V_1}} = V_1$	2.49	4.05
k_{12}	.0165	.0112
k_{21}	.0496	.0442
$V_{TOT} - V_1 = V_2$	7.51	15.95

Quantities calculated from the two compartment model of Cylindrium kinetics by fitting the experimental data (Table 20.4) to the function (20.5a)

TABLE 20.4

Time	Smaller Cylindrium		Larger Cylindrium	
	$C_1*(t)$ Fitted	$C*(0,t)$ Experimental	$C_1*(t)$ Fitted	$C*(0,t)$ Experimental
4.	.331	.383	.247	.208
5.	.316	.345	.222	.199
6.	.302	.317	.203	.191
7.	.289	.295	.188	.184
8.	.277	.276	.176	.177
9.	.266	.261	.166	.170
10.	.255	.248	.158	.163
11.	.245	.237	.151	.157
12.	.236	.227	.144	.151
13.	.227	.218	.139	.146
14.	.219	.211	.134	.141
15.	.212	.204	.129	.136
16.	.204	.197	.125	.131
17.	.198	.192	.121	.127
18.	.191	.186	.118	.123
19.	.186	.181	.115	.119
20.	.180	.177	.112	.115
21.	.175	.173	.109	.112
22.	.170	.169	.107	.108
23.	.166	.165	.105	.105
24.	.162	.162	.102	.102
25.	.158	.159	.100	.0993
26.	.154	.156	.0983	.0966
27.	.150	.153	.0965	.0941
28.	.147	.150	.0948	.0917
29.	.144	.148	.0931	.0895
30.	.141	.145	.0916	.0873
∞	.100	.100	.0500	.0500

Matching of the fitted curve to the experimental data points. In fact the experimental data can be generated by Eq. (20.1), a solution to the diffusion equation. The fitted curve has been obtained by conforming a function of the form (20.5a) to the experimental points by a least squares method.

With the evaluation of k_{12} and k_{21} the three investigators part company, each retiring to calculate the characteristic or invariant parameters of the Cylindrium - fictitium system. Investigator #1 has decided to adopt the representation of Eq. (19.1) and regards the λ's as characteristic quantities; investigator #2 adopts the mathematically equivalent form (19.4); and investigator #3 adopts the mathematically equivalent form (19.5). We may relate their calculations using (19.6) by observing that

$$k_{ij} = \lambda_{ij} = \frac{b_{ij}}{V_j} = \frac{\rho_{ij}}{M_j} \tag{20.8}$$

Investigator #1, observing that $\lambda_{12} = k_{12}$ and $\lambda_{21} = k_{21}$ differ somewhat in magnitude between the two members of the species Cylindrium, attributes the difference to statistical error in measurement and calculates an arithmetic mean value for each

$$\bar{\lambda}_{12} = 0.5(.0165 + .0112) = .0139$$

$$\bar{\lambda}_{21} = 0.5(.0496 + .0442) = .0469 \ [\text{time}^{-1}]$$

Investigator #2, using

$$b_{ij} = V_j \, k_{ij} \tag{20.9}$$

calculates

$$b_{12} = .124$$

$$b_{21} = .124$$

for the smaller Cylindrium, and

$$b_{12} = .179$$

$$b_{21} = .179$$

for the larger Cylindrium. Notwithstanding the observation that $b_{ij} = b_{ji}$ within a given individual, Investigator #2 elects to take a mean for each of b_{12} and b_{21}. That is

$$\overline{b_{12}} = 0.5(.124 + .179) = .152$$

$$\overline{b_{21}} = 0.5(.124 + .179) = .152 \, [\text{volume} \cdot \text{time}^{-1}]$$

Investigator #3 using

$$\rho_{ij} = M_j \, k_{ij} \tag{20.10}$$

and

$$\rho_{12} = \rho_{21} \tag{20.11}$$

from (15.17), obtains

$$\rho_{21} = M_1 \, k_{21}$$

$$= V_1 C \, k_{21} \tag{20.12}$$

The right hand side of (20.12) can be evaluated using Tables 20.1 and 20.3:

$$\rho_{21} = (2.49)(50.)(.0496) = 6.18$$

for the smaller Cylindrium, and

$$\rho_{21} = (4.04)(50.)(.0442) = 8.95$$

for the larger Cylindrium. Since $\rho_{12} = \rho_{21}$, he calculates a single mean value for ρ for the Cylindrium - fictitium system.

$$\bar{\rho} = 0.5(6.18 + 8.95) = 7.57 \ [\text{mass} \cdot \text{time}^{-1}]$$

Each investigator will now publish what he believes are the characteristic or invariant parameters governing the dispersion of fictitium in the Cylindrium species.

Investigator #1 will report that there are two characteristic fractional disappearance rates of .0139 and .0469 [time^{-1}].

Investigator #2 will report that the characteristic clearance is .152 [volume · time^{-1}] for each compartment.

Investigator #3 will report that the characteristic bidirectional mass flux is 7.57 [mass · time^{-1}].

——— *Etcetera*, because there are not just 3 but an indefinite number of possible compositions for the coefficients of the multicompartmental differential equations. You and I know, by Privileged Information, that the desired transport parameter here is the diffusion coefficient $D = .50$ [length^2 · time^{-1}]. But none of the 3 investigators, by applying

multicompartmental analysis, *can* obtain the "true"[†] characteristic quantity,
D. They cannot obtain the "true" picture of transport in terms of the λ's,
b's or ρ's for the same reason that copper sulphate cannot be broken down
into terms of earth, fire, air and water: *there is no consistent physical*
basis for the components into which the compound system (biological trans-
port and copper sulphate) is being analysed.

But alas, the gift of Privileged Information seldom comes. We are all in the
position of the three investigators. We can, by limited access to the system,
construct multicompartmental models. And these models are often of great
use — heuristically and for input-output studies. But we cannot, using
multicompartmental models, determine a unique, characteristic, or invariant
structure of the system.

Problems

20.1. We calculated ρ_{21} from (20.10) and (20.12), obtaining values of 6.18
and 8.95 for the two Cylindria. We then set $\rho_{12} = \rho_{21}$ by (20.11). Calculate
ρ_{12} directly from (20.10) and show that you get the same answers as before.

20.2. Do you see why $b_{12} = b_{21}$ for each of the two Cylindria?

[†]There seems to be nothing such as absolute "truth" in science. The best we
can hope for is that our studies on transport in intact organisms will
yield results which are both reproducible and contiguous with the studies of
transport in non-living systems.

SECTION 21
CURVE FITTING: EXPONENTIALS AND POWER FUNCTIONS

Exponentials

Hitherto almost nothing has been said about the process by which experimental data can be conformed to a sum of exponentials, or about the difficulties encountered in the interpretation of such curve fits. This omission has been by design. The process of curve fitting will be discussed in Appendix B, and the interpretation has been reserved until the present time. It was desirable to assume at the beginning of our studies that a unique sum of exponentials could be associated with every set of data points, and in this way concentrate our attention on the physical implications of this particular exponential function. Under these circumstances, it could be seen clearly that any uncertainty in the interpretation of the data derived squarely from uncertainty in the physical interpretation of the mathematical model. However, now that the latter point has been emphasized, we should observe that in all problems in which curve fitting is involved, further uncertainty is engendered by the lack of uniqueness of the curve fitting process itself; that is, several different mathematical functions will often fit a set of data points equally well. There may be slight variations in the sums of squares of the differences between the ordinates of the measured and the fitted points, but these deviations will not be significant. This problem is expected to be more troublesome when the functions being fitted contain more parameters — i.e. more degrees of freedom — than are required to fit the data with a prescribed level of accuracy, and when the function being fitted derives from a poor mathematical model of the physical process. A great deal has already been written on the subject of uncertainty and errors associated with exponential curve fitting (e.g. Glass and de Garreta, 1971; Myhill, 1968) but the matter is never more clearly illustrated than by Riggs (1963). This author shows for example, how different methods of numerical analysis can fit two quite different functions of the form

$$y(t) = B_1 \, e^{-\beta_1 t} + B_2 \, e^{-\beta_2 t} \tag{21.1}$$

to the same set of data points. The functions

$$y_1(t) = 42.0 \, e^{-.366t} + 21.5 \, e^{-.035t} \tag{21.2}$$

and

$$y_2(t) = 37.7\ e^{-.506t} + 28.0\ e^{-.049t} \qquad (21.3)$$

will provide equally satisfactory visual matches to the data over the range
of $0 \le t \le 24$. A comparison of the functions $y_1(t)$ and $y_2(t)$ over this
range of t is given in Table 21.1 (and is also illustrated in Fig. 6-21 of
Riggs' text).

TABLE 21.1

t	$y_1(t)$	$y_2(t)$
0.	63.5	65.7
1.	49.9	49.4
2.	40.2	39.1
3.	33.4	32.4
4.	28.4	28.0
5.	24.8	24.9
6.	22.1	22.7
7.	20.1	21.0
8.	18.5	19.6
9.	17.2	18.4
10.	16.2	17.4
11.	15.4	16.5
12.	14.6	15.6
13.	14.0	14.9
14.	13.4	14.1
15.	12.9	13.4
16.	12.4	12.8
17.	11.9	12.2
18.	11.5	11.6
19.	11.1	11.0
20.	10.7	10.5
21.	10.3	10.0
22.	9.97	9.53
23.	9.62	9.07
24.	9.29	8.64

Comparison of the values of the functions $y_1(t) = 42.0e^{-.366t} + 21.5e^{-.035t}$

and $y_2(t) = 37.7e^{-.506t} + 28.0e^{-.049t}$ over the range $0 \leq t \leq 24$. The two functions are very nearly equal-valued over this range of t.

Let us now suppose that $y(t)$ is the impulse response function of the two compartment model of the inulin system described in section 17. The input is applied to, and the output taken from compartment 1, the compartment of access. Let us evaluate K_{12}, K_{21} and K_{01} for each of the functions (21.2) and (21.3) using equations (17.30) through (17.33). The results are shown in Table 21.2. It may be seen that the K's differ appreciably in magnitude for the two functions. This situation is intolerable and often plagues the compartment modeller: two functions providing equally good matches to the experimental data generate "characteristic" interchange coefficients which differ radically in magnitude. Note, however, that the coefficient K_{01} does not differ nearly as much as do K_{12} and K_{21}; and $B_1/\beta_1 + B_2/\beta_2$ differs only by about 12%. Can the reader perceive why that should be so? We shall return to this matter subsequently.

TABLE 21.2

Function	B_1	B_2	β_1	β_2	K_{12}	K_{21}	K_{01}	$\dfrac{B_1}{\beta_1} + \dfrac{B_2}{\beta_2}$
$y_1(t)$	42.0	21.5	.366	.035	.147	.167	.0871	729.
$y_2(t)$	37.7	28.0	.506	.049	.244	.210	.102	646.

Calculation of K_{12}, K_{21} and K_{01} from the parameters of the functions $y_1(t)$ and $y_2(t)$ on the assumption that these functions describe the impulse response of a two compartment system.

Power Functions

As indicated in section 19, functions other than sums of exponentials have been found to fit tracer data quite well, using fewer ajustable parameters. This strongly suggests that the physical model on which the compartment rests is fundamentally poor. Functions (power-type functions), which have been found to fit tracer data well, include

pure powers of t
$$y = At^{-\alpha} \qquad , \qquad (21.4)$$
gamma-type
$$y = At^{-\alpha}e^{-\beta t} \qquad , \qquad (21.5)$$
and first-passage-time
$$y = At^{-\alpha}e^{-\beta t - \lambda/t} \qquad (21.6)$$
with $.1 < \alpha < 3$. The case of $\alpha = 3/2$ in (21.6) is well known in diffusion theory. The efficacy of the power function in fitting curves of disappearance of bone-seeking radioisotopes such as ^{45}Ca and ^{177}Lu was shown by Anderson, Osborn, Tomlinson and Weinbren in 1963. Further empirical evidence for the power function was adduced by Anderson, Osborne, Tomlinson and Wise (1969), who showed how excellent curve fits could be obtained for plasma activity-time curves for a variety of heavy-metal isotopes, and for the urinary excretion of tritium over a 250-day period. These results are most convincing.

These power-type functions, although originally found empirically, can be derived from probabilistic models of tracer kinetics — models which differ substantially from compartmental models. A probabilistic model which is especially suitable for understanding the kinetics of bone-seeking elements has been proposed by M.E. Wise and his colleagues in two papers (Wise et al. 1968 and Wise 1974). The gist of their argument is as follows. Suppose that radiocalcium is injected into an intact animal or human and that it finds its way to bone. A radiocalcium atom may then wander from bone to non-bone, and then back to bone. Call this one cycle. Suppose that this cycle may occur with probability p. The atom may wander from bone and be excreted, never to return to bone. This will occur with probability (1-p). Then the probability that a given atom makes just one cycle and is excreted is p(1-p). The probability of just two cycles is $p^2(1-p)$ etc. One has to introduce *time* into the system: the time taken to make exactly one cycle, exactly two cycles etc. Let these times by governed by the functions $y_1(t)$, $y_2(t)$ ... Given $y_1(t)$, all other $y(t)$'s can be deduced. Then the probability $z(t)$ governing times of residence within the system (injection to excretion) will look something like
$$z(t) = \ldots + (1-p)p\, y_1(t) + (1-p)p^2\, y_2(t) + \ldots \qquad (21.7)$$
The y's were evaluated formally in terms of their cumulants. Then for $z(t)$ to have the form of (21.5) or (21.6), $y_1(t)$ had to be a (probability density)

function for first passage times of tracer particles undergoing random walks
with drift through a mixed medium. (In a homogeneous compartment these
particles can only move randomly without drift.) For further details the
reader is referred to the original papers, or to Wise's chapter in the book
edited by Patil et al.

An equation of the form (21.5) was derived for radiocalcium by Burkinshaw et
al. (1969) on the basis of a model which postulates a continuously expanding
exchangeable calcium pool. They define M_t as the mass of a hypothetical
pool of calcium uniformly labelled at the same specific activity as the
plasma, and postulate that

$$M_t \doteqdot M_1 \, t^b \, , \; M_1, \; b \; \text{constants} > 0; \; t < 15 \; \text{hours}.$$

Then the total radioactivity in the exchangeable pool, M_t^* in our nomenclature,
is the product of tracee mass with specific activity a_t, or

$$M_t^* = M_1 t^b a_t \tag{21.8}$$

They further postulate that

$$\frac{dM_t^*}{dt} = -ka_t, \quad k > 0 \tag{21.9}$$

Eliminating M_t^* between (21.8) and (21.9) they derive for specific activity
an expression of the form (21.5) with $\alpha = b$ and βt replaced by βt^{1-b}, $b < 1$.

An expression of the form (21.5) was derived in quite a different way by
Marcus (1975) using a modification of a single compartment model. The
expected number of sites within the system which retain tracer for times
between t and $t + dt$ after injection is taken to be proportional to $\rho(t)dt$,
where $\rho(t)$ is called a "retention density". It is suggested that

$$d\rho(t) = -k(t)\rho(t) \, dt$$

or

$$\frac{d\rho(t)}{dt} = -k(t)\rho(t) \tag{21.10}$$

Here $k(t)$ is a time-varying analog of the usual rate constant k. If, for
example,

$$k(t) = \lambda + 1/t, \quad \lambda > 0 \tag{21.11}$$

then Marcus derives that $p(t)$, the fraction of the amount of material
injected remaining at time t, is given by

$$p(t) = At^{-\alpha}e^{-\lambda t} \qquad (21.12)$$

whicn is of the form (21.5).

The reader is also referred to the clock model of J.H. Marshall (1967), and in the same book, to the specific compartmental model of S.R. Bernard, where a power function results from a sum of exponential terms.

The preceding three derivations have been sketched in order to demonstrate that there are various models of the tracer transport system, both probabilistic and deterministic, which yield power functions rather than sums of exponentials for their solution. I have tried to capture the physical essence of three or four papers in just three paragraphs, and in so doing may have sacrificed some accuracy in the interests of brevity.

Because the various power functions contain only one term comprising several factors, curve fitting is facilitated by first making a log-log plot of the data. For example, if we wish to curve fit to a function of the form

$$y = At^{-\alpha}e^{-\beta t} \qquad (21.5)$$

our labour will be reduced by taking logarithms of both sides:

$$\log y = \log A - \alpha \log t - \beta t \qquad (21.13)$$

One could now proceed along the lines suggested in Appendix B.

As an example of the relative precisions of the power and double exponential functions in fitting tracer disappearance data let us return to Table 17.1, which gives plasma ^{14}C-hydroxymethyl inulin values following a single injection of this tracer. When the data were fitted to a function of the form (21.5) the result was

$$C*(t) = 48.19 \times 10^3\, t^{-.4239}\, e^{-.0120t} \qquad (21.14)$$

involving only three parameters. When fitted to a double exponential the result was

$$C*(t) = 41885.\ e^{-.278t} + 14434.\ e^{-.0198t} \qquad (17.34)$$

using four parameters. The capacities of these functions to fit the data is compared in Table 21.3. It may be seen that the two functions fit the data with comparable degrees of precision, despite the extra parameter available to the double exponential.

TABLE 21.3

Time (min)	C* measured (dpm/ml)	C* fitted (double exponential)	Sum of Squares of Residuals x 10^{-6} (double exponential)	C* fitted (power function)	Sum of Squares of Residuals x 10^{-6} (power function)
1	47700	45900	3.24	47600	.01
3	30800	31800	1.00	29000	3.24
5	22200	23300	1.21	22800	.36
8	17400	16900	.25	17900	.25
12	13600	12900	.49	14400	.64
20	10300	9900	.16	10500	.04
30	8100	8000	.01	7800	.09
45	5700	5900	.04	5500	.04
60	4000	4400	.16	4100	.01
90	2500	2400	.01	2400	.01
120	1400	1300	.01	1500	.01
			Total = 6.58×10^6		Total = 4.70×10^6

Expansion of Table 17.1. Impulse response function to an intravenous injection of 44.4×10^6 dpm ^{14}C-hydroxy-methyl inulin to a dog. The actual plasma tracer concentrations C* are shown, as well as the values of C*(t) calculated from the fitted double exponential function

$$C^* = 41885. \ e^{-.278t} + 14434. \ e^{-.0198t} \qquad \text{(4 parameters)}$$

and the values of C*(t) calculated from the fitted power-type function

$$C^* = 48190. \ t^{-.4289} e^{-.0120t} \qquad \text{(3 parameters)}$$

In this example the sum of squares of residuals is less for the power function with three adjustable parameters than for the double exponential with four adjustable parameters.

Speaking Practically

If two people are set the task of fitting the same set of data to, say, a
double exponential function, you can expect to find two entirely different
functions. They may describe similarly shaped curves over a certain range
of the independent variable, but the actual numbers which appear in the
functions may be completely different. A good alternative to the sum of
exponentials is the power function. It will often fit the data as well as or
better than the sum of exponentials. These power functions have a theoretical base which is in the process of development.

References

Anderson, J., Osborn, S.B., Tomlinson, R.W.S. and Weinbren, I., Some applications of power law analysis to radioisotope studies in man, *Phys. Med. Biol.* 8, 287-295 (1963).

Anderson, J., Osborn, S.B., Tomlinson, R.W.S. and Wise, M.E., Clearance curves for radioactive tracers — sums of exponentials or powers of time?, *Phys. Med. Biol.* 14, 498-501 (1969).

Bergner, P.-E.E., The significance of certain tracer kinetical methods, especially with respect to the tracer dynamic definition of metabolic turnover, *Acta Radiol. Scand.* 210, Suppl., 1-59 (1962).

Berman, M. and Schoenfeld, R., Invariants in experimental data on linear kinetics and the formulation of models, *J. Appl. Physics* 27, 1361-1370 (1956).

Berman, M., Shahn, E. and Weiss, M.J., The routine fitting of kinetic data to models: a mathematical formalism for digital computers, *Biophys. J.* 2, 275-287 (1962).

Berman, M., Shahn, E. and Weiss, M.J., Some formal approaches to the analysis of kinetic data in terms of linear compartmental systems, *Biophys. J.* 2, 289-316 (1962).

Bernard, S.R., in *Compartments, Pools and Spaces in Medical Physiology*, U.S. Atomic Energy Commission Monograph, 471, 1967.

Bischoff, K.B., Dedrick, R.L., Zaharko, D.S. and Longstreth, J.A., Methotrexate pharmacokinetics, *J. Pharmaceut. Sci.* 60, 1128-1133 (1971).

Bright, P.B., The volumes of some compartment systems with sampling and loss from one compartment, *Bull. Math. Biol.* 35, 69-79 (1973).

Burkinshaw, L., Marshall, D.H., Oxby, C.B., Spiers, F.W., Nordin, B.E.C. and Young, M.M., Bone turnover model based on a continuously exchangeable calcium pool, *Nature* 222, 146-148 (1969).

Burton, A.C., The basis of the master reaction in biology, *J. Cell and Comp Physiol.* 9, 1-14 (1936).

Crank, J., *The Mathematics of Diffusion*, Oxford University Press, London, 1956.

Dedrick, R.L., Zaharko, D.S. and Lutz, R.J., Transport and binding of methotrexate in vivo, *J. Pharmaceut. Sci.* 62, 882-890 (1973).

Dedrick, R.L., Animal scale-up, *J. of Pharmacokinetics and Biopharmaceutics* 1, 435-461 (1973).

Glass, H.I. and De Garreta, A.C., The quantitative limitations of exponential curve fitting, *Phys. Med. Biol.* 16, 119-130 (1971).

Harcourt, A.V. and Esson, W., On the laws of connexion between the conditions of a chemical change and its amount, *Philosophical Transactions of the Royal Society of London* 156, 193-221 (1866).

Hart, H.E., Analysis of tracer experiments in non-conservative steady-state systems, *Bull. Math. Biophys.* 17, 87-94 (1955).

Heading, J., *Matrix Theory for Physicists*, Longmans, Green & Co. Ltd., London, 1958.

Ince, E.L., *Ordinary Differential Equations*, Dover Publications, 1956.

Marcus, A.H., Power laws in compartmental analysis. Part I: a unified stochastic model, *Math. Biosci.* 23, 337-350 (1975).

Marshall, J.H., Calcium pools and the power function, in *Compartments, Pools and Spaces in Medical Physiology*, U.S. Atomic Energy Commission Monograph, 1967.

Milsum, J.H., *Biological Control Systems Analysis*, McGraw-Hill, New York, 1966.

Myhill, J., Investigation of the effect of data error in the analysis of biological tracer data from three compartment systems, *J. Theor. Biol.* 23, 218-231 (1968).

Norwich, K.H., Radziuk, J., Lau, D. and Vranic, M., Experimental validation of nonsteady rate measurements using a tracer infusion method and inulin as tracer and tracee, *Can. J. Physiol. Pharmacol.* 52, 508-521 (1974).

Perl, W., A method for curve-fitting by exponential functions, *Int. J. Appl. Radiation and Isotopes* 8, 211-222 (1960).

Plentl, A.A. and Gray, M.J., Hydrodynamic model of a 3-compartment catenary system with exchanging end compartments, *Proceedings of the Society for Experimental Biology and Medicine* 87, 595-600 (1954).

Rescigno, A. and Segre, G., *Drug and Tracer Kinetics*, Blaisdell, Waltham, Mass. 1966.

Riggs, D.S., *The Mathematical Approach to Physiological Problems*, William and Wilkins, Baltimore, 1963.

Russell, B., *Wisdom of the West*, Crescent, 1959.

SAAM Manual. Published by the U.S. Department of Health, Education and Welfare, Public Health Service.

Sheppard, C.W., The theory of the study of transfers within a multicompartment system using isotopic tracers, *J. Appl. Physics* 19, 70-76 (1948).

Soong, T.T., Pharmacokinetics with uncertainties in rate constants, *Math. Biosci.* 12, 235-243 (1971).

Soong, T.T., Pharmacokinetics with uncertainties in rate constants II: sensitivity analysis and optimal dosage control, *Math. Biosci.* 13, 391-396 (1972).

Teorell, T., Kinetics of distribution of substances administered to the body. I. Extravascular modes of administration, *Arch. Intern. Pharmacodyn.* 57, 205-240 (1937).

Wise, M.E., Osborn, S.B., Anderson, J. and Tomlinson, R.W.S., A stochastic model for turnover of radiocalcium based on the observed power laws, *Math. Biosci.* 2, 199-224 (1968).

Wise, M.E., The evidence against compartments, *Biometrics* 27, 262 (1971).

Wise, M.E., Interpreting both short- and long-term power laws in physiological clearance curves, *Math. Biosci.* 20, 327-337 (1974).

Wise, M.E., Skew distributions in biomedicine including some with negative powers of time, in *Statistical Distributions in Scientific Work*, G.P. Patil, S. Kotz and J.K. Ord (Editors), D. Reidel Publishing Co., Dordrecht, Holland, 1975.

CHAPTER V
RATES OF APPEARANCE IN STEADY STATE SYSTEMS

SECTION 22
CALCULATION OF RATE OF APPEARANCE IN STEADY STATE SYSTEMS

I. One Compartment Approximation

Under conditions of steady state (as defined rigorously by equations (13.1) to (13.3)) the rate of appearance of tracee is constant and equal to its rate of disappearance. We shall now take up the issue of how best to measure such steady state rates. Historically, the first method for making such calculations was to model the system as a single compartment. From (6.7) we have

$$R^{*}_{d,v} = a \cdot R_{d,v} \qquad (22.1)$$

For a well-intermixed region of volume, V, we may multiply both sides of this equation by V to give simply

$$R^{*}_{d} = aR_{d} \qquad (22.2)$$

Let us set up a one compartment model as illustrated in Fig. (22.1).

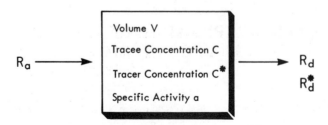

Fig. 22.1. Open single compartment model. Tracee enters and leaves at the
            constant rate $R_a = R_d$. Tracer leaves at the rate $R_d^* = aR_d$.

(a) Tracer Injection Method

Suppose that tracer is injected as a bolus at t = 0 and that no additional
tracer is added thereafter. Writing the differential equation in the usual
way

$$\frac{dM_1^*}{dt} = - k_{01}M_1^* \quad .$$
(22.3)

Dropping the subscripts and dividing through by the volume $V^\dagger$

$$\frac{dC^*}{dt} = -KC^*$$
(22.4)

We recognize that

$$R_d^* = KVC^*$$
(22.5)

$\dagger$Since both Mrs. Nish, the typist, and I became thoroughly confused by the
change in k's from lower to upper case, it is fair to assume that you, the
reader, are confused also. Here's what happened. From the defintion,
(16.5) we have $k_{11} = k_{01}$, and from the definition (17.4) we have $K_{01} = k_{01}$.
But rather than carry around two subscripts throughout the section, I have
set $K = K_{01}$. Done.

since both define the rate [mass · time^{-1}] at which tracer leaves the system (refer to Eq. (6.2)). Solving (22.4)

$$C*(t) = A e^{-Kt} ,$$ (22.6)

where

$$C*(0) = A$$

The volume may be approximated by applying the dilution principle in the usual way (Eq. (19.22)) to the conditions at zero time:

$$V = \frac{M_o*}{A}$$ (22.7)

Combining (22.2), (22.5) and (22.7) we have

$$R_d = \frac{C}{C*} R_d^* = KVC = \frac{M_o*KC}{A}$$ (22.8)

where all quantitites on the right-hand side may now be measured. The natural logarithm of the measured values of C* is plotted against t and the best straight line is drawn through all points. The slope of this straight line is –K and the intercept is the logarithm of A. V and R_d may now be measured.

Usually the measured data do not conform perfectly to a straight line on such a semi-logarithmic plot. The early points are especially recalcitrant, tending to rise above any straight line which fits the remainder of the points. Occasionally, however, the approximation is quite good (Fig. 22.2). Let us write the final result in the form

$$R_a = R_d = \frac{M_o*C}{A/K}$$ (22.9)

(b) Constant Tracer Infusion Method

Suppose that tracer is infused at a constant rate, R_a^*, until tracer achieves a steady state[†]. Since both C and C* are now constant, we obtain immediately from (22.2)

$$R_a = R_d = R_d^*/a = R_a^*/a$$ (22.10)

[†]Tracee, throughout this section and throughout this chapter, is presumed to be in steady state.

where all quantities on the right-hand side can now be measured. This is
the "tracer infusion" formula.

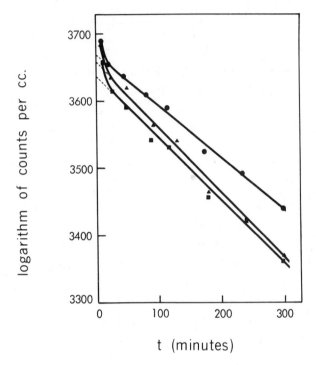

t (minutes)

Fig. 22.2.   The logarithm of the counts per c.c. of plasma phosphorus plotted
            against time. ● Dog 1B; ▲ Dog 2B; ■ Dog 3B.
From: Journal of General Physiology 26, 337, 1943 by D.B. Zilversmit et al.
Reproduced by permission of The Rockefeller University Press.

There is an important difference between the injection formula (22.9) and
the infusion formula (22.10). The injection formula is an approximation
which will be superseded when we analyse the matter more closely. The
infusion formula will be seen to hold even when we carry out the more exact
calculation. Eq. (22.9) was derived first in two papers published in 1943:
the first by Zilversmit, Entenman and Fishler, and the second by Zilversmit,
Entenman, Fishler and Chaikoff. Eq. (22.10) was derived first by Stetten,
Welt, Ingle and Morley in 1951. These are now classical equations.

Clearly, if two tracer methods can be used to measure the same physiological quantity, R_a, they should, if applied to the same subject at the same time, yield the same result. It is to the credit of Cowan and Hetenyi that they decided to test this hypothesis for the glucose system in dogs. They ascertained that the two formulae (22.9) and (22.10) did *not*, in fact, yield quite the same result, but rather (22.9) gave values about 7% higher than (22.10). This was a result of the approximation involved in (22.9).

(c) Arbitrary Tracer Infusion

There is nothing sacrosanct about tracer inputs in the form of impulses and step functions. Within the framework of a one compartment model we can calculate the rate of appearance of tracee using a tracer infusion that may be any function of time. Suppose we deal with an open one compartment model with arbitrary tracer input, $R_a^*(t)$ (Fig. 22.3). Then

$$\frac{dC^*}{dt} = \frac{R_a^*(t)}{V} - KC^* \tag{22.11}$$

Since

$$R_d = KVC \tag{22.8}$$

therefore

$$R_d = \left(\frac{C}{C^*}\right) KVC^* \tag{22.12}$$

and, eliminating K between (22.11) and (22.12)

$$R_a = R_d = \frac{1}{a}\left[R_a^*(t) - V\frac{dC^*}{dt}\right] \tag{22.13}$$

Eq. (22.13) is a general formula giving R_a (a constant) as function of $R_a^*(t)$ and $\frac{dC^*}{dt}$. If the system is not well-represented by a single compartment, then, of course, the right-hand side of (22.13) will not remain constant. The injection and constant infusion formula (22.9) and (22.10) will be seen to emerge as special cases of (22.11) (problem 22.1). Eq. (22.13) will be derived by an alternative method in section 30.

The reader is referred to section 26 for a more general treatment of the arbitrary tracer infusion. The use of (22.13) with an arbitrary tracer infusion is not very practical because the term $V\frac{dC^*}{dt}$ introduces inaccuracies, as we shall see in section 31.

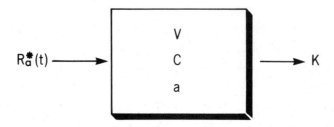

Fig. 22.3. An arbitrary tracer infusion $R_a^*(t)$, may be used to calculate the
rate of appearance of tracer, R_a. Calculations are made
assuming that the physical system may be represented by the one
compartment model depicted.

Speaking Practically

This section really contains only one applicable equation: (22.10). It
implies the following. If you deal with a system which is in steady state
with respect to tracee, and you infuse tracer at a steady rate R_a^*
(dpm/min or μCi/min) until the tracer also reaches a steady state (as
evidenced, essentially, by a constant plasma tracer concentration), then the
rate of appearance of tracee is given by the equation

$$R_a = R_a^*/a \qquad (22.10)$$

That is, the rate of appearance of tracee is given by the rate of infusion of
tracer divided by the constant plasma specific activity. For reasons which
will become apparent later, the tracer injection equation (22.9) can be
improved.

Problem

22.1. Solve Eq. (22.11) as a *linear equation* (referring, if necessary to
any text on ordinary differential equations) and show that Eqs. (22.9) and
(22.10) emerge as special cases of the general solution. You may regard

K as defined by (22.8),

$$R_a = R_d = KVC.$$

SECTION 23
CALCULATION OF RATE OF APPEARANCE
IN STEADY STATE SYSTEMS

II. Two Compartment Approximation

The use of a two compartment model permits greater accuracy of calculation
when the injection method is used. Clearly, different biological systems are
representable by different two compartment models, but since the glucose
system has been set up as a paradigm of the open system let us carry out
the calculations for glucose. The assiduous reader has already solved
Problem 17.2 and has expressed the K's as functions of the B's and β's. The
two compartment model we are dealing with is illustrated in Fig. 23.1. In
addition to the constraint

$$K_{02} = K_{01} \qquad\qquad (17.38)$$

we must stipulate that the tracee enters solely in compartment 1, the
compartment of experimental access. Any injections or infusions of tracer
are also made into compartment 1. The differential equations for tracer are
of the usual form:

$$\frac{dC_1{}^*}{dt} = -(K_{21} + K_{01})C_1{}^* + K_{12}C_2{}^* \qquad\qquad (23.1)$$

$$\frac{dC_2{}^*}{dt} = K_{21}C_1{}^* - (K_{12} + K_{01})C_2{}^* \qquad\qquad (23.2)$$

The solutions are of the form

$$C_1{}^* = B_1\, e^{-\beta_1 t} + B_2\, e^{-\beta_2 t}\,, \qquad \beta_1 > \beta_2 > 0 \qquad\qquad (23.3)$$

$$C_2{}^* = -B_3\, e^{-\beta_1 t} + B_3\, e^{-\beta_2 t} \qquad\qquad (23.4)$$

since $C_2{}^*(0) = 0$. Solving the inverse problem by substituting (23.3) and
(23.4) back into (23.1) and (23.2) provides 4 algebraic equations in the

4 unknown quantities K_{12}, K_{21}, K_{01} and B_3:

$$B_1\beta_1 = K_{21}B_1 + K_{01}B_1 + K_{12}B_3 \qquad (23.5)$$

$$B_2\beta_2 = K_{21}B_2 + K_{01}B_2 - K_{12}B_3 \qquad (23.6)$$

$$B_3\beta_1 = K_{21}B_1 + K_{12}B_3 + K_{01}B_3 \qquad (23.7)$$

$$B_3\beta_2 = -K_{21}B_2 + K_{12}B_3 + K_{01}B_3 \qquad (23.8)$$

These equations are quadratic and certain extraneous roots must be eliminated since they give rise to solutions which are not physically acceptable (e.g. negative concentrations). The solutions finally obtained are

$$K_{01} = \beta_2 \qquad (23.9)$$

$$K_{12} = \frac{B_2\,(\beta_1 - \beta_2)}{B_1 + B_2} \qquad (23.10)$$

$$K_{21} = \frac{B_1\,(\beta_1 - \beta_2)}{B_1 + B_2} \qquad (23.11)$$

$$B_3 = B_1 \qquad (23.12)$$

Adding (23.9), (23.10) and (23.11) gives the additional relationship

$$K_{01} + K_{12} + K_{21} = \beta_1 \qquad (23.13)$$

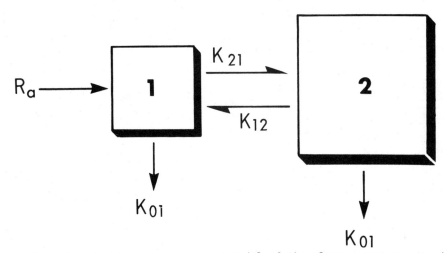

Fig. 23.1. A solvable two compartment model of the glucose system. K_{02} has been constrained to be equal to K_{01}. Compartment 1 is the sole compartment of access. The endogenous appearance of glucose is taken to be entirely within compartment 1.

(a)   Tracer Injection Method

In the problem 17.2(d), the reader was asked to calculate the rate of appearance using the assumption that $C_2 = C_1$: the constant concentration of tracee in compartment 2 is equal to that in compartment 1. Let us proceed now somewhat differently. Let us make the assumption (severely berated in earlier chapters) that

$$\text{(tracee flux)} = \text{(specific activity)}^{-1}\text{(tracer flux)} \quad (23.14)$$

We have seen that (23.14) holds for convective flux while not necessarily for diffusive flux, but let us make the assumption anyway. Then the tracer flux

$$K_{ij}C_j{}^*$$

from the jth to the ith compartment, is accompanied by the tracee flux

$$K_{ij}C_j$$

etc. Thus accompanying equations (23.1) and (23.2) we have two differential equations for tracee,

$$\frac{dC_1}{dt} = \frac{R_a}{V_1} - (K_{21} + K_{01})C_1 + K_{12}C_2 = 0 \qquad (23.15)$$

$$\frac{dC_2}{dt} = K_{21}C_1 - (K_{12} + K_{01})C_2 = 0 \qquad (23.16)$$

$\frac{dC_1}{dt}$ and $\frac{dC_2}{dt}$ have been equated to zero because the system is in steady state. The rate of appearance, R_a [mass/time], must be divided by the volume of the compartment of appearance. Equations (23.15) and (23.16) are two algebraic equations in the two unknown quantities R_a and C_2. If C_2 is eliminated, we obtain R_a as

$$R_a = \frac{K_{01}(K_{01} + K_{12} + K_{21})}{K_{12} + K_{01}} \, V_1 C_1 \qquad (23.17)$$

Simplifying the right-hand side using equations (23.9) through (23.13)

$$R_a = \frac{\beta_1\beta_2}{\dfrac{B_1\beta_2 + B_2\beta_1}{B_1 + B_2}} \, V_1 C_1 \qquad (23.18)$$

From the dilution principle,

$$V_1 = \frac{M_o{}^*}{B_1 + B_2} \qquad (23.19)$$

Hence (23.18) becomes

$$R_a = \frac{M_o {}^*C_1}{\dfrac{B_1}{\beta_1} + \dfrac{B_2}{\beta_2}} \tag{23.20}$$

Eq. (23.20) thus permits the calculation of steady state rate of appearance in terms of the mass of tracer injected, the tracee concentration in the compartment of injection, and the parameters of a double exponential curve which has been fitted to the impulse response of the system. Notice that it differs from the monoexponential formula (22.9) by the addition of an extra term in the denominator. If the rate of appearance were calculated by both formulae (22.9) and (23.20), the latter would give the smaller value, since both B_1 and B_2 will be found to be positive numbers.

(b) Tracer Infusion Method

Again, we infuse tracer at the constant rate R_a^* and await the condition of tracer steady state. We then enjoy a steady state in both tracer and tracee. In place of (23.1) and (23.2) we now have

$$\frac{R_a^*}{V_1} - (K_{21} + K_{01})C_1{}^* + K_{12}C_2{}^* = 0 \tag{23.21}$$

$$K_{21}C_1{}^* - (K_{12} + K_{01})C_2{}^* = 0 \tag{23.22}$$

while (23.15) and (23.16) remain as before. Since

$$C_1{}^* = a_1 C_1$$

and

$$C_2{}^* = a_2 C_2$$

(23.21) and (23.22) become

$$\frac{R_a^*}{V_1} - (K_{21} + K_{01})a_1 C_1 + K_{12}a_2 C_2 = 0 \tag{23.23}$$

$$K_{21}a_1 C_1 - (K_{12} + K_{01})a_2 C_2 = 0 \tag{23.24}$$

Multiplying (23.16) by a_2 and subtracting from (23.24):

$$K_{21}a_1 C_1 - K_{21}a_2 C_1 = 0$$

or

$$a_2 = a_1 = a \tag{23.25}$$

Substituting (23.25) into (23.23)

$$\frac{R_a^*/a}{V_1} - (K_{21} + K_{01})C_1 + K_{12}C_2 = 0 \tag{23.26}$$

Comparing (23.15) with (23.26) we can make the identification

$$R_a = R_a^*/a \qquad (23.27)$$

which is identical to (22.10). Thus we have obtained the same formula for
the steady state rate of appearance using the one and the two compartment
approximations.

The physical content of the results of the one and two compartment calculations is not easily recognized. Hence we now pass on to the investigation
of steady state rate of appearance in a distributed (non-compartmented)
system where we can retain some footing in physics and chemistry. It will
become apparent that all "turnover" formulae calculated hitherto are specific
instances of a general case, and we shall examine the assumptions entailed
by their use.

Speaking Practically

We have undertaken quite a bit of algebraic labour in order to calculate the
injection and infusion equations, (23.20) and (23.27) respectively, for rates
of appearance of a metabolite in a steady state system. It is neither
necessary nor desirable to go through a similar calculation for every
different metabolic system. It is much simpler to approach each system as a
distributed system, whenever possible, as we shall do in sections 25 and 26.

SECTION 24

PRIMING INFUSIONS OF TRACER IN
DISTRIBUTED SYSTEMS

When an investigator elects to measure the steady state rate of appearance
of S by means of a constant infusion of S* he will, generally, await the
condition of tracer steady state and then apply the steady infusion formula
(22.10) or (23.27),

$$R_a = R_a^*/a \qquad (24.1)$$

This formula, while derived so far only for a single compartment model and a certain type of two compartment model, will subsequently be shown to be valid in a much more general system. The problem encountered in applying a constant tracer infusion to a system which contains no tracer prior to the start of infusion is that it takes a long time to approach the tracer steady state. For example, if the impulse response function of a system is

$$C^* = A\ e^{-.01t} \tag{24.2}$$

where t is measured in minutes ($.01\ min^{-1}$ is in the usual range for intermediary metabolic reactions), then it is easily seen that upon application of a steady tracer infusion to the tracer-free system, $C^*(t)$ will approach its steady, asymptotic value with a half-time of $.69/.01$, or 69 minutes. Thus the investigator will have a long, long wait before $C^*(t)$ differs from its steady value by, say, less than 5% of the steady value. For this reason the practice has developed of speeding the approach to tracer steady state by applying, coincident with the start of the steady tracer infusion, a priming tracer injection.

(i) Priming Tracer Injection

There are two ways one might go about estimating the optimal magnitude of the priming tracer injection *provided he knew the impulse response of the system*.

(a) Cancelling the "Slow" Exponential

Suppose that the response to an impulse of M_0^* is conformed to a sum of exponentials

$$C_1^* = \sum_{i=1}^{n} B_i e^{-\beta_i t}, \quad \beta_1 > \beta_2 > \ldots > \beta_n > 0 \tag{24.3}$$

No compartments need be assumed; the sum of exponentials is just a convenient way of expressing the experimental data. Then by the usual convolution process (cf. (13.11), (13.12), (13.13)) the response to the step input R_a^* is

$$C_2^* = \frac{R_a^*}{M_0^*} \sum_{i=1}^{n} \left[\frac{B_i}{\beta_i} - \frac{B_i}{\beta_i} e^{-\beta_i t} \right] \tag{24.4}$$

The system, being in steady state with respect to tracee, is assumed to be linear. The response to a primed infusion — i.e. an injection of M_0^* coincident with the beginning of the steady infusion R_a^* — will be given by the superposition of C_1^* and C_2^*:

$$C^* = \sum_{i=1}^{n} \left[\frac{R_a^*}{M_o^*} \frac{B_i}{\beta_i} + \left(B_i - \frac{R_a^*}{M_o^*} \frac{B_i}{\beta_i} \right) e^{-\beta_i t} \right] \tag{24.5}$$

The function C^* is illustrated in Fig. (24.1). The terms $\dfrac{R_a^*}{M_o^*} \dfrac{B_i}{\beta_i}$ define the steady state level of tracer, while the terms $\left(B_i - \dfrac{R_a^*}{M_o^*} \dfrac{B_i}{\beta_i} \right) e^{-\beta_i t}$, which decay away toward zero, account for the deviation of the system from steady state and hence the *transient* behaviour of the system to a step input. The "prevailing" term in the transient portion is that exponential function which decays at the slowest rate, — i.e. the term

$$\left(B_n - \frac{R_a^*}{M_o^*} \frac{B_n}{\beta_n} \right) e^{-\beta_n t}$$

If we could "cancel" this term — i.e. arrange for its vanishing — we could then remove the primary impediment to the approach of tracer steady state. Equating the coefficient of this prevailing exponential to zero we obtain

$$B_n - \frac{R_a^*}{M_o^*} \frac{B_n}{\beta_n} = 0$$

or

$$M_o^* = \frac{R_a^*}{\beta_n} \tag{24.6}$$

If the impulse response function contains only a single exponential term, the selection of a priming injection equal to R_a^*/β_n will result in an instant tracer steady state. In the event of a multiexponential impulse response the selection of the priming injection equal to R_a^*/β_n will speed up the attainment of tracer steady state but will still leave a transient, or time-varying phase to the output. Eq. (24.6) is a very practical formula and quite commonly used. For example, if β_n were .01 min^{-1}, the priming injection would be 100 R_a^*, or equal to 100 minutes of infusion. Often the experimenter will not know the impulse response function of his system in detail, but *will* know the approximate value of the "slow" exponential. Eq. (24.6) then provides a reasonable estimate of the strength of the optimal priming injection

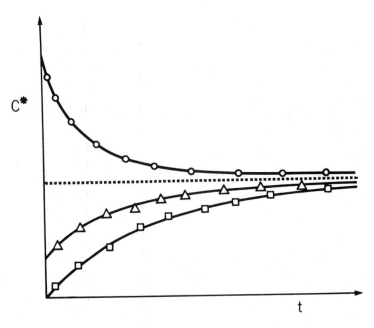

Fig. 24.1.   Illustrates the output to a primed infusion of tracer — i.e. an
             injection of tracer applied at the start of a steady tracer
             infusion. The tracee is in steady state.

             □   no priming injection given

             △   small priming injection

             ○   larger priming injection

             ···· steady state level of tracer, approached asymptotically

(b) Minimizing the Area of the Transient Offset

This method is an alternative to (a) and involves the more usual mathematical
method for finding extrema. The form of the response to a primed tracer
infusion is usually that depicted in Fig. (24.1): a monotonically rising or
falling curve asymptotically approaching the steady state level. Such a
curve is found when tracer is sampled at a point close to its injection site,
the situation observed when both input and output occur in the vascular
region.

Suppose that the response to a primed infusion is represented by

$$C* = C_1* + C_2*$$

as given by (24.5). $C*$ will deviate from the desired steady state level by an amount which might be termed the "transient offset", $C*_{trans}$, given by

$$C*_{trans} = \sum_{i=1}^{n} \left(B_i - \frac{R_a*}{M_o*} \frac{B_i}{\beta_i} \right) e^{-\beta_i t} \tag{24.7}$$

The area of the figure contained by $C*$ and the asymptote, between $t = 0$ and $t \to \infty$ (Fig. 24.2), is given by

$$\int_0^\infty C*_{trans}(t)dt = \int_0^\infty \sum_{i=1}^{n} \left(B_i - \frac{R_a*}{M_o*} \frac{B_i}{\beta_i} \right) e^{-\beta_i t} \, dt$$

$$= \left[\sum_{i=1}^{n} \left(\frac{R_a*}{M_o*} \frac{B_i}{\beta_i^2} - \frac{B_i}{\beta_i} \right) e^{-\beta_i t} \right]_0^\infty$$

$$= \sum_{i=1}^{n} \left(\frac{B_i}{\beta_i} - \frac{R_a*}{M_o*} \frac{B_i}{\beta_i^2} \right) \tag{24.8}$$

From the homogeneity of the linear system, the B's, which define the amplitude of the impulse response function, are proportional to M_o*, the strength of the impulse (Eq. (12.4)). That is

$$B_i = \gamma_i M_o* \tag{24.9}$$

where the γ's are the coefficients of the unit impulse response function. Hence

$$\int_0^\infty C*_{trans}(t)dt = \sum_{i=1}^{n} \left(\frac{\gamma_i}{\beta_i} M_o* - R_a* \frac{\gamma_i}{\beta_i^2} \right) \tag{24.10}$$

Regarding M_o* as the independent variable, let us find the value of M_o* for which $\int_0^\infty C*_{trans}(t)dt$ is an extremum. If the transient curve lies above the steady state asymptote, the extremum will be a minimum; if it lies below, the extremum will be a maximum.

$$\frac{d}{dM_o*} \int_0^\infty C*_{trans}(t)dt = \sum_{i=1}^{n} \left(\frac{\gamma_i}{\beta_i} - \frac{\gamma_i}{\beta_i^2} \frac{dR_a*}{dM_o*} \right) = 0$$

$$\frac{dR_a*}{dM_o*} = \frac{\sum\limits_{i=1}^{n} \dfrac{\gamma_i}{\beta_i}}{\sum\limits_{i=1}^{n} \dfrac{\gamma_i}{\beta_i^2}} \tag{24.11}$$

Since $M_0{}^*$ equals zero for R_a^* equals zero, (24.11) is integrated to give

$$M_0{}^* = \frac{\sum\limits_{i=1}^{n} \frac{\gamma_i}{\beta_i^2}}{\sum\limits_{i=1}^{n} \frac{\gamma_i}{\beta_i}} \cdot R_a^* \tag{24.12}$$

Eq. (24.12), then, provides the weight for an optimal priming injection of tracer, $M_0{}^*$, in terms of the constant rate of infusion of tracer, R_a^*, and the parameters γ and β of the unit impulse response function $\sum\limits_{i=1}^{n} \gamma_i\, e^{-\beta_i t}$.

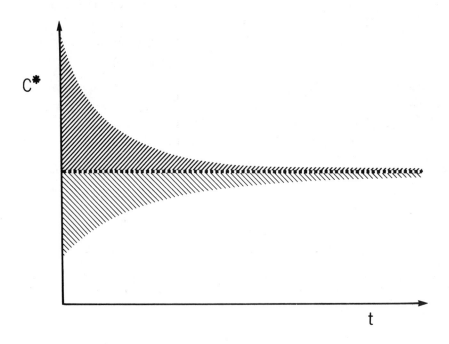

Fig. 24.2. The dashed line indicates the steady state tracer concentration
The hatched areas are two possible transient offset areas
The upper region will have a positive area and the lower a
negative area
An optimal priming injection might minimize the magnitude of the
offset area
An extremum for the upper area would be a minimum; for the lower
a maximum

In truth, the two formulae (24.6) and (24.12) optimize different things, and
it may be difficult on theoretical grounds to decide between them. In
actual practice, though, it makes very little difference. Usually the values
of β_i differ by about one order of magnitude, while the values of γ_i are of
about the same order of magnitude. Thus (24.12) will often reduce to

$$M_o{}^* \doteq \frac{\dfrac{\gamma_n}{\beta_n^2}}{\dfrac{\gamma_n}{\beta_n}} \cdot R_a^* \tag{24.13}$$

which is identical with (24.6).

(ii) A Priming Tracer Infusion for Instantaneous Steady Concentration

It is desirable to have a method for attaining an instantaneous, constant
concentration of tracer at the output. Let us represent this constant
tracer concentration by C^*. Suppose that the response to an impulse $M_o{}^*$, of
tracer is again given by (24.3) with n equal to 2 for simplicity:

$$C_1{}^*(t) = B_1 e^{-\beta_1 t} + B_2 e^{-\beta_2 t} \tag{24.14}$$

and let us represent the required tracer infusion rate by $R_a^*(t)$ — which is
the function we must find. From the convolution principle (Eq. (12.36)) we
have

$$LC^* = L \frac{C_1{}^*(t)}{M_o{}^*} \cdot L\, R_a^*(t) \tag{24.15}$$

where L designates the Laplace transform as usual. Since the Laplace transform of the step fucntion C^* is given by

$$LC^* = \frac{C^*}{s} \tag{24.16}$$

and the transform of an exponential function is given by

$$L\, e^{-\beta_i t} = \frac{1}{s + \beta_1} \tag{24.17}$$

we may expand (24.15) in the form

$$\frac{C^*}{s} = \frac{R_a^*(s)}{M_o{}^*} \left[\frac{B_1}{s + \beta_1} + \frac{B_2}{s + \beta_2} \right] \tag{24.18}$$

where

$$R_a^*(s) \equiv L R_a^*(t) \tag{24.19}$$

Hence

$$\frac{R_a^*(s)}{M_o^*} = \frac{C^*/s}{\dfrac{B_1}{s+\beta_1} + \dfrac{B_2}{s+\beta_2}} \tag{24.20}$$

Finding the desired infusion rate, $R_a^*(t)$, is now just a problem in taking
the inverse transformation of the right-hand side of (24.20). Eq. (24.20)
expands to the form

$$\frac{R_a^*(s)}{M_o^*} = \frac{C^*}{(B_1 + B_2)s} \cdot \frac{(s + \beta_1)(s + \beta_2)}{s + \dfrac{\beta_1 B_2 + \beta_2 B_1}{B_1 + B_2}}$$

which we recognize immediately as being unsuitable for inverting. Defining
a new constant

$$B \equiv \frac{\beta_1 B_2 + \beta_2 B_1}{B_1 + B_2} \tag{24.21}$$

we obtain

$$\frac{(B_1 + B_2)}{M_o^* C^*} R_a^*(s) = \frac{s + \beta_1}{s + B} \cdot \frac{s + \beta_2}{s} \tag{24.22}$$

For purposes of inversion we should like to have all rational functions
proper — i.e. the numerator of lower power in s than the denominator. Hence
(24.22) is rewritten as

$$\frac{B_1 + B_2}{M_o^* C^*} R_a^*(s) = \left(1 + \frac{\beta_1 - B}{s + B}\right) \cdot \left(1 + \frac{\beta_2}{s}\right)$$

$$= 1 + \frac{\beta_1 \beta_2}{B} \cdot \frac{1}{s} + \left(\beta_1 + \beta_2 - B - \frac{\beta_1 \beta_2}{B}\right) \cdot \frac{1}{s+B} \tag{24.23}$$

which can now be inverted using (12.23), (24.16) and (24.17):

$$\frac{B_1 + B_2}{M_o^* C^*} R_a^*(t) = \delta(t) + \frac{\beta_1 \beta_2}{B} + \left(\beta_1 + \beta_2 - B - \frac{\beta_1 \beta_2}{B}\right) e^{-Bt}$$

$$R_a^*(t) = \frac{C^* M_o^*}{B_1 + B_2} \left[\delta(t) + \frac{\beta_1 \beta_2}{B} + \left(\beta_1 + \beta_2 - B - \frac{\beta_1 \beta_2}{B}\right) e^{-Bt}\right] \tag{24.24}$$

Finally, introducing (24.21)

$$R_a^*(t) = \frac{M_o^* C^*}{B_1 + B_2} \cdot \left[\delta(t) + \frac{B_1 + B_2}{\dfrac{B_1}{\beta_1} + \dfrac{B_2}{\beta_2}} + \frac{B_1 B_2 (\beta_1 - \beta_2)^2}{(B_1 + B_2)(\beta_1 B_2 + \beta_2 B_1)} \cdot \exp\left(-\frac{\beta_1 B_2 + \beta_2 B_1}{B_1 + B_2} \cdot t\right)\right]$$

$$\tag{24.25}$$

The total tracer infusion which is expected to produce an instantaneously
constant tracer concentration at the output is, then, a combination of
                        an injection (represented by the delta function)
                        a constant infusion, and
                        a monoexponentially decaying infusion (Fig. 24.3)

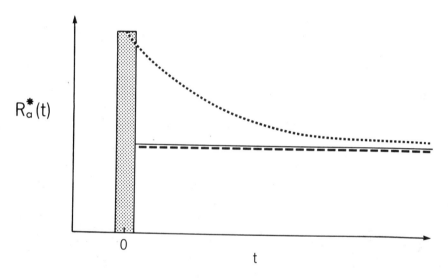

Fig. 24.3.   The total rate of infusion of tracer, $R_a^*(t)$, required to produce
             a constant tracer concentration at some output site (solid line)
             is a superposition of the 3 inputs shown:
                        the impulse (delta function)   ▓▓▓   hatched area
                        the constant infusion   ━ ━ ━
                        the monoexponentially decaying infusion •••••

The value of $R_a^*(t)$ for large t is seen to be given by the second term on the right-hand side of (24.25). That is

$$\lim_{t \to \infty} R_a^*(t) = \frac{M_o * C*}{\frac{B_1}{\beta_1} + \frac{B_2}{\beta_2}} \qquad (24.26)$$

which is the steady state relationship expected (24.4). (Show that the argument of the exponential function must have a negative value for t > 0.)

Daniel, Donaldson and Pratt (1974,5) devised a method for producing a rapid, constant concentration of certain metabolites in the plasma. Their method utilizes a two compartment model of the metabolic system, and determination of the impulse response function by curve fitting to a double exponential function. These investigators designed an electrical analogue of the physiological system, determined how current must be varied (in the model) to produce a constant voltage, and then, analogously, how $R_a^*(t)$ should be varied to produce a constant C* or sometimes, C. The infusate often contained appreciable quantities of tracee as well as tracer. Their calculated optimal infusion rate is identical in form to Eq. (24.25), consisting of three components. They then constructed an electronically controlled, motor-driven syringe to deliver the infusion at the appropriate rate automatically. Their results are very impressive (e.g. Fig. 24.4) and their papers are certainly recommended reading. The value of the theory developed in this section is to show how the optimal $R_a^*(t)$ can be obtained without recourse to a compartmental model or an electrical analogue, but directly from the measurement of the impulse response function.

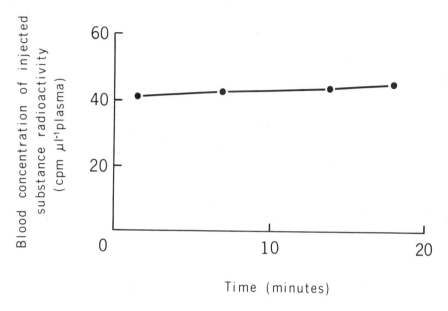

Fig. 24.4.   Shows the level of radioactivity found over a 20 minutes period
            in the blood plasma of a rat during the injection of ^{14}C L-
            leucine. The injection was started at zero time.

From: Medical and Biological Engineering 13, 223, 1975 by P.M. Daniel,
J. Donaldson and O.E. Pratt. Reproduced by permission of the International
Federation for Medical and Biological Engineering.

Speaking Practically

If you want to produce a constant, elevated blood level of some substance
rather rapidly, you are assured that there is a fairly simple way of doing
so if the system is a linear one. Suppose that the substance is a tracer
whose tracee is in steady state. If the approximate half-life of the tracer
in plasma is, say, $t_{1/2}$, and the desired constant infusion rate of tracer is
R_a^*, then the best, single "priming" injection you can give is one consisting
of $R_a^* \cdot t_{1/2}/.69$ dpm or μCi. If you are short of time or patience, are dealing
with an unstable system, or are just plain compulsive, you can achieve a
constant plasma level instantaneously, having fitted the impulse response of

the system to a double exponential function of the form of Eq. (24.3), with
n = 2. You now give a combination of three tracer inputs, whose magnitudes
are given by the three terms on the right-hand side of Eq. (24.25):
The first is an injection of magnitude $M_o*C*/(B_1+B_2)$ at t = 0, C* arbitrary;
The second is a constant infusion of magnitude ... ;
The third is an infusion declining progressively with time, of magnitude ...
This third infusion could be given by a programmable infusion apparatus
if you are lucky enough to have one. If not you can give it manually using
a Sage infuser, which permits smooth adjustment of the infusion rate, or a
Harvard infuser, which permits smooth adjustment from 0% to 100% of the
nominal rate at any gear setting. Changing the setting every minute or two
is often enough to simulate a continuous change when dealing with many
metabolic systems.

Problems

24.1 Verify that (24.25) is the correct solution by convolving $R_a^*(t)$,
given by (24.25), with the unit impulse response function (obtained by
dividing Eq. (24.14) by M_o*) to obtain the constant output, C*.

24.2 ^{14}C-glycerol has a half-life in plasma of about 3 minutes. From a
practical point of view, what would you recommend as the optimal priming
injection?

24.3 The techniques described in part (ii) above theoretically assure that
the output at some selected sampling point will be a step function of
desired amplitude. But do they assure an instantaneous steady state in
tracer? Is the output at all possible sampling points a step of the same
amplitude?

24.4(a) In a hypothetical physiological system in steady state, the
response to a unit impulse input is accurately described by
$$C* = A\,t^{-1/2} \ [dpm/ml], \ A \text{ is a constant} > 0$$
This decay curve is qualitatively similar to the usual exponential decay;
but calculate the step response of the system. Does it plateau for large t?
Suppose an infusion is given at the rate
$$R_a^*(t) = B\,t^{-1/2} \ [dpm/min], \ \ B \text{ constant}$$
Calculate the response. Compare the infusion rate $R_a^*(t) = B\,t^{-1/2}$ to the
rate given by Eq. (24.25).

24.4(b)   In a hypothetical physiological system in steady state, the response to a unit impulse input is accurately described by

$$C^* = \sqrt{.1}\ t^{-.5}\ e^{-.1t}$$

Calculate the unit step response in the limit as $t \to \infty$.

Functions of the type given here have been suggested by M.E. Wise (1975) for 45- or 47-calcium. These functions emerge from a convincing physical model for bone-seeking elements such as calcium, strontium, barium and radium.

SECTION 25

STEADY STATE RATE OF APPEARANCE IN DISTRIBUTED SYSTEMS BY THE TRACER INFUSION METHOD

We have already seen, using a one and (selected) two compartment approximation that the steady state rate of appearance of S can be obtained as the ratio of the steady rate of tracer infusion to the constant specific activity in the compartment of access, when tracer (as well as tracee, of course) is maintained in a steady state (Eqs. (22.10) and (23.27)). The tracer is generally presumed to be in a steady state when it has been infused for some time at a constant rate, R_a^*, and its concentration at some sampling site remains constant. In section 24, we discussed a number of methods for hastening the arrival of constant C^*, the simplest being the application of a priming injection of tracer. The various theorems of section 24 were developed for a distributed (non-compartmentalized) system and depended only upon the existence of a linear tracer system. We shall now continue our explorations of distributed systems and examine the validity of Eq. (25.1) below (same as (23.27)) in the more general system where transport occurs by convection and diffusion.

$$R_a = R_a^*/a \qquad\qquad (25.1)$$

In section 14 we combined the equations of convective diffusion for tracer and tracee to give the result (Eq. (14.21))

$$C^2 \frac{\partial a}{\partial t} = C R^*_{a,v} - C^* R_{a,v} - \vec{u} \cdot (C^2 \nabla a) + D\nabla \cdot (C^2 \nabla a) \quad (25.2)$$

For the case where (Eq. (14.49))

$$\nabla_4 \, a = 0 \qquad\qquad\qquad (25.3)$$

where the boundaries of the distribution space are impermeable to carrier fluid, to S and to S*, and where there is a single source point for each of S and S* ($\vec{r}_1$ and $\vec{r}_2$ respectively), we derived the result (14.53)

$$R_a = \frac{C(\vec{r}_2)}{C^*(\vec{r}_1)} \cdot R^*_a \qquad\qquad (25.4)$$

If we now apply the additional constraint

$$\vec{r}_1 = \vec{r}_2 = \vec{r}_0 \qquad\qquad\qquad (25.5)$$

i.e. that S and S* share a single, common source point (corresponding to the state where tracee and tracer enter the system via the same compartment), then

$$R_a = R^*_a/a_0 \qquad\qquad\qquad (25.6)$$

Eq. (25.6) is the counterpart of (25.1) in a distributed system. Because Eq. (25.1), or equivalently (25.6), is of fundamental importance, let us derive it (a) more generally, and

(b) more rigorously than before.

More General Boundary Conditions

We recall (section 9) that the divergence theorem, which was employed in the development of the equation of convective diffusion for a system with rigid boundaries, allowed for the existence of an exterior surface and any number of interior surfaces. The interior surfaces can now be taken to be cellular surfaces, and we shall permit the substance S to disappear by reacting chemically at these surfaces. The impermeability of the enclosing surfaces to the carrier fluid will be retained. These boundary conditions seem more realistic for intermediary metabolites which may, effectively, vanish from the system at the cellular surfaces. The model so constructed is still not perfectly general, but it is decidedly more general than that used in section 14.

Uniqueness

Let us suppose, again, that we deal with the case of a single source point of S at $\vec{r}_1$. Hence $R_{a,v}$ will be given by (14.51)

$$R_{a,v} = R_a \cdot \delta(\vec{r} - \vec{r}_1) \tag{25.7}$$

The equation of convective diffusion or convection-diffusion-reaction for S (11.21) then assumes the form

$$\frac{\partial C}{\partial t} = D\nabla^2 C - \vec{u} \cdot \nabla C + R_a \cdot \delta(\vec{r}-\vec{r}_1) - R_{d,v}(C,t) \tag{25.8}$$

where R_a may be a function of time. Let us now attempt to demonstrate the theorem that *corresponding to any given point source of S of strength $R_a(t)$ there exists one and only one solution, $C(\vec{r},t)$, to (25.8), the equation of convective diffusion.*

The method of proving this uniqueness theorem is quite similar to that used by Maxwell to demonstrate the uniqueness for Laplace's equation. Suppose that corresponding to some value of R_a applied at $\vec{r} = \vec{r}_1$ there are two solutions to (25.8), $C_a(\vec{r},t)$ and $C_b(\vec{r},t)$, each satisfying the initial and boundary conditions in C. Hence

$$\frac{\partial C_a}{\partial t} = D\nabla^2 C_a - \vec{u} \cdot \nabla C_a + R_a \cdot \delta(\vec{r}-\vec{r}_1) - R_{d,v}(C_a,t) \tag{25.9}$$

and

$$\frac{\partial C_b}{\partial t} = D\nabla^2 C_b - \vec{u} \cdot \nabla C_b + R_a \cdot \delta(\vec{r}-\vec{r}_1) - R_{d,v}(C_b,t) \tag{25.10}$$

Define

$$w = C_a - C_b \tag{25.11}$$

where $w = 0$ at $t = 0$, since C_a and C_b satisfy the same initial conditions in C. Subtracting (25.10) from (25.9) leaves

$$\frac{\partial w}{\partial t} = D\nabla^2 w - \vec{u} \cdot \nabla w + R_{d,v}(C_b,t) - R_{d,v}(C_a,t)$$

Multiplying this equation by w we have

$$\frac{1}{2}\frac{\partial w^2}{\partial t} = wD\nabla^2 w - \frac{1}{2}\vec{u} \cdot \nabla w^2 + w\left[R_{d,v}(C_b,t) - R_{d,v}(C_a,t)\right] \tag{25.12}$$

Integrating (25.12) over the entire volume of the system

$$\frac{1}{2}\int_V \frac{\partial w^2}{\partial t}\,dV = D\int_V w\nabla^2 w\,dV - \frac{1}{2}\int_V \vec{u} \cdot \nabla w^2\,dV + \int_V w\left[R_{d,v}(C_b,t) - R_{d,v}(C_a,t)\right]dV$$

$$\tag{25.13}$$

We shall now set about simplifying the right-hand side of this equation.

Let us define $R_{d,s}(C,t)$ as the rate of disappearance of S per unit of surface area on the boundary. Since we are assuming no flow of carrier fluid through the boundary, the substance S must diffuse toward the surface at points close to the surface. That is

$$R_{d,s}(C,t) = -D\nabla C \cdot \vec{n} \qquad (25.14)$$

where $\vec{n}$ is the unit, outward-drawn normal at the surface. Thus

$$R_{d,s}(C_a,t) = -D\nabla C_a \cdot \vec{n}$$

and

$$R_{d,s}(C_b,t) = -D\nabla C_b \cdot \vec{n}$$

for the two supposed solutions to the convective diffusion equation. Subtracting the above two equations, multiplying the difference by w, and integrating over the entire surface, we obtain

$$\int_S w\nabla w \cdot \vec{n} \, dS = \frac{1}{D} \int_S w \left[R_{d,s}(C_b,t) - R_{d,s}(C_a,t) \right] dS \qquad (25.15)$$

If U and W are any two scalar point functions, then from Green's Theorem (see Appendix C),

$$\int_V U\nabla^2 W \, dV = \int_S U\nabla W \cdot \vec{n} \, dS - \int_V \nabla U \cdot \nabla W \, dV \qquad (25.16)$$

If we then set

$$w = U = W$$

(25.16) becomes

$$\int_V w\nabla^2 w \, dV = \int_S w\nabla w \cdot \vec{n} \, dS - \int_V (\nabla w)^2 \, dV \qquad (25.17)$$

Substituting the value of $\int_S w\nabla w \cdot \vec{n} \, dS$ from (25.15) into (25.17),

$$D\int_V w\nabla^2 w \, dV = \int_S w \left[R_{d,s}(C_b,t) - R_{d,s}(C_a,t) \right] dS - D\int_V (\nabla w)^2 \, dV \qquad (25.18)$$

which we shall reserve for later use.

Since fluid does not cross the bounding surface we can write

$$\vec{u} \cdot \vec{n} = 0 \qquad (25.19)$$

and since the fluid is incompressible

$$\nabla \cdot \vec{u} = 0 \tag{11.16}$$

Using the vector identity ((10.26), (C.19))

$$\nabla \cdot (w^2\vec{u}) = w^2\nabla \cdot \vec{u} + \vec{u} \cdot \nabla w^2$$

$$= \vec{u} \cdot \nabla w^2$$

as a consequence of (11.16). Hence

$$\int_V \vec{u} \cdot \nabla w^2 \, dV = \int_V \nabla \cdot (w^2\vec{u}) dV = \int_S w^2\vec{u} \cdot \vec{n} \, dS \tag{25.20}$$

using the divergence theorem. Introducing (25.19) leaves

$$\int_V \vec{u} \cdot \nabla w^2 \, dV = 0 \tag{25.21}$$

We may now replace the first two integrals on the right-hand side of (25.13) by their values given by (25.18) and (25.21):

$$\frac{1}{2}\int_V \frac{\partial w^2}{\partial t} \, dV = \int_S w\left[R_{d,s}(C_b,t) - R_{d,s}(C_a,t)\right] dS - D\int_V (\nabla w)^2 dV$$

$$- 0 + \int_V w\left[R_{d,v}(C_b,t) - R_{d,v}(C_a,t)\right] dV \tag{25.22}$$

If we now make the assumption that the rate of a chemical reaction is a monotone increasing function of the concentration, then the two expressions

$$R_{d,s}(C_b,t) - R_{d,s}(C_a,t)$$

and

$$R_{d,v}(C_b,t) - R_{d,v}(C_a,t)$$

each have the sign of $(C_b - C_a)$. This assumption is probably well-taken in steady state circumstances. It implies that increasing the concentration of some substance, S, will never result in a decrease in the rate of chemical reaction, nor will decreasing the concentration result in an increase in the rate of reaction[†]. Introducing this monotonicity assumption, we see that each of the integrals on the right-hand side of (25.22) must have a value equal to or less than zero, so that

---

[†]The assumption is not valid in at least one metabolic system out of steady state.

$$\frac{1}{2} \int_V \frac{\partial w^2}{\partial t} \, dV \leq 0 \qquad\qquad (25.23)$$

If the boundaries are fixed then

$$\frac{d}{dt} \int_V w^2 \, dV \leq 0 \qquad\qquad (25.24)$$

Suppose that this derivative has a value less than zero. Now $w = 0$ at $t = 0$, since C_a and C_b each satisfy the initial conditions, and therefore, $\int_V w^2 \, dV$ is equal to zero. Since the time derivative of this integral is negative, therefore a short time, Δt, later the integral must assume a negative value. But this is not possible because the integrand is the square of a real number and must therefore remain positive. A contradiction has been reached, and so we must adopt the alternative possibility

$$\frac{d}{dt} \int_V w^2 \, dV = 0$$

implying

$$w = a \text{ constant} = 0$$

since $w = 0$ at $t = 0$. Hence

$$C_a = C_b \qquad\qquad (25.25)$$

and only one possible solution exists.

We have now demonstrated the theorem that *corresponding to any given point source of strength* R_a *there exists one and only one solution,* $C(\vec{r},t)$, *to the equation of convective diffusion* for a given set of initial conditions, for a selected set of boundary conditions, and subject to the constraint that chemical reaction is a monotone increasing function of concentration. It follows as a subcase that when S is in steady state, so that $\frac{\partial C}{\partial t} = 0$, there is only one solution, $C(\vec{r})$, possible etc.

Having proved this theorem for the unlabelled substance, S, we may now prove the corresponding theorem for the labelled substance, S*. Since the term $R^*_{d,v}(\vec{r},t)$, which appears in the equation of convective diffusion for S*, depends in the usual way upon $R_{d,v}$, hence the solution, $C*(\vec{r},t)$, depends upon $R_{d,v}$ and therefore upon $C(\vec{r},t)$. The theorem for S* will then be stated: *Corresponding to any given point source of strength* R^*_a *there exists one and only one solution,* $C*(\vec{r},t)$, *to the equation of convective diffusion — for each function* $C(\vec{r},t)$. In other words, for each state of tracee in a system, tracer will behave in a unique fashion. This theorem may be proved in exactly the same way as the previous one. And it follows as a subcase that when S* is in steady state so that $\frac{\partial C*}{\partial t} = 0$, there is only one solution, $C*(\vec{r})$, possible etc.

Conditions for a Zero Gradient in Specific Activity

Suppose we have a condition where both tracee and tracer are in a steady state; i.e. we have achieved the condition of tracer steady state by means of a primed or unprimed tracer infusion at the point source $\vec{r}_1$. Then the equations of convective diffusion assume the form

$$D\nabla^2 C - \vec{u} \cdot \nabla C + R_a \cdot \delta(\vec{r}-\vec{r}_1) - R_{d,v}(C,t) = 0 \qquad (25.26)$$

and

$$D\nabla^2 C* - \vec{u} \cdot \nabla C* + R_a^* \cdot \delta(\vec{r}-\vec{r}_1) - a \cdot R_{d,v}(C,t) = 0 \quad (25.27)$$

Suppose now that $C = \sigma$ is an admissible solution to (25.26) satisfying the boundary conditions. Then from the uniqueness theorem, σ is the *only* solution. Let us define a constant scalar, α, by the equation

$$\alpha \equiv R_a^*/R_a \qquad (25.28)$$

Since α is constant,

$$\frac{\partial \alpha}{\partial t} = 0 \qquad (25.29)$$

and

$$\nabla \alpha = 0 \qquad (25.30)$$

Let us now examine the function

$$C* = \alpha\sigma \qquad (25.31)$$

as a possible solution to (25.27). Substituting (25.31) into the left-hand side of (25.27) and bearing in mind (25.28), (25.29) and (25.30), we obtain

$$\alpha\left[D\nabla^2\sigma - \vec{u} \cdot \nabla\sigma + R_a \cdot \delta(\vec{r}-\vec{r}_1) - R_{d,v}(\sigma,t)\right]$$

But the quantity in square parentheses is equal to zero, by virtue of (25.26). Therefore

$$C* = \alpha\sigma$$

is an admissible solution to (25.27) if

$$C = \sigma$$

is an admissible solution to (25.26). For the double steady state condition (in S and S*) then, specific activity

$$C*/C = \alpha \qquad (25.32)$$

and the gradient of specific activity (from (25.30)) is everywhere equal to zero.

A General Proof for Eq. (25.6)

Under steady state conditions we have by definition (13.1)

$$R_a = R_d$$

and

$$R_a^* = R_d^*$$

$$(25.33)$$

Under any circumstances, by definition (5.1)

$$R_d = \int_V R_{d,v}(\vec{r},t)dV$$

$$= \int_V \frac{R_{d,v}^*(\vec{r},t)}{a} \, dV \qquad (25.34)$$

by (6.7). For the double steady state condition

$$a = \alpha \qquad (25.32)$$

Thus

$$R_d = \frac{1}{\alpha} \int_V R_{d,v}^*(\vec{r},t)dV$$

$$= R_d^*/\alpha \qquad (25.35)$$

by definition (5.2). Hence, using (25.33) and (25.35)

$$R_a = R_a^*/\alpha \qquad (25.36)$$

which is identical with (25.6). This equation has now been derived somewhat more generally and somewhat more rigorously than before. It may now be seen to hold both in a compartmentalized and in a distributed system. It has not been necessary to assume *a priori* that $\nabla_4 a$ is equal to zero.

This material has been developed by the author (1973).

Some Practical Considerations

In the above derivation it has been assumed that there is a single, common source point for tracer and tracee. Suppose that there are multiple source points or even a distributed source for tracee. It is not hard to see that *if* it could be arranged that the tracer source, $R_a^*(\vec{r},t)$, were such that

$$R_a^*(\vec{r},t) = \alpha \cdot R_a(\vec{r},t)$$

(cf. Bergner's *equivalent tracer supply*) then the derivation of (25.36) is unchanged.

In practice, it may be difficult or impossible to arrange a single, common source point for tracer and tracee. Usually the two sources are located at different points within the vascular system, or there may be several tissues secreting tracee into the capillary blood and a single source for tracer in peripheral venous blood. Under such circumstances, one can probably proceed as if the tracer and tracee entered through the same source point. Blood intermixes rapidly enough within the vascular system so that to a good approximation one might take the rate of appearance of tracee to be $R_{a,v}$ — a uniform value over the entire vascular system — and equal to zero outside this region. The proof of (25.36) then proceeds as before.

In the event that the sources of tracee are unknown or that there are many sources and the experimenter cannot arrange an equivalent tracer supply, no simple formula such as (25.36) exists.

Eq. (25.4)(designed for use with different source points for tracer and tracee) is based upon the *a priori* validity of

$$\nabla_4 a = 0$$

and it is difficult to see how this requirement can be attained experimentally under conditions of multiple tracee sources.

There is no universal formula for R_a by tracer infusion techniques.

Speaking Practically

By all means, calculate R_a from the equation $R_a = R_a^*/$specific activity, but do remember that this equation depends for its validity on the constancy of specific activity everywhere in the system.

SECTION 26

STEADY STATE RATE OF APPEARANCE IN DISTRIBUTED SYSTEMS BY THE TRACER INJECTION METHOD

In this section we shall develop other tracer equations for computing steady state rate of appearance of tracee *in systems for which (25.36) is theoretically valid.*

$$R_a = R_a^*/\alpha \qquad (26.1)$$

R_a^* is the constant rate of tracer infusion and α the constant specific activity in steady state. Eq. (25.36) (or (26.1)) is a pivotal equation. We must be assured that all conditions for its validity are satisfied, and we shall always employ (26.1) in the development of subsequent equations for rate of appearance. It will be assumed that there is a single, common source point for tracer and tracee. (The rapidly intermixing blood pool may aid in approximating this condition.)

Sudden Injection or "Bolus" of Tracer

Suppose that a mass of tracer, M_0^* is injected suddenly at the common tracer-tracee inlet. The concentration of tracer at the sampling site, $C_I^*(t)$, is divided by M_0^* to give the unit impulse function, $h(t)$ (in units of volume^{-1}). Using the convolution integral (12.25) we can calculate the response to the step input, R_a^*:

$$C_s^*(t) = \int_0^t R_a^* h(\tau) d\tau \qquad (26.2)$$

where $C_s^*(t)$ is the step response at the sampling site. Since R_a^* is constant we obtain for $t \to \infty$

$$\frac{R_a^*}{C_s^*} = \frac{1}{\int_0^\infty h(t) dt} \qquad (26.3)$$

But from (26.1)[†]

[†] C signifies the tracee concentration at the sampling point while C_s^* signifies the tracer concentration at the sampling point. The extra subscript is necessary to avoid ambiguity later in this section.

$$\frac{R_a^*}{C_s^*} = \frac{R_a}{C} \tag{26.4}$$

Equating the right-hand sides of (26.3) and (26.4)

$$R_a = \frac{C}{\int_0^\infty h(t)dt} \tag{26.5}$$

Equivalently,

$$R_a = \frac{M_o^* C}{\int_0^\infty C_I^*(t)dt} \tag{26.6}$$

Equation (26.6) is a form of the *occupancy principle* as enunciated by Orr and Gillespie (1968). However their derivation of the principle is not fundamental enough for our purposes here. For example, they state that if a fraction f(t) of injected tracer is found in some defined region t seconds after injection, then the same fraction f(t) of the quantity of tracee $R_a\delta t$ which entered the system during time δt will be in the defined region t seconds later. We have seen earlier that such arguments are a little too general. The argument is, for example, clearly not true when tracer and tracee are transported by pure diffusion.

Equations (26.5) and (26.6) provide a means of calculating the steady state rate of appearance, R_a, in terms of the impulse response function of the system. If $C_I^*(t)$ could be expressed as (i.e. fitted to) a double exponential function,

$$C_I^*(t) = B_1 e^{-\beta_1 t} + B_2 e^{-\beta_2 t} \tag{26.7}$$

then (26.6) would assume the form

$$R_a = \frac{M_o^* C}{\dfrac{B_1}{\beta_1} + \dfrac{B_2}{\beta_1}} \tag{26.8}$$

which is identical in form with (23.20). If $C_I^*(t)$ could be expressed as a monoexponential function,

$$C_I^*(t) = A e^{-kt} \tag{26.9}$$

then (26.6) would assume the form

$$R_a = \frac{M_o^* C}{A/k}$$

(26.10)

which is identical with (22.9).

The bolus injection method is very commonly used in the measurement of steady rates of appearance. When carefully tested in a suitable system (Norwich and Hetenyi, 1971), the bolus injection and steady infusion methods are found to give nearly identical values for R_a. When R_a is calculated from (26.10) in a system where $C_I^*(t)$ is not monoexponential, the value so-obtained will differ from the value of R_a obtained using (26.1) — as indeed it should.

While the sudden injection method is quite a practical method, it is certainly not the only type of injection which can be used. The tracer mass, M_o^*, can also be infused over a somewhat longer period of time, with perfectly usable results.

Infusion of Tracer over a Short Time Interval

Suppose that instead of injecting M_o^* suddenly, the same tracer mass is infused according to some rather arbitrary function $R_a^*(t)$ such that

$$M_o^* = \int_o^\infty R_a^*(t)dt$$

(26.11)

The present case is more general than the previous one and reduces to it for the particular infusion

$$R_a^*(t) = M_o^* \, \delta(t)$$

Digression. In order to complete the derivation of the formula for R_a in terms of $R_a^*(t)$, we must first prove a theorem: *If $C^*(t)$ is the response to the general input $R_a^*(t)$, then*

$$\int_o^\infty C^*(t)dt = M_o^* \int_o^\infty h(t)dt = \int_o^\infty R_a^*(t)dt \cdot \int_o^\infty h(t)dt$$

(26.12)

Proof. (Essentially the derivation given by Perl, Effros and Chinard (1969)) From the physical nature of the quantities, all three functions $C^*(t)$, $R_a^*(t)$ and $h(t)$ have the value zero for $t = 0$ and for $t = \infty$, and we may assume that the time integral of these functions between the limits $t = 0$ and $t = \infty$ is finite. That is, all three functions will look something like the sketch in

Fig. 26.1. Using the convolution integral as in section 13, we have the relationship

$$C*(t) = \int_0^t R_a^*(\sigma) \ h(t - \sigma)d\sigma \qquad (26.13)$$

Integrating both sides between the times 0 and T,

$$\int_0^T C*(t)dt = \int_0^T\int_0^t R_a^*(\sigma) \ h(t - \sigma)d\sigma \ dt \qquad (26.14)$$

The double integral is evaluated (Fig. 26.2) by first fixing the strip of width dt and integrating with respect to σ between $\sigma = 0$ and $\sigma = t$ (i.e. between the t-axis and the straight line $\sigma = t$), and then integrating with respect to t over all such strips between t = 0 and t = T. The order of integration may be reversed by first fixing a strip of width $d\sigma$ and integrating with respect to t etc. Thus we have, equivalently

$$\int_0^T C*(t)dt = \int_0^T\int_\sigma^T R_a^*(\sigma) \ h(t - \sigma)dt \ d\sigma \qquad (26.15)$$

Making a transformation of variables in the inner integral from t to τ

$$t - \sigma = \tau$$

$$\int_0^T C*(t)dt = \int_0^T\int_0^{T-\sigma} R_a^*(\sigma) \ h(\tau)d\tau \ d\sigma \qquad (26.16)$$

Removing $R_a^*(\sigma)$ to the outer integral (since it is not a function of τ) and taking the limit as $T \to \infty$,

$$\lim_{T\to\infty} \int_0^T C*(t)dt = \lim_{T\to\infty} \int_0^T R_a^*(\sigma)d\sigma \int_0^{T-\sigma} h(\tau)d\tau \qquad (26.17)$$

For very large values of T, the upper limit of integration in the integral on the right may be regarded as T rather than T - σ because as $\sigma \to T$, $R_a^*(\sigma) \to 0$ from the nature of the function $R_a^*(\sigma)$ (Fig. 26.1). Hence

$$\int_0^\infty C*(t)dt = \int_0^\infty R_a^*(\sigma)d\sigma \cdot \int_0^\infty h(\tau)d\tau \qquad (26.18)$$

which demonstrates (26.12). It should be noted that the same result follows directly from (12.36) replacing the Laplace operator by $\int_0^\infty e^{-st}dt$ and letting $s \to 0$.

End of Digression

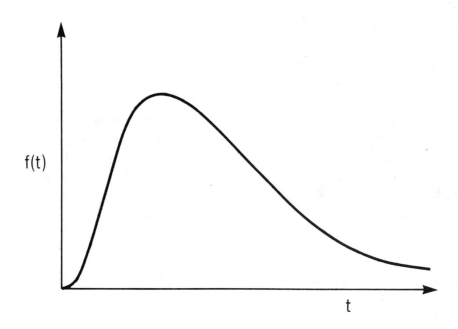

Fig. 26.1. All three of the functions, C*(t), $R_a^*(t)$ and h(t) will have the *general appearance* of f(t) above. $R_a^*(t)$ is arbitrary; h(t) will be shifted to the left; C*(t) will be shifted to the right.

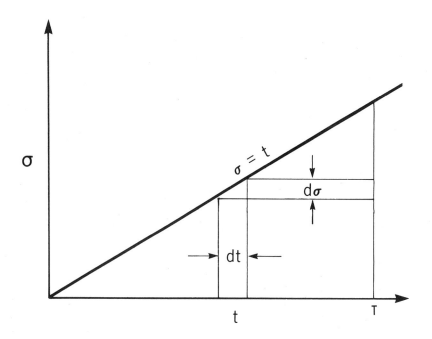

Fig. 26.2. The double integral (26.14) is evaluated over the triangular
            region bounded by the t-axis and the straight lines t = T and
            σ = t. The integration may be carried out by fixing dt first and
            integrating with respect to σ, or vice versa.
From Journal of Theoretical Biology 25, 313, 1969 by W. Perl, Effros and
Chinard. Reproduced by permission of Academic Press.

We may now calculate the required formula for rate of appearance from an
arbitrary tracer infusion, $R_a^*(t)$. R_a is calculated from the bolus or instantaneous injection using

$$R_a = \frac{C}{\int_0^\infty h(t)\,dt} \tag{26.5}$$

From (26.12)

$$\frac{1}{\int_0^\infty h(t)\,dt} = \frac{M_0^*}{\int_0^\infty C^*(t)\,dt}$$

which, upon introduction into (26.5) yields

$$R_a = \frac{M_o *C}{\int_0^\infty C*(t)dt} \tag{26.19}$$

Thus the response to an arbitrary infusion of tracer, $C*(t)$, is used in exactly the same fashion as the response to a bolus injection of tracer, $C_I*(t)$, in calculating the steady state rate of appearance (cf. Eq. (26.6)).

Finally, let us combine the arbitrary tracer infusion formula, (26.19), with the steady tracer infusion formula (26.4):

$$\frac{R_a}{C} = \frac{R_a^*}{C_s^*} = \frac{M_o^*}{\int_0^\infty C*(t)dt}$$

or

$$C_s^* = R_a^* M_o^{*-1} \int_0^\infty C*(t)dt \tag{26.20}$$

The above equation was derived by Perl et al. (1969), and termed by the authors *the indicator equivalence theorem*, since it relates the two types of tracer or indicator experiments.

It should always be borne in mind that the derivations of all equations for the rate of appearance depend upon the validity of the steady infusion equation (26.1) or (26.4).

For the sake of completeness, let me remark explicity that by dividing both the numerator and denominator of the right-hand sides of (26.6) or (26.19) by C   we obtain R_a in terms of specific activity:

$$R_a = \frac{M_o^*}{\int_0^\infty a(t)dt} \tag{26.21}$$

Speaking Practically

One of the assumptions under which we are proceeding is that so little tracer is being added to the system that R_a, the rate of appearance of tracee, is virtually unaffected by it. Therefore all tracer methods for calculating R_a must be consistent one with the other, and all methods must give the correct result if any one method does. R_a can be calculated by injecting a known quantity of tracer, either rapidly or slowly, and dividing the quantity of tracer injected by the area under the specific activity-time curve measured thereafter (26.21). But this method will give the correct result only if the constant infusion method would have given the correct result *had it been used*. Thus the conditions for the validity of (26.21) are the same as the conditions for the validity of (25.36).

Problem

26.1 Eq. (26.6) has been established for a single source point of S. Devise an injection technique theoretically applicable in a system with multiple source points.

SECTION 27
EXPERIMENTAL VALIDATION IN VIVO OF THE EQUATIONS GOVERNING STEADY STATE RATE OF APPEARANCE

The equation for calculating the rate of appearance of S in steady state using the constant tracer infusion method is (25.36):

$$R_a = R_a^*/\alpha \tag{27.1}$$

where α is the constant specific activity at the sampling site. The equation for calculating this same quantity using the tracer injection method is (26.6)

$$R_a = \frac{M_0 {}^*C}{\int_0^\infty C_I{}^*(t)dt} \tag{27.2}$$

where C is the constant tracee concentration and $C_I{}^*(t)$ the tracer concentration at the sampling site. We now ask: what is the experimental evidence for (i) the equivalence and (ii) the validity

of these equations for R_a? We can answer these questions for the glucose system.

(i)   Eq. (27.2) was derived from (27.1) using the theory of linear systems, and it has been shown experimentally (section 13) that the steady state glucose system is, indeed, a linear one with respect to tracer (equivalent).

(ii) Since we now know that the equivalence has been demonstrated experimentally, it is necessary to demonstrate the validity for only one of the equations. Strong experimental evidence can be adduced for the glucose system as follows. Under normal circumstances, glucose is produced endogenously by the liver primarily. When glucose is infused intravenously into a dog, the result is a suppression of the endogenous glucose production. As the rate of exogenous glucose infusion and the plasma glucose concentration rise progressively, the endogenous glucose production is suppressed more and more. When plasma glucose concentration is raised *and held* above a level of 1.3 - 1.5 mg/ml (the level varying between dogs), the endogenous glucose output tends to approach zero — which seems quite reasonable. (Why should the liver produce what you are already providing?) When the total rate of appearance — equal to the endogenous plus exogenous rate — is calculated for steady states in the region $C \geq 1.3$ mg/ml using a tracer injection method, this total rate is found to be very nearly equal to the rate of exogenous glucose infusion (Table 27.1). That is, in steady state

$$R_a + R_{inf} = R_d = R_{a_{total}} \qquad (27.3)$$

where R_a represents the endogenous rate of appearance and R_{inf} the exogenous rate (the rate of infusion). When an injection of ^{14}C - or ^3H - labelled glucose is given, one can calculate (27.2)

$$R_{a_{total}} = \frac{M_o {}^* C}{\int_0^\infty C_I {}^*(t)dt} \qquad (27.4)$$

In the high plasma glucose region it is found that

$$R_{a_{total}} = R_{inf} \qquad (27.5)$$

which of course means (27.3) that

$$R_a = 0 \qquad (27.6)$$

The endogenous supply is completely suppressed. Both the left- and right-hand sides of (27.5) are measured independently and are quite consistently found to be equal (within about 6%).

TABLE 27.1

Dog Number	Plasma glucose	R_{inf}	$R_{a_{total}}$	$(R_{a_{total}} - R_{inf})/R_{inf}$
	[mg/ml]	[mg/min]	[mg/min]	
13	2.91	17.89	17.31	-.0324
15	1.32	14.00	13.78	-.0157
17	1.52	7.09	7.67	+.0818
19	1.58	5.97	5.64	-.0553
22	1.45	6.22	6.17	-.00804

Data obtained from dogs maintained in various glucose steady states above the fasting level. (G. J. Hetenyi Jr., K. H. Norwich and S. Zelin, American Journal of Physiology 224, 635, 1973.) The data corresponding to the greatest rate of intravenous glucose infusion, R_{inf}, is shown for each dog. One paradoxical case has been omitted. In theory, the total rate of appearance, $R_{a_{total}}$, should exceed R_{inf} by the amount of the endogenous rate of glucose production. At high plasma glucose levels, the endogenous glucose supply may be totally suppressed, so that $R_{a_{total}}$ equals R_{inf}. As may be seen from the final column of the table, the tracer-calculated value of $R_{a_{total}}$ (obtained from (27.2) or (27.1)) usually fell slightly *short* of R_{inf} - which, strictly speaking, is not possible - illustrating the limitations of the method. The greatest deviation was 5.53% for dog 19. Endogenous supply was not totally suppressed in dog 17.

Such experiments provide very strong, although indirect, evidence that the various tracer equations are valid for the steady state glucose system. The experimental data has been taken from a paper by Hetenyi and colleagues, 1973. Direct experimental evidence that the appropriate tracer equations govern *nonsteady* rate of appearance of S will be given in a later section.

Speaking Practically

There is good reason to believe that tracer infusion and injection methods
should give the same value for R_a. If you try both methods and obtain
different values, something other than the theory is probably wrong. Ask
yourself such questions as:

Is the label recycling?

Is all the measured radioactivity originating with the tracer molecules?

Is there an error in the biochemical technique?

Are the early minutes of the impulse response curve sampled? (Failure to record the rapid fall of tracer activity in plasma immediately following injection has been known to produce an error in excess of 100% in calculating
R_a.)

SECTON 28

THE DECLINE IN TRACER MASS AS A
MONOEXPONENTIAL PROCESS

The impulse response (response to a sudden injection of a mass, $M_0{}^*$, of
tracer), when recorded at a sampling point close to the injection site, is
usually a monotonically decreasing function of time (Fig. 13.2). Very seldom
can such a decay curve (tracer concentration versus time) be conformed, over
its entire duration, to a monoexponential function, but rather requires a
linear combination of two or more exponential functions (Fig. 22.2). And
there is no reason to stick slavishly to exponentials. We have seen that one
can often do equally well or better using functions of the form $at^{-\alpha}$ or
$at^{-\alpha}e^{-\beta t}$ (section 21).

We have seen (section 22) that if the distribution space of S can be treated
like the contents of a tank which are being stirred vigorously, then the
impulse response curve will be truly monoexponential, and the steady state
rate of appearance of S will be given by (22.9)

$$R_a = \frac{M_o{*}C}{A/k} \tag{28.1}$$

But, since the observed impulse response is very seldom monoexponential, we may safely assume that the single, well-stirred tank model is just a first order approximation.

We have seen (section 26) that with a single, common source point for S and S*, the impulse response need not be monoexponential, but may be conformed to a sum of many exponentials — n say; and the steady state rate of appearance will be given by [(26.6), (26.8)]

$$R_a = \frac{M_o{*}C}{\sum\limits_{i=1}^{n} \dfrac{B_i}{\beta_i}} , \qquad \beta_1 > \beta_2 > \ldots > \beta_n \tag{28.2}$$

Eq. (28.1) will usually provide a good estimate of (28.2) when A/k is taken to be equal to B_n/β_n — the contribution of the "slow" or "prevailing" exponential term.

We now ask the question: *Are there any circumstances under which the impulse response function, C*(t), will not be monoexponential in character, — say*

$$C{*}(t) = \sum\limits_{i=1}^{n} B_i e^{-\beta_i t} , \qquad n > 1 \tag{28.3}$$

— and yet the steady state rate of appearance will still be given accurately (not just as a first approximation) by (28.1) in the form

$$R_a = \frac{M_o{*}C}{B_n/\beta_n} \quad ? \tag{28.4}$$

Suppose that we have a distributed system (not well-mixed)[†] in which the concentration of S is, effectively, uniform at steady state. That is, at steady state

---

[†]There may be many sources and sinks dispersed throughout the system so that, effectively, $\nabla C = 0$.

$$\nabla C = 0 \tag{28.5}$$

Suppose moreover, that $R_{d,v}$ is also uniform so that

$$\nabla R_{d,v} = 0 \tag{28.6}$$

The boundaries will be taken to be rigid and impermeable to S. We need make no supposition here about the mode of transport in the system, but we shall suppose that the dispersion of material within the system is "rapid" in comparison with metabolism — an expression which we shall invest with some mathematical meaning in a short while. Suppose that an impulse of $M_o{}^*$ is applied at an arbitrary inlet site and that C^* is measured at an arbitrary sampling site. Prior to the injection of tracer, such a system is very much like a single, well-stirred tank in that S has a uniform concentration throughout its distribution space. But the system is *not* well-mixed and the injected tracer will require a finite time to disperse.

Introducing the equation of continuity in the form (10.6) without regard to the mode of dispersion (convection or diffusion),

$$\frac{\partial C^*}{\partial t} = -\nabla \cdot (C^*\vec{v}) - R^*_{d,v} \tag{28.7}$$

Since

$$R^*_{d,v} = C^* \frac{R_{d,v}}{C} \tag{28.8}$$

we have, as a consequence of (28.5), (28.6) and (28.8)

$$R^*_{d,v} = kC^* \tag{28.9}$$

where

$$k = R_{d,v}/C \quad , \tag{28.10}$$

a positive constant. Introducing (28.9) into (28.7)

$$\frac{\partial C^*}{\partial t} = -\nabla \cdot (C^*\vec{v}) - kC^* \quad , \tag{28.11}$$

and integrating over the entire distribution volume, V,

$$\int_V \frac{\partial C^*}{\partial t} \, dV = -\int_V \nabla \cdot (C^*\vec{v}) dV - \int_V kC^* dV \tag{28.12}$$

That is

$$\frac{d}{dt} \int_V C^* dV = -\int_S C^*\vec{v} \cdot \vec{n} \, dS - k \int_V C^* dV$$

using the divergence theorem. The surface integral vanishes due to the impermeability of the surface, and $\int_V$ C*dV is equal to the total mass of tracer, M*(t), present in the system. Thus

$$\frac{dM*}{dt} = -kM* \qquad (28.13)$$

and

$$M* = M_0* \, e^{-kt} \qquad (28.14)$$

It is seen that the *total mass* (not concentration) of tracer declines mono-exponentially in this system regardless of the initial spatial distribution of tracer, as long as the tracer was entered as an impulse in *time*. This *theorem on heterogeneous tracer distribution* was introduced by Norwich and Hetenyi (1971) for a general metabolic system, and earlier by Perl (1963) for a linear system.

Now the assumption has been made that the rate of dispersion "exceeds" the rate of metabolism, which will be interpreted to mean that tracer will dis-tribute itself fairly uniformly in space before, say, 66% of the injected tracer mass has been degraded metabolically. In the strict mathematical sense, tracer will never become uniformly distributed. The assumption is made that uniformity will have been reached for all intents and purposes (within the limits of experimental resolution) before the arbitrary fraction, two-thirds, of the injected tracer has vanished. If and when tracer has been distributed uniformly in this fashion, the tracer concentration, C*, will be given simply by dividing the total tracer mass M* by the volume V. From (28.14)

$$C* = \frac{M_0*}{V} \, e^{-kt}, \qquad t > t_u \qquad (28.15)$$

where t_u is the time of effective uniform tracer distribution. If C* has been fitted to a sum of exponentials, then $t > t_u$ corresponds to the time when the data may be fitted by the single, prevailing exponential function,

$$C* = B_n \, e^{-\beta_n t} \qquad (28.16)$$

comparing (28.15) and (28.16)

$$\beta_n = k \qquad (28.17)$$

and

$$B_n = M_0*/V \qquad (28.18)$$

Since (28.15) would have been valid for *any* initial tracer distribution
applied as an impulse in time, it would therefore have been valid if tracer
had been distributed uniformly at zero time. But under the latter conditions
the initial uniform tracer concentration, $C_0{}^*$, would be given by

$$C_0{}^* = M_0{}^*/V = B_n \qquad (28.19)$$

Thus by plotting a graph of the logarithm of C^* against t and extrapolating
the prevailing monoexponential portion back, to obtain a y-intercept, $\log C_0{}^*$,
we obtain

$$V = \frac{M_0{}^*}{C_0{}^*} = \frac{M_0{}^*}{B_n}{}^\dagger \qquad (28.20)$$

Clearly, the volume of the system is independent of the initial tracer distribution, so that V will always be given by (28.20): the mass of tracer injected divided by the intercept (Fig. 28.1).

Now from (28.10)

$$R_{d,v} = kC \qquad (28.10)$$

and since k and C are constants

$$R_d = V \cdot R_{d,v} = kVC \quad, \qquad (28.21)$$

or equivalently, using (28.17), and (28.20)

$$R_d = \beta_n \frac{M_0{}^*}{C_0{}^*} C \qquad (28.22)$$

Finally, introducing (28.19)

$$R_d = \frac{M_0{}^*C}{B_n/\beta_n}{}^\dagger \qquad (28.4)$$

which is (28.4) since $R_a = R_d$.

---

†The validity of this equation rests upon the hypothesis that there will come
a time when tracer is fairly uniformly distributed spatially while an appreciable amount of tracer still remains in the system. A computer simulation of a system of rather arbitrary structure, arbitrary convection currents etc. has shown that this equation gives a very close approximation of
the measured value (V, R_d) (Spiegel and Norwich, unpublished).

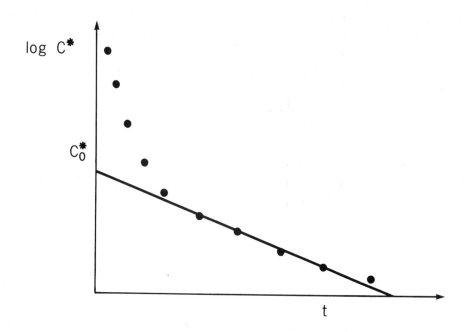

Fig. 28.1. The logarithm of the impulse response is plotted against time and
            a straight line is drawn through the prevailing portion, if it
            exists. The volume of the system is estimated by dividing M_0^*,
            the mass of injected tracer, by the antilogarithm of the inter-
            cept, C_0^*.

Thus we have examined a system which is not well-mixed, in which the monoexponential equation theoretically provides the correct value of steady state
R_a. In such a system the tracer was not infused "equivalently".

The desirability of comparing the speed of metabolic with that of dispersive
processes suggest that we should seek a dimensionless number which would contain this information. for example, in chemical reactor theory the extent of
diffusion of material in comparison with bulk flow is given by a dimensionless number, D/vL, called the *reactor dispersion number* (or its reciprocal,
the *axial Peclet number*). L is some characteristic length. The quantity *c*
defined by

$$c = R_d/C \qquad\qquad (28.23)$$

has the dimensions of [volume / time] and is known in physiological studies
as the *clearance* or *metabolic clearance rate*. It represents the (virtual)
volume "cleared" of a metabolite per unit time and is, therefore, an index of
the extent of metabolic activity.

We can then define a dimensionless number,

$$\frac{uc}{D^2}$$

relating the three quantitites, clearance, convective velocity magnitude and
diffusion coefficient. Since the convective speed, u, will be constant for
many substances of varying molecular weight and metabolic activity, the number uc/D^2 would reflect the ratio of metabolic to diffusive activity. Since
it is anticipated that u will be difficult to evaluate, one might, alternatively, define the ratio D^2/c as a characteristic velocity. The greater
this velocity, the greater the ratio of diffusive to metabolic activity.

Speaking Practically

In this section we derived two practical equations, (28.21) for calculating
R_a and (28.20) for calculating V, the volume of the system. If you are sure
that tracee, the unlabelled substance, is quite uniform in its concentration
throughout the system, then you may proceed very simply, as follows. Inject
a certain mass of tracer and measure C* for as long a period as is prac-
ticable. Plot $\ln$ C* vs t using semi-log paper, draw the best straight line
through the later points (omitting the earlier ones), and extrapolate this
straight line backwards (to the left) until it intersects the C*-axis. The
value of the intercept, C_0^*, may then be read from the semi-log paper. The
volume, V, is then obtained by dividing the mass of tracer injected by the
value of the intercept, C_0^* (28.20). By taking any two points on the straight
line, determine its slope, k ($time^{-1}$). Representing the concentration of the
unlabelled substance, as usual, by C, $R_a = R_d = kVC$ (28.21). This is one of
the oldest of tracer procedures. It is usually not the most accurate way of
determining either V or R_a because C is not usually perfectly uniform.

SECTION 29

DETERMINING THE VOLUME OF A DISTRIBUTED METABOLIC SYSTEM

A Problem Only Partially Solved

How does one determine the volume through which a given substance, S, is distributed? We could, perhaps, answer this question more easily if we knew just what was meant by the term *distribution volume of S*. Is this the volume within a living system which will contain *all* the molecules of S? If this is the case, the distribution volumes of effectively all molecules or species in the system will equal the total volume of the system. For example, the distribution volume of haemoglobin would generally be taken to be the intravascular volume; but who would state that not a single molecule of haemoglobin penetrates the extravascular space or the urine? It is clearly not useful to regard the distribution volume of S as that volume which contains all the molecules of S. Turning for a moment to physics, we might ask: what is the volume occupied by an atom of hydrogen? Applying Schrödinger's equation we conclude that there is a finite probability of locating this atom at any point in space[†]. Thus a much more useful question is: what is the volume in which we should expect to find the hydrogen atom 99% of the time? Such a formulation of the question leads to a more useful "volume" for the hydrogen atom. The analogy between the living system and the atom is quite plain, and we should probably do well to define the distribution volume of S as the minimum volume in which 90 to 99% of the molecules of S can be contained. If we invoke an *ergodic* hypothesis we could equivalently define the distribution volume of S as that volume where a given molecule of S within the system will spend 95 or 99% of its time. Definitions such as the latter

---

[†]For the reader unfamiliar with this view of modern physics we might state the wave mechanical view rather briefly. The atom is assigned a wave function, ψ, such that the probability of finding the atom between r and r + dr is given by $\psi\psi^*dr$, where ψ^* is the complex conjugate of ψ. For a free hydrogen atom, $\psi\psi^*dr > 0$ for all r, dr > 0.

will permit us at least to estimate the distribution volume of materials *in vivo*.

Before approaching a general method, let us review some of the methods which have been employed in earlier sections of the text, implicitly or explicitly, for finding volumes.

(i) Special Methods for Particular Systems

The distribution space for glucose was taken to be equal to that of the extracellular fluid because numerous assays of plasma and interstitial fluids have shown them to contain substantial quantities of glucose, while measurements made on intracellular water, hepatocytes excluded, have indicated much lower concentrations of glucose (Morgan et al., 1961). Labelling components of extracellular fluid and applying the dilution principle (section 19) have shown that extracellular fluid has a measured value equal to about 20% of body volume etc. This was a special method for measuring the glucose space, and it has been partially confirmed by tracer studies. The division of total body water into an extracellular and an intracellular component has been carried out by utilizing the known electrolyte distribution between components (Manery and Hastings, 1939). Etcetera. Special methods exist for certain substances.

(ii) Direct Mesurement using a Measuring Rod

Obviously, the volume of urine contained within the bladder can be measured by evacuating the bladder and measuring, with a graduated cylinder, the volume of expelled urine. Similarly, the volume of fluid filling a cystic container can be estimated by measuring the dimensions of the container directly, (Fig. 29.1). In this way the volume of the chambers of the heart may be calculated by filling them with radio-opaque material and making radiographs at proper angles. In the "Cylindrium" example (section 20) we assumed that the total distribution volume of fictitium could be obtained from the external measurements of the organism. However, the method of direct measurement is not of great use in finding the distribution space of most substances *in vivo*.

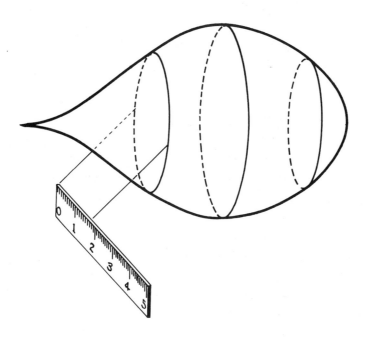

Fig. 29.1. The volume of an irregular container may be estimated by means of
            a series of measurements made with a measuring rod, provided
            that the container is accessible.

(iii) The Dilution Principle: Zero-Gradient Systems

In a system where the gradients in C and $R_{d,v}$ are equal to zero (see section
28) we may apply the dilution principle. Injecting a mass M_o^* of S we obtain
the volume V from the intercept of the monoexponential portion of the C*
decay curve. From (28.20)

$$V = \frac{M_o^*}{C_o^*}$$

where C_o^* is the intercept (Fig. 29.2). In this way we can find the volume
of the blood using ^{32}P - labelled red blood cells (Lawson, 1962)[+]. Since the

---

[+]We did not invoke a particular mode of transport such as convection in the
  derivation of (28.20) and therefore the failure of plasma to entrain blood
  cells (see section 6) is not of importance here.

concentration of red blood cells is fairly constant (i.e. hematocrit fairly uniform), since there are multiple sites of red cell destruction, and since blood cell dispersion is rapid in comparison with blood cell destruction, probably the criteria for validity of (28.20) are fulfilled (i.e. conditions (28.5) and (28.6) are met).

In dealing with multicompartment systems with a single compartment of access, we used (28.20) to calculate the volume, V_1, of this compartment:

$$V_1 = \frac{M_o^*}{C_1^*(0)} \qquad (19.22)$$

Usually (19.22) was used as the definition of V_1 so that no risk of experimental confrontation was run.

The above three techniques: the dilution technique, the measuring rod and special *ad hoc* techniques have, therefore, been introduced in earlier sections. Let us now turn to a final, and probably most useful, approach to determining the distribution volume of a substance *in vivo*.

(iv) Approximate Distribution Volumes using the Mean Residence Time in Steady State Systems

a) Mean residence time. The *mean residence time* or *mean sojourn time* of a molecule of S in a system is the mean or average period of time spent by such a molecule within the system. In order to calculate the mean residence time (represented by $\bar{t}$) we must know the times of entrance of a large number of molecules into the system and the times of exit of each of these molecules from the system. By subtracting the entrance time from the exit time for each molecule, we obtain the residence time for that molecule, and by averaging the residence times for all molecules we obtain the mean residence time $\bar{t}$. However, such complete data is very difficult to obtain in general, and we must be content to measure $\bar{t}$ in special systems, or to estimate $\bar{t}$ in more general systems.

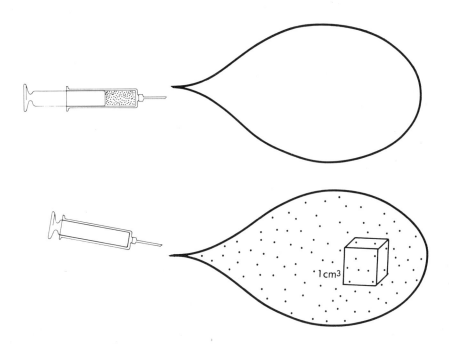

Fig. 29.2. The dilution principle. A syringe containing 100 particles is
            emptied into an irregular container of unknown volume. After
            mixing, a sample of one cubic centimetre within the container is
            found to contain 4 particles. The volume of the container is
            then 100/4 = 25 cubic centimetres.

Suppose that a bolus of tracer of mass M_o^* is injected into a steady state
system at zero time. Then the rate of disappearance of tracer per unit volume is given by

$$R_{d,v}^* = C*\left(\frac{R_{d,v}}{C}\right)$$ (29.1)

as usual. Hence the mean residence time for particles leaving the system via
some particular volume element, dV, may be designated $\bar{t}_v$, where

$$\bar{t}_v = \frac{\int_0^\infty t R_{d,v}^* \, dt}{\int_0^\infty R_{d,v}^* \, dt} \quad , \qquad R_{d,v}^* \neq 0$$ (29.2)

if the integrals converge, using the usual statistical representation for a mean. Substituting (29.1) into (29.2)

$$\bar{t}_v = \frac{\int_0^\infty t\left(\frac{R_{d,v}}{C}\right) C* \, dt}{\int_0^\infty \left(\frac{R_{d,v}}{C}\right) C* \, dt}$$

Since the tracee is in steady state, the factor $R_{d,v}/C$ is constant in time, comes out in front of the integral sign, and is cancelled, leaving

$$\bar{t}_v = \frac{\int_0^\infty t C* \, dt}{\int_0^\infty C* \, dt} \qquad (29.3)$$

where the arguments of the functions are $\bar{t}_v(\vec{r})$ and $C*(\vec{r},t)$. If the integrals converge, we can now, formally anyway, integrate (29.3) over all elements of volume to obtain the mean residence time. That is

$$\bar{t} = \frac{\int_V \bar{t}_v R_{d,v} \, dV}{\int_V R_{d,v} \, dV} \qquad (29.4)$$

$\bar{t}$ may now be evaluated for the case described in section 28, where the gradients in C and in $R_{d,v}/C$ were equal to zero ((28.5) and (28.6)) and the initial bolus of tracer had an arbitrary spatial distribution. Eq. (28.15)

$$C* = \frac{M_0^*}{V} e^{-kt} \qquad (29.5)$$

is still valid for $t > t_u$, where t_u is the time when tracer assumed an effectively uniform spatial concentration. Now $\bar{t}$ is a property of unlabelled molecules of the species of S and, therefore, cannot vary with the initial distribution of molecules of S*. Therefore, let us take the case where S* was uniformly distributed at zero time; i.e. $t_u = 0$. Then (29.5) is valid for $0 \leq t < \infty$. Substituting for C* from (29.5) into (29.3)

$$\bar{t}_v = \frac{\int_0^\infty t \, e^{-kt} \, dt}{\int_0^\infty e^{-kt} \, dt} = \frac{1}{k} \qquad (29.6)$$

But S and S* are now uniform throughout all V. Therefore

$$\bar{t}_v = \bar{t} = \frac{1}{k} \qquad (29.7)$$

Eq. (29.7) should, then, be valid regardless of the initial spatial distribution of tracer. Regardless of the sampling site, e^{-kt} will be the prevailing exponential in the decay of C* with time, and thus $\bar{t}$ from (29.7) is unique.

Let us turn, now, to the general or unrestricted system. An impulse of tracer is applied at some point in the system and C* is measured at some arbitrary sampling site, $\vec{r}_s$. Using (29.3) we may calculate $\bar{t}_v(\vec{r}_s)$ as

$$\bar{t}_v(\vec{r}_s) = \frac{\int_0^\infty tC^*(\vec{r}_s,t)dt}{\int_0^\infty C^*(\vec{r}_s,t)dt} \tag{29.8}$$

but unfortunately, we cannot in general determine $\bar{t}_v$ at other points. In other words, we cannot obtain $\bar{t}_v(\vec{r})$ in general, so that $\bar{t}$ may not be obtained from (29.4). Hence for such a system we have two choices:

   (1) Refrain from calculating V by tracer methods
   (2) Assume that the sampling site is a *representative* point, so that

$$\bar{t} \doteq \bar{t}_v(\vec{r}_s) \tag{29.9}$$

If we are fortunate enough to deal with a system with a single site of disappearance of S, and if we can select the sampling point very close to the site of disappearance, then

$$\bar{t} = t_v(\vec{r}_s) \tag{29.10}$$

exactly. As we shall see later, the inulin system, having the kidneys as a "single" site of disappearance, is a good example of this convenient type of system.

Let us now try to get the gist of the idea of determining distribution spaces using mean transit times. Suppose that a substance enters a certain system at the constant rate R_a mass units per unit time. Suppose, moreover, that on the average, each of these mass units dwells within the system for $\bar{t}$ seconds before leaving it. The system will then contain $R_a \cdot \bar{t}$ mass units of material. If there is only one point of entrance to and exit from the system it will contain a "string" of $R_a \cdot \bar{t}$ mass units (Fig. 29.3). If $\bar{C}$ is a mean or average concentration taken over the entire system, then the volume of

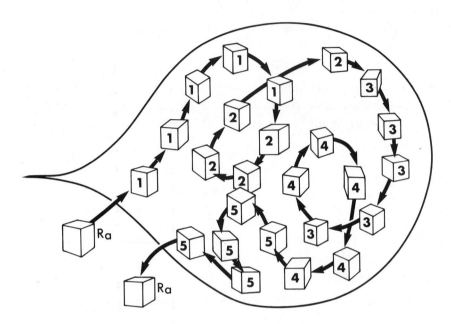

Fig. 29.3.   The principle of mean transit time illustrated by a system with
             a single site of input and of output. If a substance of mean
             concentration C [mass/volume] enters a system at the steady rate
             R_a [mass/time], it is *as if* a volume R_a/C entered the system in
             each unit of time. If the mean or average time of passage from
             inlet to outlet is $\bar{t}$, then each volume R_a/C occupies all possible
             positions within the container during the time $\bar{t}$. The total
             volume of the container is then $(R_a/C)\bar{t}$. In the diagram, each
             cubic centimetre contains one particle so that C = 1. Five par-
             ticles enter the system in one second so that R_a = 5. The 5 cc's
             which entered 5 seconds ago are marked *5*; the 5 cc's which
             entered 4 seconds ago are marked *4* etc. Each cubic centimetre
             occupies all possible positions within the container during its
             sojourn, and the mean residence or sojourn time is 5 seconds.
             Therefore the volume of the container is 5/1[cc's per second] ·
             5 [seconds] = 25 cubic centimetres. The diagram is, of course,
             schematic and the little cubes are supposed to fill the entire
             container.

the system will be approximated by the mass of material it contains divided by the mean concentration, or

$$V \doteq \frac{R_a \cdot \bar{t}}{\bar{c}}$$

(29.11)

This is an intuitive derivation and we shall now attempt to improve upon it.

b) Approximate pool sizes and distribution volumes. The *pool size* is a term sometimes applied to the total mass of S within a system. We shall now see in more detail how this quantity may be approximated by estimating the mean residence time of S. The distribution volume can then be obtained as a further approximation by estimating the mean concentration of S within the steady state system. This demonstration for the volume of a metabolic system is parallel to the method of Meier and Zierler (1954) or Sheppard (1962) for the volume of blood in a capillary labyrinth. The latter, however, can often be estimated with greater accuracy.

As we discussed above, calculating $\bar{t}_v(\vec{r}_s)$, the mean residence time for particles of S departing from a sampling point with coordinates designated by $\vec{r}_s$, is not difficult. But obtaining $\bar{t}$, the mean residence time for particles in the entire system, with exactness, is often not possible. We shall meet this problem by pretending that it does not exist. Let us estimate $\bar{t}$ by $\bar{t}_v(\vec{r}_s)$; that is let us assume that the sampling point is a representative point (29.9).

Let us represent the distribution of *exit* times for molecules of S entering the system at zero time at its only source point by g(t). That is, if a molecule of S enters the system at t = 0, the probability that it will disappear (by excretion, degradation etc.) between the times t and t+dt is given by g(t)dt. Define f(t) by

$$f(t) \equiv \int_0^t g(\tau)d\tau$$

(29.12)

Then f(t) gives the probability that a molecule which enters the system at t = 0 will disappear *by* the time, t. Since all molecules of S entering the system must ultimately leave,

$$\lim_{t \to \infty} f(t) = 1$$

(29.13)

The system is in steady state and S enters via its single inlet point at the constant rate R_a. Thus,

the mass of S entering between times 0 and Δt which disappears by time t

$$= R_a \cdot \Delta t \cdot f(t);$$

the mass of S entering between times Δt and $2\Delta t$ which disappears by time t

$$= R_a \cdot \Delta t \cdot f(t-\Delta t);$$

.

.

.

the mass of S entering between times $t-\Delta t$ and t which disappears by time t

$$= R_a \cdot \Delta t \cdot f(\Delta t)$$

The mass of S which enters the system between the times zero and t and which disappears from the system by time t is then given by

$$\int_0^t R_a f(\tau)d\tau$$

The total mass of S which has entered the system in time, t, is $R_a \cdot t$. Therefore the mass of S which entered the system after zero time and which remains in the system at time t, is equal to the total mass which entered minus that mass which has disappeared during the time, t, or

$$R_a \cdot t - \int_0^t R_a f(\tau)d\tau$$

Eventually this "new" S (that which entered after t = 0) will constitute all the substance S which is in the system. That is, the "new" mass will equal the pool size M. Hence M is given by

$$M = \lim_{t\to\infty} R_a[t - \int_0^t f(\tau)d\tau] \qquad (29.14)$$

We can write

$$\frac{df(\tau)}{d\tau} = g(\tau) . \qquad (29.15)$$

That is, since f(0) is effectively equal to zero, integrating (29.15)

$$f(\tau) = \int g(\tau)d\tau + a \text{ constant} \qquad (29.16)$$

and

$$f(t) = \int_0^t g(\tau)d\tau \qquad (29.16a)$$

as indicated by the definition (29.12). Therefore, integrating by parts using (29.16)

$$\int_0^t \tau g(\tau) d\tau = \left[\tau f(\tau) \right]_0^t - \int_0^t f(\tau) d\tau$$

$$= t\, f(t) - \int_0^t f(\tau) d\tau \qquad (29.17)$$

Substituting for $\int_0^t f(\tau) d\tau$ from (29.17) into (29.14)

$$M = \lim_{t \to \infty} R_a \left[t - t f(t) + \int_0^t \tau g(\tau) d\tau \right]$$

Introducing (29.13) and changing the dummy variable from τ to t leaves

$$M = R_a \int_0^\infty t\, g(t) dt \qquad (29.18)$$

Since g(t) gives the distribution of exit times from the system, $\int_0^\infty t\, g(t) dt$

by definition is equal to the mean residence time for all molecules of S in the system. That is

$$\bar{t} = \int_0^\infty t\, g(t) dt \qquad (29.19)$$

so that

$$M = R_a \cdot \bar{t} \qquad (29.20)$$

Eq. (29.20) states that the pool size, or mass of unlabelled material in a steady state system is equal to the product of the rate of appearance of S and its mean residence time. It has been proved for a system with a single source point. The right-hand side of (29.20) may be evaluated explicitly in terms of the impulse response function determined at some representative sampling point[†]. Since

$$R_a = \frac{M_0 {*} C}{\int_0^\infty C{*} dt} \qquad , \qquad (26.6)$$

[†] C and C* in equations used from here to the end of the section will expressly mean tracee and tracer concentrations respectively at the designated sampling point. That is, they are C_s and $C_s^*(t)$.

and we are approximating $\bar{t}$ by $\bar{t}_v(\vec{r}_s)$ so that

$$\bar{t} = \frac{\int_0^\infty tC^* dt}{\int_0^\infty C^* dt} , \qquad (29.8)$$

therefore

$$M = \frac{M_0^* \, C \int_0^\infty tC^* dt}{\left[\int_0^\infty C^* dt\right]^2} \qquad (29.21)$$

The total mass, M, of the substance S in the system may be expressed as

$$M = \bar{C}V \qquad (29.22)$$

that is, as the product of the distribution volume, V, and some mean concentration of S, $\bar{C}$. Because it is usually not possible to estimate $\bar{C}$ accurately, the best one can do is set

$$\bar{C} = C \qquad (29.23)$$

that is, set the mean concentration of S equal to the concentration at the sampling site. Thus the sampling site is taken as a *representative* point for determining $\bar{t}$ and $\bar{C}$. Usually this cannot be justified rigorously. However, introducing the approximation (29.23) into (29.22) and referring to (29.20) we obtain

$$V \doteq \frac{M}{C} = \frac{R_a \cdot t}{C} \qquad (29.24)$$

Finally, introducing (29.24) into (29.21)

$$V = \frac{M_0^* \int_0^\infty tC^* dt}{\left[\int_0^\infty C^* dt\right]^2} \qquad (29.25)$$

Suppose that one localized region of a larger system maintains uniform concentrations C and C*, and that this region serves as the sole port of entry and exit for S and S* to and from the whole system. Then a sampling point in this region is a representative point and Eq. (29.21) is expected to hold exactly. If, in addition, C is uniform throughout the entire system so that (29.23) is valid, then Eq. (29.25) is expected to hold true exactly.

Bright (1973), using multicompartmental analysis, demonstrated that (29.25) was valid for measurements made in the sole entry-exit compartment. We have demonstrated the validity of (29.21) for a somewhat more general system which need not consist of compartments except for the entry-exit region, which sustains no gradients, and (29.25) for a system with $\nabla C = 0$.

When plasma serves both as the sampling site and the only sink (approximated, for example, by the case of inulin discussed below), then the sampling site is truly representative of $\bar{t}$. If $\vec{r}_s$ designates the sampling point, then from (29.4)

$$\bar{t} = \frac{\bar{t}_v \cdot R_{d,v}(\vec{r}_s) \cdot \Delta V}{R_{d,v}(\vec{r}_s) \cdot \Delta V} = \bar{t}_v$$

The primary assumption remaining in the use of (29.25) is, then, the equating of plasma concentration with mean concentration (29.23), which will usually not introduce a great error. The calculation of the pool size from (29.21) is free of even this approximation, and so should be quite accurate.

c) Applications to the analysis of physiological systems. The calculation of volumes using the tracer technique described above involves a number of assumptions: e.g. representative sampling points, zero concentration gradients, etc. So the whole matter of tracer-calculated volumes is not on firm footing. Nevertheless, it is possible to make some interesting observations about the nature of certain systems by applying the techniques such as they are.

Both glucose and inulin are believed to be distributed essentially throughout the extracellular space, based upon special (non-tracer) methods (see (i) above). The extracellular fluid in a dog may be estimated by calculating 20% of the total body volume; that is extracellular volume (millilitres) approximately equals 20% of total body weight (grams).

Glucose is a substance entering its distribution space primarily through one inlet (the hepatic vein) and metabolized by very many tissues disseminated throughout the body. Thus for the glucose system we might venture to make the approximations ((28.5) and (28.6))

$$\nabla C = 0$$

$$\nabla R_{d,v} = 0$$

Hence for glucose we should expect that V calculated from (28.20)

$$V = M_0^*/C_0^* \qquad\qquad (29.26)$$

should be reasonably accurate. M_0^* is the mass of tracer injected as an
impulse, and C_0^* is the y-intercept of the prevailing exponential function in
the impulse response curve. However, the fact that there are multiple sites
of degradation in no way disqualifies V as calculated from (29.25). Therefore the distribution volumes of glucose calculated using (29.25) and (29.26)
should be nearly equal and both values of V should match the extracellular
volume quite closely (Table 29.1).

Inulin is not a metabolite, but can be infused intravenously and is excreted
nearly totally in the kidney by a process of glomerular filtration. Thus
inulin has a single source point but a relatively isolated site of excretion.
Moreover, the inulin "molecule" is large (molecular weight $\sim$ 2500) and con-
sequently its diffusion is slow. Thus it will have non-zero gradients in
both C and $R_{d,v}$. For example it is not expected that inulin in steady state
would be found in the same concentration in both plasma and interstitial
fluids (Schachter et al., 1950). Therefore V calculated from (29.26) is not
expected to be accurate. V should be calculated for inulin only by use of
(29.25), and when calculated by (29.25) should match the extracellular volume
quite closely (Table 29.1)[†].

Calculations were made on the glucose and inulin systems in dogs. Both
systems were in steady state with respect to tracee. To establish a steady
state in inulin, the substance was infused intravenously for a long period of

---

[†]The use of (29.25) is fraught with pitfalls. Eq. (29.25) implies that the
sampling site is a representative point with $\bar{t}_v$ calculated from (29.2). And
(29.2) is valid only if $R_{d,v}^* \neq 0$; i.e. if the sampling site is "representative" of a sink. But the glomerulus is the only real sink for extracellular
inulin. Can we then be certain *a priori* that the use of (29.25) is
justified?

TABLE 29.1

Substance	Expt. No.	M_0^* [dpm]	B_1 [dpm/ml]	B_2 [dpm/ml]	β_1 [min^{-1}]	β_2 [min^{-1}]	Body wt [g]	V = 20% body volume [ml]	V = $R_a \cdot \bar{t}/C$ [ml]	V = M_0^*/B_2 [ml]
Inulin	1A	4.56×10^7	41600.	15700.	0.18	0.022	9700.	1940.	1720.	2900.
Inulin	4A1	4.04×10^7	18700.	7590.	0.114	0.015	15200.	3040.	3160.	5320.
Inulin	4A2	7.34×10^7	37200.	12400.	0.10	0.015	15200.	3040.	3010.	5920.
Glucose	14D	3.25×10^7	6710.	8770.	0.282	0.0108	17100.	3420.	3510.	3710.
Glucose	15C	3.23×10^7	13800.	11700.	0.274	0.0133	12400.	2480.	2480.	2760.
Glucose	19A	3.32×10^7	8380.	8450.	0.290	0.0112	16500.	3300.	3650.	3930.

Plasma impulse responses to inulin and to glucose were conformed to the function $C^* = B_1 e^{-\beta_1 t} + B_2 e^{-\beta_1 t}$. The amount of tracer injected, M_0^*, and the four parameters of the impulse response function are tabulated. The distribution volumes of inulin and glucose are calculated (a) as equal to the extracellular space, 20% of the body volume; (b) as the product of clearance R_a/C with mean residence time, $\bar{t}$; (c) by the distributed sink equation, equal to M_0^*/B_2. The three calculations give nearly the same value of V for glucose, while calculation (c) differs from the other two for inulin. Most of the numbers in the table have been rounded off to 3 significant figures.

time. The experiments were performed at different times; by different
groups; on different dogs. The glucose experiments were reported by Hetenyi
et al. (1973) and the inulin experiments by Norwich et al. (1974). ^{14}C was
used as a label for glucose, and either ^{3}H or ^{14}C for inulin. The tracer was,
of course, injected as an impulse, and the impulse response in plasma was
found to be adequately fitted over a two-hour period by a double exponential
function. Multiexponential curve fitting was not absolutely necessary for
these calculations but facilitated the analysis. The impulse responses were,
therefore, represented by a function of the form

$$C^* = B_1 e^{-\beta_1 t} + B_2 e^{-\beta_2 t} \qquad \beta_1 > \beta_2 > 0 \ . \qquad (29.27)$$

For purposes of calculating V by (29.25) the evaluation of two integrals was
required:

$$\int_0^\infty C^* dt = \frac{B_1}{\beta_1} + \frac{B_2}{\beta_2} \qquad\qquad (29.28)$$

$$\int_0^\infty t C^* dt = \frac{B_1}{\beta_1^2} + \frac{B_2}{\beta_2^2} \qquad\qquad (29.29)$$

These integrations can be readily verified using (29.27). In calculating V
from (29.26),

$$C_0^* = B_2$$

since B_2 is the coefficient of the prevailing exponential ($\beta_2 < \beta_1$). The
results of the calculation of the distribution volumes by three techniques
are assembled in Table 29.1.

Examining the three right-hand columns in Table 29.1 reveals the expected
result. V calculated by (29.25) always agrees reasonably well with the volume of the total extracellular fluid as estimated by taking 20% of the body
volume. The agreement is equally close for both inulin and glucose. However, the value of V calculated from (29.26) is greater than the other two
values of V by a factor of nearly 2 for inulin, while it nearly agrees with
the other two values for glucose.

Now, it may be seen by substituting (29.28) and (29.29) into (29.25) that

$$V = \frac{M_o^* \left(\dfrac{B_1}{\beta_1^2} + \dfrac{B_2}{\beta_2^2} \right)}{\left(\dfrac{B_1}{\beta_1} + \dfrac{B_2}{\beta_2} \right)^2}$$ (29.30)

so that when B_1 is greater than zero and of the order of magnitude of B_2

$$V \to M_o^*/B_2$$

when

$$\beta_2 \ll \beta_1$$

That is, V calculated from (29.25) will approach V calculated from (29.26) as
the ratio of β_1 to β_2 increases. A glance at Table 29.1 shows this ratio to
be much higher for glucose than for inulin, as it must be, since (29.25) and
(29.26) nearly agree in the case of glucose. The probable physiological
explanation for the discrepancy in the kinetics of the two substances is the
larger size and hence slower diffusion of the inulin molecule, coupled with a
paucity of portals of exit for inulin. These conditions conspire to produce
a gradient in inulin concentration under steady state conditions which pro-
hibits the use of (29.26).

As a final example, let us consider the case of glycerol. When C* is the
concentration of labelled glycerol in plasma of a dog, the impulse response
C*(t) is fitted far more satisfactorily by a power function of the form

$$C^* = At^{-\alpha}, \quad 2 > \alpha > 1$$ (29.31)

than by a sum of exponentials (M.A. Kallai, K.H. Norwich and G. Steiner,
in publication). A graph of log C* against log t is linear for t between 15
seconds and 75 minutes following tracer injection. It is seen, then, that the
integral

$$\int_0^\infty tC^* dt$$

fails to converge, and hence $\bar{t}_v$ is not obtainable by Eq. (29.3). Thus the
distribution volume of glycerol does not seem to be obtainable by the
methods of this section using samples of plasma only.

And now, how close have we come to the concept of volume with which this
section was introduced — the idea that a volume of distribution is that
volume in which a molecule will spend 95% or 99% of its time? Actually we
have utilized this concept, but rather insidiously. We measured the impulse
response function of a tracer over some finite period of time, and fitted the

data to a double exponential function (29.27). We then calculated $\bar{t}$ (and thence the mass, M) by extrapolating the double exponential to infinity, and in this way trying to guess when those few recalcitrant molecules which were hiding away in remote reaches of the organism were going to emerge and exit past the sampling site. Finally, in order to calculate the volume V from (29.24) we had to estimate some average value of the concentration, C. We usually did this by accepting as C the tracee concentration at the sampling site, reasoning (implicitly) that this concentration prevailed (approximately) throughout the "effective" distribution volume — i.e. concentrations in regions where a molecule spent   1% of its time were not important enough to affect the average. And so, in a qualitative sense, we have embodied the wave mechanical concept of volume in our equations.

Speaking Philosophically

The variables associated with the kinetics of a substance could be classified into two groups: *intrinsic* and *extrinsic* variables. The intrinsic variables are related to the transport of a substance within its distribution space. They may be rate constants, transition probabilities, velocities, diffusion coefficients etc., depending on what model of transport is being employed. The extrinsic variables are related to the structure and function of the system as a whole. There are only a few of such variables discussed in this book: total mass, M, total volume, V, total rates of appearance and dis-appearance, R_a and R_d. It transpires that we calculate values for the extrinsic variables based upon the mathematical relationships between intrinsic variables which characterize the model being used. Thus we can calculate a value for R_a (extrinsic) based upon relationships between rate constants K_i (intrinsic) within the framework of a multicompartmental model; and we can calculate a value for R_a based upon relationships between velocities $\vec{v}$ within the framework of a distributed model. The results obtained by the two methods may agree very nearly. Volumes etc. may also be calculated using different models, with similar results. This phenomenon is well-known in physics. For example, various equations pertaining to the propagation of light can be derived equally well, for all intents and purposes, using the concept of a "luminiferous ether", or using the newer view of space which does not demand an ether. When one particular set of intrinsic variables tends to confer advantage, a model which it comprises may be promoted to the status of *the* model of the system.

Speaking Practically

It is difficult to set forth hard-and-fast rules for the calculation of distribution volumes, and the following are only rules of thumb.

(1) If the system is closed with respect to S, use the dilution principle,

(2) If the system is open — i.e. if S is in a dynamic steady state — estimate the distribution volume as the product of the clearance $Ra/\bar{C}$ and the mean residence time $\bar{t}$. Use any convenient sampling point to approximate $\bar{C}$ and $\bar{t}$.

(3) If the system is open, if the sites of degradation of S are well-distributed in space and/or if dispersion of S throughout the distribution volume is rapid, then V may be estimated by dividing the mass of tracer injected, M_0^*, by the y-intercept of the prevailing exponential function.

The systems decribed by (3) comprise a sub-set of the systems described by (2). Therefore, within the sub-set, either technique (2) or (3) should work. The discrepancy between the results of (2) and (3) may provide information about the distribution of sinks and/or the rapidity of dispersion.

Problems

29.1 The following measurements were made of the plasma *unit* impulse response of ^{131}I-labelled insulin (not *inulin*, but the hormone *insulin*) in a normal dog by Campbell and Rastogi (1969). The tracer concentrations are given in units of [counts per minute] per [litre] per [counts per minute] injected per unit body weight i.e. $\frac{cpm/\ell}{cpm}$. The cpm is used here rather than the dpm as the unit of radioactivity.

Time [minutes]	Unit impulse response $\left[\frac{cpm/\ell}{cpm}\right]$ per unit body weight
4.5	0.371
10.0	0.188
16.5	0.124
23.0	0.0956
31.0	0.0784
39.5	0.0540
51.0	0.0454
60.0	0.0406
92.0	0.0262

The data were fitted to the double exponential function

$$h(t) = 0.487\ e^{-.141t} + 0.101\ e^{-.0149t}$$

The weight of the dog was 16.9 kg

Check the quality of the curve-fitting by evaluating $h(t)$ for a few values of t.

Calculate V by modelling the system with grad $R_{d,v} = 0$.

Calculate V using the mean residence time of ^{131}I-insulin with the plasma as a representative sampling site.

Based upon these calculations what might be guessed about the number and anatomical locations of the sites of insulin degradation in the dog?

29.2 Returning to the purely hypothetical example of the Cylindrium in section 20, let us regard fictitium as a substance with an indefinitely small but nonetheless uniform value of $R_{d,v}$ throughout its distribution space. Then the impulse response to fictitium at the tail canduit (disregarding compartments completely) may be fitted to a function of the form

$$C^*(t) = B_1 e^{-\varepsilon t} + B_2 e^{-\beta_2 t}$$

where ε is very nearly equal to zero (cf. Eq. (20.5a)). Recall that $M_o^* = 1$. Calculate V for the larger and smaller Cylindria using the data of Table 20.3 and compare these values with the actual volume which is given in Table 20.1.

References

Bergner, P.E.E., The significance of certain tracer kinetical methods. Especially with respect to the tracer dynamic definition of metabolic turnover, *Acta Radiol. Scand.* 210 Suppl., 1-59 (1962).

Bright, P., The volumes of some compartment systems with sampling and loss from one compartment, *Bull. Math. Biophys.* 35, 69-79 (1973).

Campbell, J. and Rastogi, K.S., Actions of growth hormone: enhancement of insulin utilization with inhibition of insulin effect on blood glucose in dogs, *Metabolism* 18, 930-944 (1969).

Cowan, J.S., Schachter, D. and Hetenyi, Jr., G., Validity of a tracer injection method for studying glucose turnover in normal dogs, *J. Nucl. Med.* 10, 98-102 (1969).

Daniel, P.M., Donaldson, J. and Pratt, O.E., The rapid achievement and maintenance of a steady level of an injected substance in the blood plasma, *J. Physiol.* 237, 8-9p(1974).

Daniel, P.M., Donaldson, J. and Pratt, O.E., A method for injecting substances into the circulation to reach rapidly and to maintain a steady level, *Med. Biol. Eng.* 13, 214-227 (1975).

Hetenyi, Jr., G., Norwich, K.H. and Zelin, S., Analysis of the glucoregulatory system in dogs, *Am. J. Physiol.* 224, 635-642 (1973).

Lawson, H.C., *The Volume of Blood - A Critical Examination of Methods for its Measurement*, Chapter 3, Handbook of Physiology, Section 2, Circulation Volume I, W.F. Hamilton, Editor, Published by American Physiological Society, 1962.

Manery, J.F. and Hastings, A.B., The distribution of electrolytes in mammalian tissues, *J. Biol. Chem.* 127, 657-676 (1939).

Maxwell, J. Clerk, *A Treatise on Electricity and Magnetism*, Vol. I, Clarenden Press, 1891.

Meier, P. and Zierler, K.L., On the theory of the indicator-dilution method for measurement of blood flow and volume, *J. Appl. Physiol.* 6, 731-744 (1954).

Morgan, H.E., Henderson, M.J., Regen, D.M. and Park, C.R., Regulation of glucose uptake in muscle, *J. Biol. Chem.* 236, 253-261 (1961).

Norwich, K.H., Measuring rates of appearance in systems which are not in steady state, *Can. J. Physiol. Pharmacol.* 51, 91-101 (1973).

Norwich, K.H. and Hetenyi, Jr., G., Basic studies on metabolic steady state. Incompletely mixed systems, *Bull. Math. Biophys.* 33, 403-412 (1971).

Norwich, K.H., Radziuk, J. and Vranic, M., Experimental validation of non-steady rate measurements using a tracer infusion method and inulin as tracer and tracee, *Can. J. Physiol. Pharmacol.* 52, 508-521 (1974).

Orr, J.S. and Gillespie, F.C., Occupancy principle for radioactive tracers in steady-state biological systems, *Science* 162, 138-139 (1968).

Perl, W., An extension of the diffusion equation to include clearance by capillary blood flow, *Ann. N.Y. Acad. Sci.* 108, 92-105 (1963).

Perl, W., Effros, R.M. and Chinard, F.P., Indicator equivalence theorem for input rates and regional masses in multi-inlet steady-state systems with partially labeled input, *J. Theor. Biol.* 25, 297-316 (1969).

Schachter, D., Freinkel, N. and Schwartz, I.L., Movement of inulin between plasma and interstitial fluid, *Am. J. Physiol.* 160, 532-535 (1950).

Sheppard, C.W., *Basic Principles of the Tracer Method,* p. 178, Wiley, New York, 1962.

Stetten, Jr., D.W., Welt, I.D., Ingle, D.J. and Morley, E.H., Rates of glucose oxidation in normal and diabetic rats, *J. Biol. Chem.* 192, 817-830 (1951).

Wise, M.E., *Skew distributions in biomedicine including some with negative powers of time*, in Statistical Distributions in Scientific Work, Vol. 2, G.P. Patil, S. Kotz and J.K. Ord, Eds., D. Reidel Publishing Co., Dordrecht, Holland, 1975.

Zilversmit, D.B., Entenman, C. and Fishler, M.C., On the calculation of "turnover time" and "turnover rate" from experiments involving the use of labeling agents, *J. Gen. Physiol.* 26, 325-331 (1943).

Zilversmit, D.B., Entenman, C., Fishler, M.C. and Chaikoff, I.L., The turnover rate of phospholipids in the plasma of the dog as measured with radioactive phosphorus, *J. Gen. Physiol.* 26, 333-340 (1943).

CHAPTER VI
RATES OF APPEARANCE IN
NONSTEADY STATE SYSTEMS

SECTION 30
CALCULATION OF RATES OF APPEARANCE IN
NONSTEADY STATE SYSTEMS:
ONE COMPARTMENT APPROXIMATIONS

In this section and the following ones we shall deal with systems in which the substance S appears (is released, secreted or produced) at the time-varying rate $R_a(t)$ (mass/time). The discussion will generally be restricted to the case where there is a single source point or inlet for S and S*. All the arguments and derivations to be advanced would hold equally well for the case of equivalent tracer supply as shown in section 25.

The straightforward one compartment representation is depicted in Fig. 30.1. The simplest way to derive the desired equation giving $R_a(t)$, and possibly the most satisfying one from a physicochemical viewpoint, is to apply appropriate constraints to the combined equation of convective diffusion (14.21):

$$C^2 \frac{\partial a}{\partial t} = CR^*_{a,v} - C^* R_{a,v} - \vec{u} \cdot (C^2 \nabla a) + D\nabla \cdot (C^2 \nabla a) \qquad (30.1)$$

Gradients vanish by definition in a single compartment model; and we must set

$$R_{a,v} = R_a/V$$

$$R^*_{a,v} = R^*_a/V$$

and

$$\frac{\partial a}{\partial t} = \frac{da}{dt}$$

since spatial dependence of variables does not enter explicitly into the compartmental equations. Hence (30.1) becomes

$$C^2 \frac{da}{dt} = \frac{C}{V} R^*_a - \frac{C^*}{V} R_a \qquad (30.2)$$

239

or

$$R_a(t) = \frac{R_a^*(t)}{a(t)} - \frac{VC}{a(t)}\frac{da(t)}{dt} \tag{30.3}$$

Eq. (30.3) is identical to (22.13) which gave the steady state rate of appearance. When derived in this way (30.3) can be seen to be valid for a real system which intermixes rapidly; it does not depend for its validity on the concept of a rate coefficient.

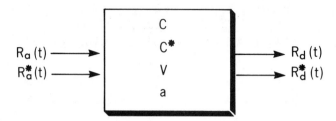

Fig. 30.1 A "physicochemical" single compartment model. The system is well-mixed so that no gradients may exist in C, C* and a. Tracer and tracee enter and leave at time-varying rates $R_a^*(t)$, $R_a(t)$ etc. No rate coefficients are required.

Eq. (30.3) can be derived using the ordinary differential equations of compartmental analysis postulating the existence of a rate coefficient K(t) or K(C,t) (see Fig. 30.2). We make the assumptions that (a) when the system is not in a steady state with respect to tracee the ordinary differential equation used in steady state analysis continues to hold; and (b) that when the system in not in steady state, the kinetics of tracee are described by the same differential equation as tracer, with the starred quantities replaced by the unstarred. These are new assumptions and do not follow necessarily from physicochemical principles.

$$\frac{dC}{dt} = \frac{R_a}{V} - KC \tag{30.4}$$

$$\frac{dC^*}{dt} = \frac{R_a^*}{V} - KC^* \qquad (30.5)$$

Multiplying (30.4) by C^* and (30.5) by C, we eliminate K by subtracting

$$C^* \frac{dC}{dt} - C\frac{dC^*}{dt} = C^* \frac{R_a}{V} - C \frac{R_a^*}{V} \qquad (30.6)$$

Since

$$C^* \frac{dC}{dt} - C \frac{dC^*}{dt} = -C^2 \frac{da}{dt}$$

(30.6) becomes

$$C^2 \frac{da}{dt} = \frac{C}{V} R_a^* - \frac{C^*}{V} R_a \qquad (30.2)$$

which again gives (30.3). Since the first derivation does not require assumptions beyond those of rapid mixing and convective diffusion, it is, perhaps, to be preferred.

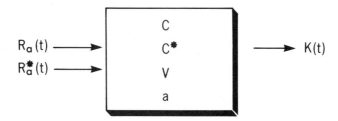

Fig. 30.2 A conventional single compartment model. Disappearance is governed by the time-varying rate coefficient, K(t).

Eq. (30.3) is, then the one compartment approximation for $R_a(t)$. It is simplest to apply when R_a^* is constant — i.e. with a steady infusion of tracer. The fundamental problem with this equation is that it doesn't work. As we shall see in the next section, when results of experiments for validation are presented, this *unmodified* one compartment formula gives a poor approximation of the actual nonsteady rate of appearance.

Eq. (30.3), in just this form, was probably never seriously considered as a means of measuring nonsteady rates of appearance. After all, in its derivation we have assumed that any newly-added materials will mix instantaneously with the entire pool of S. In 1959, R. Steele postulated that (30.3) could be used to measure the nonsteady rate of appearance of glucose, provided that the total volume, V, was replaced by a smaller quantity, pV, where p, the *pool fraction* was some number less than unity. The idea was that material newly-added to the system may be taken to intermix rapidly with only a certain fraction, p, of the total glucose pool. The problem is, of course, that the value of p is unknown. For lack of any definitive information, Steele and his colleagues proposed that a value of 0.5 be taken for p, and, as we shall see later, their guess was not far from the optimal value.

The single compartment equation (30.3) can then be replaced by the modified single compartment equation,

$$R_a(t) = \frac{R_a^*(t)}{a(t)} - \frac{pVC}{a(t)} \frac{da(t)}{dt} \qquad (30.7)$$

Since the rate of change of concentration in a one compartment model is simply the difference between the rates of appearance and disappearance per unit volume, one might also calculate the time-varying rate of disappearance from

$$R_d(t) = R_a(t) - pV \frac{dC}{dt} \qquad (30.8)$$

Problems of Application

These equations for $R_a(t)$ and $R_d(t)$ involve the evaluation of derivatives, and this may be achieved by a number of numerical methods of which some of the simpler are given in Appendix A.

In applying these equations, the experiment was usually begun in the steady state condition. R_a^* consisted of a constant tracer infusion accompanied by an initial priming injection. By plotting the logarithm of C* against time, the total volume, V, of the glucose system was determined in the manner described in the previous chapter. After a steady state interval of suitable duration, the system was displaced from steady state (by the exogenous infusion of glucose or insulin) and calculations for $R_a(t)$ and $R_d(t)$ begun.

As may be appreciated immediately, the primary problem associated with the application of this modified one compartment model is the determination of the pool fraction, p. An ingenious method for evaluating a mean or average value of p for the glucose system was devised by Cowan and Hetenyi (1971), and their evaluation has since been confirmed by other investigators using other methods (v.i.). Cowan and Hetenyi injected a bolus of unlabelled glucose (i.e. tracee) of mass M_0 into a fasted dog and thereafter measured the time-varying rates of appearance and disappearance using (30.7) and (30.8) until steady state was, effectively, regained. Taking the time of injection as zero and utilizing the principle of mass balance one must have

$$\int_0^\infty [R_d(t) - R_a(t)]dt = M_0 \qquad (30.9)$$

But from (30.8)

$$\int_0^\infty [R_d(t) - R_a(t)]dt = -pV \int_0^\infty \frac{dC}{dt}\, dt$$

$$= -pV[\bar{C} - C(0)] \qquad (30.10)$$

where $\bar{C}$ is the steady state value of C, and $C(0)$ its value at $t = 0^+$. Equating the right-hand sides of (30.9) and (30.10) permits the calculation of an average value for p over the duration of this experiment. The investigators measured a value of 0.65 for p — not far from the original estimate of 0.5.

Speaking Practically

I do not recommend that the equations developed in this section be applied in the analysis of data. They involve the use of an unknown factor, $p(t)$, whose properties one can only surmise.

Problems

30.1. Suppose you are studying two unknown metabolic systems, S and S', and suppose, for simplicity, that each has a single source point for the substance being studied. In the case of S, the sampling site is very close to the source point, while in the case of S', the sampling site is remote from the source point.

Compare the magnitudes of p and p', the pool fractions expected for the two systems.

CALCULATION OF RATES OF APPEARANCE
IN NONSTEADY STATE SYSTEMS:
TWO COMPARTMENT APPROXIMATIONS

It seems reasonable that if a one compartment model permits an approximate
calculation of the nonsteady rate of appearance, then a two compartment model
should permit a closer approximation. Just as in the steady state case, there
are many possible two compartment models, and in this section we shall select
just one illustrative example: the inulin system. Inulin is a nonmetabolite,
a substance which does not occur naturally in the body. It is an ideal substance to choose for illustrative purposes because the investigator can infuse this material at known, nonsteady rates. He may calculate the nonsteady
rate of appearance using various tracer techniques and then verify his calculations. A two compartment model of the inulin system is depicted in Fig.
31.1. The substance is infused intravenously into the plasma, which is
assumed to form a part of compartment 1. The elimination of inulin is nearly
but not totally (Ruch and Patton, 1965), by a process of filtration of plasma
in the glomerulus of the kidney and so the disappearance of inulin is also
restricted to compartment 1. The model is identical with that selected for
inulin in section 17 but there is a difference in the way we shall handle it
in the nonsteady state case.

By a *nonsteady state* system we mean, strictly, one which violates the conditions of steady state expressed by equations (13.1), (13.2) and (13.3).
There is no *a priori* reason for believing that the K's, the rate coefficients,
are constant in time as they were for the steady state case. Ideally, we
should like to leave all K's as functions of time, but for practical reasons
this is just not possible — as will soon be apparent. And so two of the
three rate coefficients must be fixed at some constant value and only the
third permitted to vary with time; the interchange coefficients, K_{12} and K_{21},
will be fixed, and the coefficient governing disappearance, K_{01}, will vary
with time. Any change in the glomerular filtration rate *will* reflect itself
in a change in the value of $K_{01}(t)$; any change in the concentration of unlabelled inulin *may* reflect itself in a change in the value of $K_{01}(t)$.

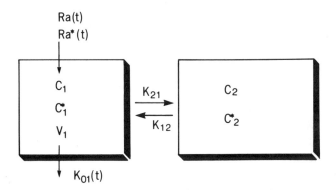

Fig. 31.1 A two compartment model describing inulin kinetics in the nonsteady
state. The interchange coefficients, K_{12} and K_{21}, are held constant while K_{01}, the coefficient governing disappearance of inulin
from the system, may vary with time.

Tracer may be infused intravenously into the system in the same manner as
tracee. A set of four differential equations is now required: one equation
for each compartment for each of tracer and tracee. The same assumption is
required here as was required in the formulation of (30.4) — i.e. that differential equations for tracee are obtained from the differential equations
for tracer by replacing the starred by the unstarred quantities. In other
words, all fluxes of tracer are equal to the corresponding fluxes of tracee
multiplied by the specific activity of the compartment of origin. The
physiocochemical limitations to this oversimplified model of transport have
been discussed earlier (Eq. (15.13)). The four differential equations may
then be written down directly from Fig. 31.1:

$$\frac{dC_1{}^*}{dt} = \frac{R_a^*(t)}{V_1} - (K_{21} + K_{01}(t))C_1{}^* + K_{12}C_2{}^* \qquad (31.1)$$

$$\frac{dC_2{}^*}{dt} = K_{21}C_1{}^* - K_{12}C_2{}^* \qquad (31.2)$$

$$\frac{dC_1}{dt} = \frac{R_a(t)}{V_1} - (K_{21} + K_{01}(t))C_1 + K_{12}C_2 \qquad (31.3)$$

$$\frac{dC_2}{dt} = K_{21}C_1 - K_{12}C_2 \qquad (31.4)$$

Preliminary Steady State Experiment

In order to measure the nonsteady rate $R_a(t)$ using the above set of four equations, we must perform a preliminary steady state experiment in order to evaluate K_{12} and K_{21}. During such a preliminary experiment, unlabelled inulin is infused at some arbitrary constant rate, R_a, until a steady state is achieved. The simplest infusion to administer is the one where R_a is equal to zero, thus dispensing totally with the unlabelled substance during this preliminary phase. With the system thus in steady state, an impulse of tracer is applied to compartment 1, and samples from this compartment are taken at intervals thereafter. The kinetics of the labelled inulin are then described by Equations (31.1) and (31.2) with K_{01} now a constant (since tracee does not vary in concentration) and $R_a^*(t)$ equal to zero. Equations (31.1) and (31.2) are now identical in form with (17.23) but carry a source term R_a^*/V_1. K_{12} and K_{21} may then be evaluated using (17.31) and (17.32), and we shall assume that they retain these values not only in the steady state but also when the system is displaced from steady state. The preliminary steady state experiment thus completed, the nonsteady state experiment may be conducted.

Nonsteady State Experiment

The system is presumed to be in steady state prior to zero time. Beginning at zero time unlabelled inulin is infused into the system at the rate $R_a(t)$ while labelled inulin (tracer) is infused at the rate $R_a^*(t)$. There is no reason why R_a^* should not simply be a steady infusion. Concentrations of tracer and of tracee, C_1^* and C^*, are measured at intervals thereafter. Equations (31.1) through (31.4) may now be used to calculate the "unknown"[†] rate of appearance, $R_a(t)$, at any time, t.

[†]In this illustrative example, of course, we know $R_a(t)$ *a priori* since we are infusing the substance S. But under more usual circumstances, the organism is delivering S at the unknown rate, $R_a(t)$.

Calculations

There are well-developed numerical techniques for solving (31.1) ... (31.4)
for $R_a(t)$, and for a discussion of these methods the reader is referred to the
original paper (Norwich et al., 1974). These techniques require either some
facility with the digital computer or access to the original computer programs. But the investigator can obtain a very good numerical solution for
$R_a(t)$ by the most elementary of techniques: by solving the differential
equations (31.1) ... (31.4) using the method of finite elements (Appendix A).

Suppose that samples have been taken at intervals of about 10 minutes, so that
tables of values of C_1 and $C_1{}^*$ are available. The method of finite element
analysis implies that some small but finite time interval, Δt, is taken to
represent the differential dt. Possibly the changes in C_1 and $C_1{}^*$ are rapid
enough that $\Delta t = 10$ would be a poor approximation, but let us suppose that
$\Delta t = 1$ would be more acceptable. If the investigator can fit a smooth curve
through his C_1 and $C_1{}^*$ data points, this is fine. But if such a curve fitting
process is too difficult he may dispense with it altogether and simply inter-
polate smoothly between data points. The interpolation may be done by a
Lagrangian 2- or 3-point process (found on computer library tapes routinely),
or failing that, simply by drawing a curve between data points by hand. When
the interpolation has been carried out, a value for C_1 and $C_1{}^*$ may be read
for each value of t from zero time to the end of the experiment *at 1-minute
intervals*. These pairs of values of (t,C_1) and $(t,C_1{}^*)$ are then read into or
calculated by the computer. An example of a smoothed curve is shown in Fig.
31.2.

We can now proceed with the finite element analysis, starting with (31.2)
which we write in the form (see Appendix A)

$$C_2{}^*(t+\Delta t) - C_2{}^*(t) = [K_{21}C_1{}^*(t) - K_{12}C_2{}^*(t)]\Delta t \qquad (31.5)$$

Since K_{12} and K_{21} are known we may solve for $C_2{}^*(t+\Delta t)$:

$$C_2{}^*(t+\Delta t) = \Delta t\, K_{21}C_1{}^*(t) + [(1-\Delta t K_{12})C_2{}^*(t)] \qquad (31.6)$$

Since $C_1{}^*(t)$ is known for values of t at intervals of Δt, and since $C_2{}^*(0)$ is
known to be equal to zero, we can calculate $C_2{}^*(\Delta t)$, $C_2{}^*(2\Delta t)$ ... by suc-
cessive applications of (31.6) and in this way compile a table of values for
$C_2{}^*$ as we have for C_1 and $C_1{}^*$.

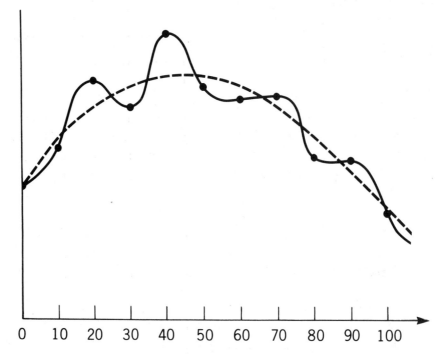

Fig. 31.2 Smoothing the data. The data contain a good deal of "noise". A smooth curve (solid line) has been drawn through adjacent points without any attempt to retrieve the "true" curve (interpolation). Alternatively, one may attempt to obtain the true, or more likely, curve around which the points scatter (broken line). Either curve will do for purposes of analysis.

Eq. (31.4) can be handled in exactly the same manner as (31.2) except that the initial condition is obtained by setting $\frac{dC_2}{dt}$ equal to zero giving

$$C_2(0) = \frac{K_{21}}{K_{12}} C_1(0) \qquad (31.7)$$

The next step is the evaluation of $K_{01}(t)$ from (31.1). Expressing (31.1) in terms of finite elements

$$C_1{}^*(t+\Delta t) - C_1{}^*(t) = \left[\frac{R_a{}^*(t)}{V_1} - (K_{21} + K_{01}(t))C_1{}^* + K_{12}C_2{}^* \right] \Delta t \qquad (31.8)$$

and solving for $K_{01}(t)$:

$$K_{01}(t) = \frac{C_1*(t) - C_1*(t+\Delta t)}{C_1*(t)\Delta t} + \frac{R_a*/V_1 - K_{21}C_1*(t) + K_{12}C_2*(t)}{C_1*(t)} \qquad (31.9)$$

The values of $C_2*(t)$ required on the right-hand side have been provided by
(31.6), and values of $K_{01}(t)$ can now be tabulated.

Finally we express (31.3) in terms of finite elements and solve for $R_a(t)$,
which is the nonsteady rate we set out to discover. The quantities $K_{01}(t)$
and $C_2(t)$ which are required for the evaluation of $R_a(t)$ have already been
calculated. The expression is long and there is not much gained by writing
it out fully here.

All of the above steps can be incorporated very simply into a computer program. It is not necessary even to store all intermediary values of C_2*, C_2
or K_{01}; only the final product, $R_a(t)$, need be retained. Thus, in Fortran,
(31.6) becomes

```
          REAL K21, K12, KO1
          C2STAR = DELTAT*K21*C1STAR(J)
          + ((1.  -DELTAT*K12)*C2STAR)                    (31.10)
```

The new value of C_2* appears on the left-hand side of the equation, and is derived from the old value on the right-hand side; K21 is a REAL variable;
C1STAR(J) is the Jth member of the array C1STAR and gives (by interpolation)
the value of C_1* at time, t. Etc.

```
          C2 = ...
          KO1(J) = ...
          RA(J) = ...
          IF... = , 1
    1     WRITE ... RA(J)
          CONTINUE
```

The procedure is really very simple and the result is a table of values giving
$R_a(t)$ at selected time intervals.

Since the original C_1* and C_1 data may not have been pre-smoothed, and it was
necessary, using this method, to take derivatives numerically (a process
which accentuates noise), the final graph of R_a against t may not be smooth.
This final graph may, if desired, be smoothed by hand or by a process of digital filtering. The result will probably be quite acceptable   (Fig. 31.5).

An alternative method is to integrate equations (31.1) ... (31.4) to obtain an analytical expression for $R_a(t)$. For example, if (31.4) is multiplied by the integrating factor $e^{K_{12}t}$ it can then be solved to give C_2 as a function of the measured variables C_1 and t:

$$e^{K_{12}t}\left(\frac{dC_2}{dt} + K_{12}C_2\right) = K_{21}e^{K_{12}t} C_1$$

$$\frac{d}{dt}\left(e^{K_{12}t} C_2\right) = K_{21}e^{K_{12}t} C_1$$

$$e^{K_{12}t} C_2 - C_2(0) = K_{21} \int_0^t e^{K_{12}\tau} C_1(\tau)d\tau$$

$$C_2(t) = C_2(0)e^{-K_{12}t} + K_{21}e^{-K_{12}t} \int_0^t e^{K_{12}\tau} C_1(\tau)d\tau \qquad (31.11)$$

And a similar expression may be obtained for $C_2*(t)$ from (31.2). By a continuation of this process we may obtain analytical expressions for $K_{01}(t)$ and $R_a(t)$ (Norwich et al., 1974). The expression for $R_a(t)$ is again rather long and not much would be gained here by writing it out.

Comparison of the One and Two Compartment Methods

The issue is quite clear. It is much easier to use a single compartment model because one simply has to evaluate the right-hand side of (30.7) using the measured values of C and a. But, one must know *a priori* the correct value of the pool fraction, p. The use of a two compartment model necessitates a preliminary experiment (or a concurrent experiment using a different tracer) to measure the impulse response of the system in order that the interchange coefficients K_{12} and K_{21} may be evaluated. But, no problem arises about the selection of a pool fraction. In general, the one compartment model gives results as good as the two compartment model if and only if the best value of p is known *a priori*.

Validation of the One and Two Compartment Methods

You will recall that the inulin system was chosen as an example because inulin is not a metabolite and therefore, it could be infused at known, nonsteady rates. We are now in a position to verify our calculation of $R_a(t)$ for inulin using (30.7) for the one compartment method and (31.1) ... (31.4) for the two compartment method. This experiment was performed on a dog resting comfortably in a Pavlov harness.

Following precisely the protocol outlined in this section and in section 30, the figures (31.3), (31.4) and (31.5) were obtained. Fig. (31.3) was obtained by calculating $R_a(t)$ from Eq. (30.7) taking p equal to 0.25, which in this case, was found to be the best *constant* value of the pool fraction. (Admittedly, if p were permitted to vary with time, one could always find *a posteriori* a function p(t) which would give the value of $R_a(t)$ precisely, but this is highly unrealistic.) The calculated value of $R_a(t)$ is compared with the actual rate at which inulin was infused. No attempt whatever was made to smooth the calculated $R_a(t)$ curve. Smoothing would clearly improve the result aesthetically. For a fuller treatment of both the theory and the experiments involved, the reader is referred to the original paper or to the review paper by Radziuk et al., 1974. In these papers, the data has been "pre-smoothed"; in the example just given, the output has been "post-smoothed". The validation process has been carried out somewhat more extensively for the former than for the latter type of smoothing. A series of experiments validating the one and two compartment methods for the glucose system has also been carried out, but the results have not yet been published.

Summary and Conclusion

The most accurate method of calculating nonsteady rates of appearance using a compartmental model in the systems studied is the use of a two compartment model. (More complex multicompartmental models have been studied; the results have not been presented here, but they are not superior.) The simplest treatment of the two compartment model is to solve the associated differential equations by the method of finite elements, and to smooth the final $R_a(t)$ curves by hand. The methods advanced have been validated for inulin and for glucose and are probably generally valid. But their general validity remains to be demonstrated.

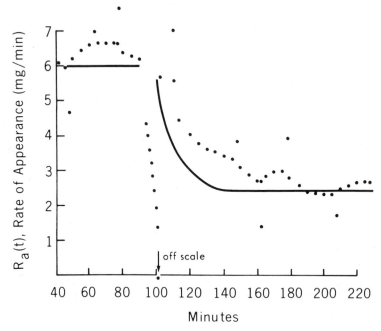

Fig. 31.3. Validity of the calculation of nonsteady rate of appearance of
          inulin using a single compartment model.

——————   Actual rate of intravenous infusion of inulin

••••••   Rate of infusion of inulin calculated using a single compart-
          ment model taking the value of p to be 0.25 (about the best
          constant value as determined *a posteriori*).

          A constant intravenous infusion of inulin was given for a period
          of 90 minutes at the rate of 6 mg/min. The infusion rate was zero
          between 90 and 100 minutes. Between 100 and 140 minutes the in-
          fusion was decreased smoothly to 2.4 mg/min, and between 140 and
          230 minutes it was held at 2.4 mg/min. The measured data, C_1 and
          C_1^*, were not smoothed so that the calculated $R_a(t)$ curve is noisy,
          but the tendency of the calculated values to follow the actual
          values is apparent. Calculated values for the first 40 minutes are
          not shown because of the excessive noise associated with the intro-
          duction of step functions in C and C* at the beginning of the ex-
          periment. Notice that the abrupt discontinuity at 100 minutes
          associated with a sudden change in rate of appearance has given
          rise to high frequency oscillation in the calculated $R_a(t)$.

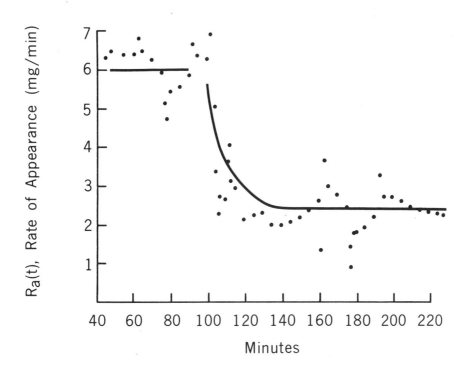

Fig. 31.4. Validity of the calculation of nonsteady rate of appearance of in-
            ulin using a two compartment model.
            ───────   Actual rate of intravenous infusion of inulin.
            ••••••   Rate of infusion of inulin using a two compartment model.
            For a fuller explanation of the experimental process see the le-
            gend to Fig. 31.3. The measured C_1 and C_1* data were not smoothed
            and so the calculated $R_a(t)$ curve is noisy; i.e. it probably con-
            tains sizable random errors. Nevertheless, the calculated points
            cluster closely about the actual curve of infusion.

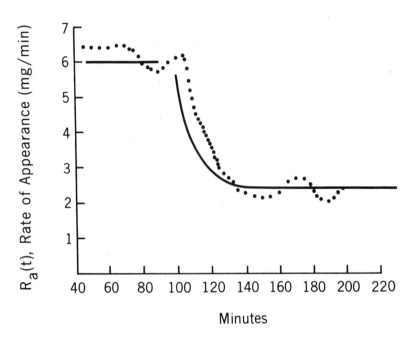

Fig. 31.5 The effect of smoothing the calculated $R_a(t)$ curve. Using a very
simple digital computer program, the calculated $R_a(t)$ curve shown
in Fig. 31.4 was (effectively) passed through a filter which at-
tenuates high frequencies. Hence all very rapid changes in the
calculated $R_a(t)$ curve tend to vanish. The result is this quite
smooth, very acceptable curve which closely approximates the actual
curve of infusion. An equally acceptable result would have been
obtained by manually smoothing the calculated curve of Fig. 31.4.

Speaking Practically

It is not difficult to measure nonsteady rates of appearance using a two compartment model. You can do it by yourself with only the most elementary knowledge of digital computer programming (e.g. a smattering of Fortran, Basic or
Focal). Measure the impulse response of the system, i.e. the response to a
bolus of tracer, when the system is in a steady state with respect to the

tracee. Fit the response to say, a double exponential function, either by
hand, by means of a computer library program, or using the very simple program given in Appendix B. Decide upon the configuration of a two compartment
model which best matches the known structure of the biological system. It
probably doesn't matter much which configuration you choose, because (recalling the steady state case) the compartmental model is just being used as a
vehicle for calculating an input-output relationship. Calculate K_{12} and K_{21}
using an inverse precedure (17.31, 17.32). Proceed with the nonsteady state
experiment or test, infusing tracer throughout, say, at a constant rate for
simplicity. Measure C and C* at intervals throughout the experiment or test.
Smooth the graphs of C and C* vs time by hand, giving you numerical values of
C and C* at all relevant times. Prepare a table of values for t, C, C*.
Solve 4 equations corresponding to (31.1 - 31.4) like simple algebraic equations using the method of finite elements, giving $R_a(t)$ in terms of the coefficients K_{12} and K_{21}. The latter coefficients are known from the preliminary impulse response measurement. Enter your table of values, your
evaluations of K_{12} and K_{21}, and your algebraic solution for $R_a(t)$; the program can then supply a numerical solution for $R_a(t)$ at specified times.
Smooth the graph of $R_a(t)$ vs t by hand.

SECTION 32
RATES OF APPEARANCE IN DISTRIBUTED, NONSTEADY STATE SYSTEMS

Suppose, first of all, that we deal with the very restricted class of systems
which is linear and stationary with respect to *tracee*. That is, if we regard
the system in which the input is the rate of appearance of tracee, $R_a(t)$, and
the output is the concentration of tracee at some sampling point, C(t), then
this system is linear and stationary by the criteria of section 13. Suppose,
moreover, that we can obtain the unit impulse response function, h(t), of this
system. Then by definition

$$C(t) = \int_0^t R_a(t - \tau)h(\tau)d\tau \qquad (32.1)$$

The desired nonsteady rate may then be obtained by solving (32.1) for $R_a(t)$. The solution of this integral equation is termed *deconvolution*.

Unfortunately, one cannot usually assume *a priori* that the system he is dealing with is a linear one, and so it is necessary to search for a more general method for determining nonsteady rates.

Herman Branson (1952a, b) introduced a more general integral equation capable of treating non-linear metabolic systems. It was applied not to a strictly distributed system but to one where a metabolite might react through a number of possible paths or routes. If the mass, $M(t)$, of the metabolite is such that

$$M_o = 0$$

then

$$M(t) = \int_0^t R(\tau)F(M_\tau, t - \tau)d\tau$$

where R corresponds to our R_a and F, the *metabolizing function,* corresponds to our h. But let us return now to a system of convective and diffusive transport.

Suppose we deal with a system (not necessarily linear) having a single, common source point for tracer and tracee. Constrained only by the usual assumptions of convective diffusion, and taking the fairly general boundary conditions introduced in section 25, let us derive an equation for the non-steady rate of appearance. If this is to be something more than an academic exercise, such an equation must, of course, give the rate in terms of quantities measurable in the laboratory.

The object here is to measure changing appearance rates, but let us suppose that the experiment is divided into two phases, the first of which is a steady state phase.

Phase 1

During this initial period, the tracee is held in steady state. An infusion (or primed infusion) of tracer is given (sections 24 and 25) such that after some time has elapsed, tracer too, is in steady state. By the end of the first phase the specific activity at all parts of the system will have assumed a constant value α, and the steady state rate of appearance will be given by (25.36)

$$R_a = R_a^*/\alpha$$

Phase 2

At the beginning of phase 2, the system is displaced from steady state by any means deemed appropriate by the investigator; e.g. by infusing tracee exogenously or by injecting a drug or hormone which affects the kinetics of the tracee.

Without, for the moment, discussing the details of the experiment, let us suppose that *during this second phase the investigator infuses tracer at some nonsteady rate $R_a^*(t)$ such that the specific activity at the sampling site is held constant at α — it's value at the end of the first phase.* Will it now be generally true that

$$R_a(t) = R_a^*(t)/\alpha \quad ? \tag{32.2}$$

Let us explore this possibility by setting up the equations of convective diffusion for tracer and tracee [(11.21, (11.23), (25.8)]:

$$\frac{\partial C}{\partial t} = D\nabla^2 C - \vec{u} \cdot \nabla C + R_a(t) \cdot \delta(\vec{r} - \vec{r}_1) - R_{d,v}(C,t) \tag{32.3}$$

$$\frac{\partial C^*}{\partial t} = D\nabla^2 C^* - \vec{u} \cdot \nabla C^* + R_a^*(t) \cdot \delta(\vec{r} - \vec{r}_1) - a(\vec{r},t)R_{d,v}(C,t) \tag{32.4}$$

where $\vec{r}_1$ is the common entry point for tracer and tracee. Taking first the *converse* of the proposition we wish to prove, let us suppose that $R_a^*(t)$ is selected in such a way that

$$R_a^*(t)/R_a(t) = \alpha = \text{a constant} \tag{32.5}$$

That is

$$\frac{\partial \alpha}{\partial t} = 0 \tag{32.6}$$

Suppose as well that

$$C^*(\vec{r},t)/C(\vec{r},t) = \alpha$$

with

$$\text{grad } \alpha = 0 \tag{32.7}$$

so that

$$C* = \alpha C \tag{32.8}$$

is a solution to (32.4). If (32.8) is a solution to (32.4), then substituting the former into the latter should demonstrate an identity:

$$\alpha \frac{\partial C}{\partial t} = \alpha [D\nabla^2 C - \vec{u} \cdot \nabla C + R_a(t) \cdot \delta(\vec{r} - \vec{r}_1) - R_{d,v}(C,t)] \tag{32.9}$$

using (32.5). Equation (32.9) is identically true as a consequence of (32.3). Moreover, the solution (32.8) satisfies the initial conditions

$$C*(\vec{r},0) = \alpha C(\vec{r},0)$$

because these initial conditions of phase 2 of the experiment are identical with the final conditions of phase 1. Equation (32.8) is, therefore, an acceptable solution to (32.4) satisfying the inital conditions, and satisfying the boundary conditions as long as $C(\vec{r},t)$ does. It has been shown in section 25 that solutions are unique provided that $R_d(C,t)$ conforms to certain conditions and it will be assumed here that these conditions are met. We may conclude, therefore, that *if* the ratio of $R_a^*(t)$ to $R_a(t)$ is set at the constant value α, *then* specific activity will everywhere assume the same constant value. Let us call this *Proposition A*.

We are really interested in the converse of this proposition: *if* the specific activity is everywhere set at the constant value α, *then* the ratio of $R_a^*(t)$ to $R_a(t)$ will then assume the same constant value. Let us call this *Proposition B*. Proposition B can be proved by appealing to (14.21), the combined equation of convective diffusion:

$$C^2 \frac{\partial a}{\partial t} = CR*_{a,v} - C*R_{a,v} - \vec{u} \cdot (C^2\nabla a) + D\nabla \cdot (C^2\nabla a) \tag{32.10}$$

Setting

$$\nabla_4 a = 0$$

(see section 14) and expressing the single, common source point by the equations

$$R^*_{a,v} = R_a^*(t)\delta(\vec{r} - \vec{r}_1)$$

$$R_{a,v} = R_a(t)\delta(\vec{r} - \vec{r}_1)$$

we have from (32.10)
$$CR_a^*(t)\delta(\vec{r} - \vec{r}_1) - C^*R_a(t)\delta(\vec{r} - \vec{r}_1) = 0$$
Integrating over the volume of distribution
$$C(\vec{r}_1,t)R_a^*(t) - C^*(\vec{r}_1,t)R_a(t) = 0$$

or

$$R_a(t) = \frac{C(\vec{r}_1,t)}{C^*(\vec{r}_1,t)} R_a^*(t) \tag{32.11}$$

Now, specific activity has everywhere the value α, so that (32.11) becomes
$$R_a(t) = R_a^*(t)/\alpha \tag{32.2}$$

as required.

During phase 2 of the experiment, the investigator has held the specific
activity *at the sampling site* constant at the value α. But it is not generally possible to monitor the entire distribution space to assure that specific
activity has this value everywhere. Nevertheless, we know from Proposition A
that the only *constant* value which the ratio of $R_a^*(t)$ to $R_a(t)$ may assume is
α, the constant value at the single sampling site. The remaining possibility
— that $R_a^*(t) : R_a(t)$ may assume some time-varying value while leaving specific activity at the sampling site fixed at α — seems remote, and will not
be pursued here. We shall, then, accept the truth of Eq. (32.2). If the
investigator can, by some means or other, fix the specific activity at the
sampling site during phase 2 at the same constant value achieved during phase
1, then the ratio of the rate of appearance of tracer to that of tracee will
assume the same value.

Experimental Considerations

By what means can the experimenter keep specific activity constant despite
changes in the rate of appearance of tracee? There are three possible means:
instantaneous measurement of specific activity, sequential guesses, and
multiple tracers.
(1) If specific activity can be measured nearly instantaneously, then the
rate of infusion of tracer may be varied manually, continuously in such a way
that the specific activity is kept nearly constant. The problem is analogous
to driving an automobile along a straight path. Visual information about the
position of the automobile is available nearly instantaneously so that the
position of the steering wheel can be adjusted continuously etc.

(2) If, as is usually the case, the value of specific activity cannot be
known until sometime later, then the problem is more acute. The method we
have resorted to is the method of sequential or "verified" guesses. This
method is applicable only if the experiment for discovery of nonsteady $R_a(t)$
can be repeated under identical conditions several times. Usually the investigator has some idea about the behaviour of $R_a(t)$ when the system is displaced
from steady state in a particular manner. He can, therefore, make an educated
guess about $R_a(t)$ and adjust $R_a^*(t)$ in such a way that he expects specific
activity in phase 2 to remain constant. He will either succeed or he will
fail. He will then repeat the experiment under identical conditions, but will
vary the infusion rate $R_a^*(t)$ in order to retain constancy of specific activity
in phase 2 for a longer period of time. And so on. Each iteration brings
the experimenter closer to the correct value of $R_a(t)$.
(3) In the event that several tracers for the same tracee are available, and
that facilities for mulitple isotope counting are available, an experimenter
may "hedge his bets" by infusing each of the available tracers at a slightly
different rate. Thus the probability that one of the tracers will yield a
constant specific activity is enhanced.

The only one of these methods which we have attempted in our laboratories is
(2), the method of sequential guesses. The results of one such experiment
has been published (Norwich 1973) and the success of this experiment is largely the result of the experimental skills of Dr. Geza Hetenyi Jr. Briefly,
this is a description of the experiment.

The purpose was to measure the decrease in the rate of release of glucose (by
the liver essentially) consequent upon the commencement of an intravenous
glucose infusion into a fasted dog. The dog, who was trained to stand quietly
in a Pavlov harness, was fasted for 14 hours prior to the experiment. About
30 minutes before beginning the experiment, two cannulae were inserted under
local anaesthesia, one double lumen catheter into the cephalic vein and one
single lumen catheter into the vena cava via the saphenous vein. Infusions
and injections were given via the cephalic cannula, tracer through one lumen
and tracee through the other. Blood samples were drawn through the saphenous
catheter. During the steady state phase, phase 1, a primed infusion of
^{14}C-1-glucose was given. The magnitude of the priming dose was calculated by
the method of section 24. At the end of 80 minutes, it was presumed that

specific activity was constant and phase 2 was begun. At the beginning of
phase 2, a steady infusion of unlabelled glucose was begun, in this way dis-
placing the system from steady state.

Fig. 32.1 illustrates how C* was increased in a manner nearly parallel with
the increase in C. Fig. 32.2 illustrates that specific activity was held very
nearly constant for a period of 7 minutes following the start of the infusion
of unlabelled glucose despite the rapid rise in glucose concentration. That
is, the "guess" for $R_a^*(t)$ was accurate for about 7 minutes. $R_a^*(t)$ was decreased by manually changing the setting on the infusion apparatus once each
minute. The resulting "descending staircase" approximated a smooth reduction
in the rate of infusion of tracer. The fall in endogenous glucose release,
essentially the hepatic glucose output, from about 32 to 5 mg/min is shown in
Fig. 32.3. It should be appreciated that the experimenter had virtually no
knowledge of the magnitude of the specific activity until several hours after
the experiment was completed. The correct guess for $R_a^*(t)$ during these 7
minutes was facilitated by many earlier experiments carried out in the steady
state condition showing how R_a diminished with increasing C.

A Blend of Techniques
It may be readily observed that the single compartment equation for $R_a(t)$,
Eq. (30.3) or (30.7), reduces to Eq. (32.2) when the derivative da/dt
vanishes. It may similarly be shown that the two compartment equation for
$R_a(t)$ reduces to (32.2) when da/dt vanishes (Radziuk, 1974). It therefore
seems reasonable, for two reasons, for the investigator to endeavour always
to maintain a constant specific activity by varying $R_a^*(t)$ in a manner which
seems appropriate to him. First, numerical evaluation of the derivative da/dt
adds a great deal of noise to the calculated values of $R_a(t)$. Any reduction
in the rate of change of specific activity will, therefore, improve the precision of the calculations. Even if the guess of the appropriate $R_a^*(t)$ is not
exact, so that da/dt does not vanish completely, a reduction in its magnitude
is useful. Second, as the magnitude of da/dt is reduced, the calculation of
$R_a(t)$ becomes less dependent on the validity of a particular compartmental
model. The calculation of $R_a(t)$ approaches that given by the general
equation, (32.2), which depends only on the convection-diffusion assumptions.[†]

[†]Eq. (32.2) may be valid for a more general system, but it has only been
demonstrated for a system of convection-diffusion-reaction.

My advice to other experimenters is, therefore, to make a guess at the rate of
tracer infusion which you think will leave specific activity constant. Since
you undoubtedly understand something about the qualitative behaviour of the
system out of steady state, quite probably your efforts will lead at least to
a diminution in the magnitude of the rate of change of specific activity —
and this will be all to the good.

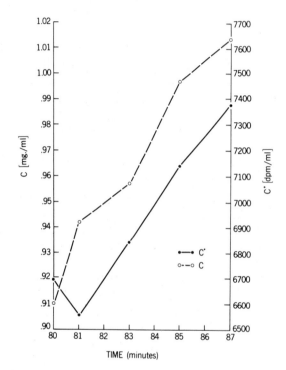

Fig. 32.1   The glucose system is driven out of steady state by infusion of unlabelled glucose (see text). The exogenous infusion *increased* C
more rapidly than the induced reduction in hepatic glucose output
and the increased peripheral uptake of glucose could *reduce* it.
The concentration of tracer increased because the infusion rate of
tracer was increased.
C and C*, the concentrations of tracee and tracer respectively, are
plotted against time, for the second phase of the experiment.
● Concentration of tracer, C*
○ Concentration of tracee, C

From Can. J. Physiol. Pharmacol., 51, 91-101, 1973 by K.H. Norwich. Reproduced by permission of National Research Council of Canada.

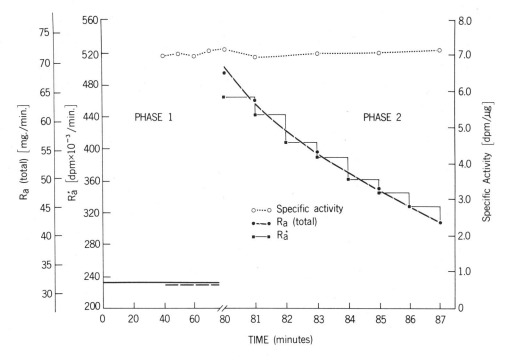

Fig. 32.2   Specific activity, nonsteady rate of appearance of glucose and rate
of infusion of tracer are plotted against time (see text).

o--------o   specific activity

●——————● $R_a(t)$

■————————■ $R_a^*(t)$

$R_a(t)$ has been calculated using Eq. (32.2):

$$R_a(t) = R_a^*(t)/\alpha \doteqdot R_a^*(t)/7.1$$

From Can. J. Physiol. Pharmacol. 51, 91-101, 1973 by K.H. Norwich.   Reproduced by permission of the National Research Council of Canada.

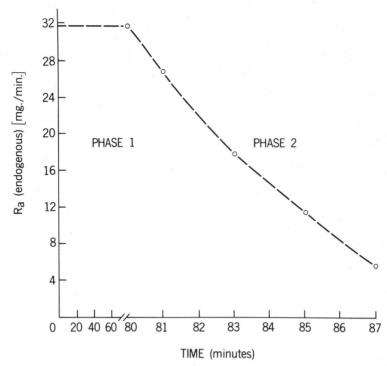

Fig. 32.3 Endogenous rate of appearance of glucose plotted against time dur-
           ing the minutes just preceding and just following the beginning of
           a steady, intravenous glucose infusion (see text).

$$R_a(t)_{TOTAL} \text{ (from Eq. (32.2))} = R_a(t)_{INFUSED} + R_a(t)_{ENDOGENOUS}$$

Hence the endogenous (essentially hepatic) glucose release rate is
obtained as the difference between the calculated total rate of
glucose infusion and the known rate of intravenous glucose infusion.

From Can. J. Physiol. Pharmacol. 51, 91-101, 1973 by K.H. Norwich. Reproduced by permission of the National Research Council of Canada.

Conclusions

For a full comparison of the accuracy of the various techniques for computing
nonsteady rate of appearance, the reader must be referred to the original
papers where validation for the (linear) inulin system and the (non-linear)
glucose were presented. The results, however, can be summarized quite
succinctly.

(1) The modified one compartment model is quite satisfactory if and only if one knows *a priori* the best value for the pool fraction. For the systems studied, this value varied between the limits of 0.25 and 0.8 approximately   (But see problem 30.1). No method was found for determining the best pool fraction *a priori*.

(2) The two compartment model yields results as good as the one compartment model with best pool fraction. No *a priori* information is required.

(3) If specific activity can be held constant, the convective diffusion equation, (32.2), must provide the most accurate value of $R_a(t)$.

New and very general methods for computing nonsteady rate of appearance are being developed, but the results have not yet been published.

Methods for measuring nonsteady rate of disappearance are lacking. Certainly $R_d(t)$ can be obtained using compartmental models (e.g. Eq. (30.8)). A general equation for $R_d(t)$ in a distributed system has not yet been developed.

I believe that the development and *validation* of very general methods for measuring rates of appearance and disappearance of metabolites in intact biological systems is the most important task confronting the biokineticist today. I believe that accurate measurement of these rates is a necessary prerequisite to the mathematical description of metabolic control systems.

Speaking Practically

When carrying out a tracer experiment (test) for the measurement of $R_a(t)$, the nonsteady rate of appearance, always try to infuse tracer at such a rate that specific activity within the system remains constant. If you do not know the correct rate of infusion, guess at it. You have nothing to lose and everything to gain. If you attain the constant specific activity, α, the nonsteady rate is obtained simply as the quotient

$$R_a(t) = \frac{R_a^*(t)}{\alpha} \qquad (32.2)$$

If you do not attain it, you can still carry out a two compartment calculation as shown in section 31, and your calculation will probably be more accurate for the attempt you have made to attain it. As is the case when measuring any of the extrinsic variables, you do not need a compartmental model to make the measurement. However, the more general methods are just being prepared for publication at the time of writing, and are best pursued at the level of the journals.

SECTION 33

LOOKING TO THE FUTURE

In this brief section I should like to express some of my thoughts and venture some opinions about the role of the determination of the fluxes of metabolites in the elucidation of physiological mechanisms, and to speculate somewhat about methods for flux measurement which may be developed in the future.

When the "flux of a metabolite" refers to the rate at which a substance arrives at some biosystem of *well-specified geometric or anatomic boundaries*, or refers to the rate at which the substance leaves such a biosystem, we have been referring to such a flux as the rate of appearance or the rate of disappearance respectively. I believe that the accurate measurement of such fluxes — rates of appearance and disappearance — is a problem of fundamental importance in physiology. One method for its solution, which we have examined in some depth, is postulation of an internal compartmentalized structure for the system. In the application of such compartmentalized models to the calculation of rates of appearance and disappearance we have specified that material exchanges at definite rates between the hypothesized compartments comprising the system. Since the compartments themselves need have no real, unique existence, these *internal fluxes* need not be real nor unique entities. But, in the case of steady state systems we have shown (sections 23 and 26) that the correct values for the input and output rates (rates of appearance and disappearance) may be obtained, regardless of the reality and uniqueness of the internal fluxes[†]. Probably this is also the case in nonsteady systems,

---

[†]There is the implicit assumption here that if a constant tracer infusion is used, the tracer steady state can be attained, and if the tracer injection is used, the C* vs t curve can be extrapolated reliably to infinite time.

which is the basis for the compartmental calculations of nonsteady rates of
appearance and disappearance offered in sections 30 and 31.

An approach to the calculation of such rates which is theoretically stronger
is based upon the convective and diffusive nature of the transport of materials *in vivo*. This approach is, I believe, more satisfactory than the compartmental approach because it is more firmly grounded in physics and chemistry. A technique has been described for measuring nonsteady rates of
appearance by adjusting the tracer infusion rate in such a way that specific
activity remains constant (section 32). The theory underlying this technique
has been derived for a distributed system where transport occurs solely by
convection and diffusion. Quite possibly the derivation can be generalized.
If improvements in the technology of measuring the specific activity of a substance in plasma are forthcoming, such that specific activity can be measured
very rapidly, then this technique could become the technique of choice. One
can imagine a servodevice employing a simple algorithm which is capable of
maintaining plasma specific activity at a constant level automatically. In
such an instance nonsteady rates of appearance could be read very nearly *on
line*. Some preliminary work in this area has been done by Hetenyi et al.
(1971), who used a lithium-drifted semiconductor detector to provide an instantaneous measure of ^{42}K activity in the ambient tissue.

A method of great generality for measuring nonsteady rates of appearance has
been developed by J. Radziuk (1974). Radziuk's method is theoretically valid
for any system consisting of a single, well mixed compartment which exchanges
material with a second "region". This second region may be, for example, a
system of n well mixed compartments, or a subsystem with convective and diffusive transport. This *impulse response* method, at its current level of
development, requires for its application some facility with mathematics and
digital computers.

One can envisage the development of a set of integral equations involving nonsteady rates of appearance which will be valid in any system involving linear
transport mechanisms. For example, in a system devoid of S and S* for all
t < 0

$$C^*(t) = \int_0^t h(\tau,\ t-\tau)R_a^*(\tau)d\tau, \quad t > 0 \qquad (33.1)$$

$$C(t) = \int_o^t h(\tau, t-\tau)R_a(\tau)d\tau, \qquad t > 0 \qquad\qquad (33.2)$$

which might be solved simultaneously for $R_a(t)$ (and $h(t,u)$ if desired). These
equations do not, in themselves, contain sufficient information for complete
solution of the problem. The challenge here is twofold:

(a) to develop the theory for a system of the greatest possible generality,
and

(b) to simplify the numerical analysis so that the experimental physiologist
     can readily employ it.

I believe that the accurate determination of rates of appearance and disappearance is mandatory if we are to gain insight into the function of metabolic control systems. To appreciate my view it is only necessary to recall
the close parallel between the laws of heat flow and the laws of material
flow. For example, the diffusion of heat is governed by Fouriers' law which
states that heat flux (in *calories* · $area^{-1}$ · $time^{-1}$) is proportional to
temperature gradient; and the diffusion of matter is governed by Fick's law
which states that material flux (in *mass* · $area^{-1}$ · $time^{-1}$) is proportional
to *concentration* gradient. Thus there is an analogy between quantity of heat
(calories) and quantity of mass on the one hand; between temperature and concentration on the other. Similarly there is an analogy between *specific heat*,
quantity of heat required to raise the temperature of an element of some system by one unit, and *dilution*, which specifies the quantity of material required to raise the concentration of an element of some system by one unit.
Just as one cannot understand the changes in temperature in a body in simple
terms without the concept of caloric flux, so too one cannot understand the
changes in concentration in a biosystem in simple terms without the concept
of material flux. The quantity

$$R_a(t) - R_d(t)$$

is the net material flux, which is the determinant of concentration change.
Fig. 33.1 compares a simple thermoregulator, such as the one which regulates
the temperature of a room, with a linearized representation of the glucoregulator, the system which regulates plasma glucose levels (Zelin et al.,
1975). The transfer functions of the glucoregulator, W, H and G have all
been evaluated by the tracer techniques which were discussed in preceding
sections. W and H are functions of the rates of appearance and disappearance.

To discuss metabolic control systems without measurements of rate of appearance and disappearance is to omit an entire dimension.

Finally, it must be stressed that despite the strenuous efforts which were
made in the preceding chapters to be very general in the approach to transport
problems in intact organisms, a great number of restrictions crept in nonetheless. It was assumed that we were dealing with simple diffusion, with a
single diffusion coefficient, D, and that concentrations were of the order of
magnitude such that $D^* = D$. The boundaries of the intact organism were taken
to be either impermeable to the substance S or absorbent of S in the simplest
possible way. The reality of bounding, semi-permeable membranes was totally
ignored. Boundary membranes contain pores whose diameters are of the order of
magnitude of the diameters of the molecules of S, and the true boundary
conditions for our differential equations are not nearly as simple as those
which have been assumed. The solvent, or carrier fluid for S has been taken
to be wholly confined within some distribution volume within the intact
organism. In fact, these fluids themselves pass through the boundary membranes and have associated with themselves a turnover rate. Concepts of
hydrostatic versus osmotic pressures, and of reflection coefficients at membranes have been omitted from our discussion. We have done the best we could
with the present level of understanding of fundamental processes, and with
the present capabilities for making measurements with minimal perturbations in
intact organisms. But looking to the future, it is hoped that we will see a
more complete blending of the approaches of the membrane biophysicist, the
physical biochemist and the systems physiologist — to the advantage of all
three.

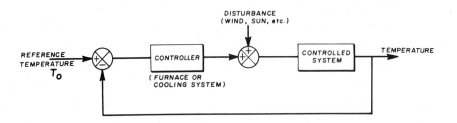

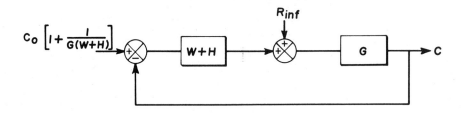

Fig. 33.1

(Caption on next page)

Fig. 33.1   The glucoregulator vis à vis the thermoregulator. In the classical
            thermoregulator (upper diagram) the existing temperature, T, (say
            of a room) is "fed back" and subtracted from the constant reference
            temperature T_0, to generate an error signal which activates a con-
            troller. The controller which is a heating and/or cooling device,
            generates a heat current. This generated heat is augmented by the
            disturbing heat current and the net heat current acts upon the con-
            trolled region, e.g. a room. The specific heat of the controlled
            region must come into play here, as the heat current (calories ·
            time^{-1}) induces a temperature change. The primary difference ob-
            served between the thermoregulator and the glucoregulator below is
            that the reference input to the glucoregulator is "prefiltered" —
            i.e. is not a simple constant, C_0,. The significance of this dif-
            ference need not concern us here. The difference between the
            existing concentration, C, and the reference input again gives
            rise to an error signal, which is led to the controller block. The
            transfer function of the controller is composed of two terms: W
            and H. The transfer function W deals with the partial glucose flux
            R_a, which gives the rate of production of glucose by liver and kid-
            ney. The transfer function, H, deals with the partial glucose flux
            R_d, which gives the rate of glucose utilization by the peripheral
            cells. The net glucose flux, $R_a - R_d$, is augmented by the disturb-
            ing flux (say from the gastrointestinal tract) and the resulting
            flux acts upon the controlled region, say the extracellular fluid.
            The analogue of the specific heat, namely the dilution effect, must
            act at this point to transduce the resultant glucose flux into a
            concentration change (transfer function G).

            The lower diagram is *not* based upon an hypothesis of the mechanism
            of glucoregulation. It is a synthesis of numerous experimental
            measurements. A better representation of this regulator is obtain-
            ed by using a somewhat more complicated diagram, but all the basic
            elements are present in this figure.

From Medical and Biological Engineering 13, 506, 1975 by S. Zelin, K.H.
Norwich and G. Hetenyi, Jr. Reproduced by permission of the International
Federation for Medical and Biological Engineering.

References

Branson, H., The kinetics of reactions in biological systems, *Arch. Biochem. Biophys.* 36, 48-59 (1952).

—————, Metabolic pathways from tracer experiments, *Arch. Biochem. Bio-Phys.* 36, 60-70 (1952).

Cowan, J.S. and Hetenyi Jr., G., Glucoregulatory responses in normal and diabetic dogs recorded by a new tracer method, *Metabolism* 20, 360-372 (1971).

Hetenyi Jr.,G., Dalziel, P.D., Studney, D.R. and Howell, W.D., The processing and interpretation of data by a semiconductor radiation needle-detector, *Int. J. Appl. Rad. Isot.* 22, 706-710 (1971).

Norwich, K.H., Measuring rates of appearance in systems which are not in steady state, *Can. J. Physiol. Pharmacol.* 51, 91-101 (1973).

> Note: Although this paper was written by a single author, it owes its strength in large part to the superlative experimental work of Dr. G. Hetenyi Jr.

Norwich, K.H., Radziuk, J., Lau, D. and Vranic, M., Experimental validation of nonsteady rate measurements using a tracer infusion method, *Can. J. Physiol. Pharmacol.* 52, 508-521 (1974).

Radziuk, J., *Measurement and Validation of Nonsteady Turnover Rates with Applications to the Inulin and Glucose Systems,* Ph.D. Thesis, University of Toronto, 1974.

Radziuk, J., Norwich, K.H. and Vranic, M., Measurement and validation of nonsteady turnover rates with applications to the inulin and glucose systems, *Federation Proceedings* 33, 1855-1864 (1974).

Ruch, T.C. and Patton, H.D., *Physiology and Biophysics,* 19 edition; page 887, W.B. Saunders Co., Philadelphia, 1965.

Steele, R., Influences of glucose loading and of injected insulin on hepatic glucose output, *Ann. N.Y. Acad. Sci.* 82, 420-430 (1959).

Zelin, S., Norwich, K.H. and Hetenyi Jr., G., The glucose control mechanism viewed as a regulator, *Med. & Biol. Eng.* 13, 503-508 (1975).

CHAPTER VII
INDICATORS AND BLOOD FLOW

SECTION 34
HYDRO- AND HAEMOKINEMATICS: STEADY FLOW

The term *kinematics* is usually used to refer to motion without respect to the forces which cause the motion, while *dynamics* treats the relationships between forces and motion. *Haemokinematics* is, then, a reasonable word to use to describe the study of blood flow without respect to the associated forces. In this book we shall be concerned largely with the aspects of haemokinematics which can be studied with the use of tracers[†]. While it is possible to use labelled red blood cells as tracer, we shall be concerned essentially with those tracers which affix themselves to some component of the blood plasma, or interstitial fluid. In studies of blood flow, tracers are usually called *indicators*.

The literature on the subject of indicators and fluid flow is bountiful. But as usual, we shall confine our studies to those aspects of the subject which are:

(a) more firmly fixed in physics and

(b) amenable to verification or validation in the laboratory.

[†]It is a moot point whether the term *tracer* should be applied in flow studies, since there is not really a tracee *in the sense that we have previously used*. The fluid component to which the tracer affixes itself could be called the tracee, but this is not the meaning previously imported to the term.

273

Thus purely mathematical or theoretical studies on this subject will be excluded, which reduces the reference literature considerably. It will also become apparent that we are not starting *de novo*; many of the equations which we have already developed for metabolic systems can be readily applied in the study of flow.

There are many types of indicators from which an investigator may choose. They may be classified usefully as follows (examples in parentheses):

    I   stable (saline)
        unstable (ascorbic acid)

   II   short biological half-life (indocyanine green)
       long biological half-life (tagged red blood cells)

 III   thermal
       chemical

  IV   optical (dyes)
       non-optical (anything else)

   V   radioactive
      non-radioactive

  VI   measurable continuously (dyes)
       not measurable continuously (radioactive)

The above 6 categories are not mutually exclusive.

For all the profusion of available indicators, the ones currently used most frequently are optical and thermal indicators. The concentration or density of both of these indicators may be recorded continuously, so that a smooth concentration-time curve may be obtained. The temperature of flowing blood may be measured nearly instantaneously by inserting a small heat sensor into the stream; the concentration of an optical indicator in flowing blood may be measured nearly instantaneously by the use of fibre-optic devices. The most common way at present of measuring the concentration of an optical indicator in the blood stream is to sample the flowing stream continuously by means of a sampling catheter. The sampled blood is then led to a densitometer where its optical density at an appropriate wavelength is measured. By means of a series of standards, the optical density measurements are converted to indicator concentrations. The concentration - time curve recorded at the densitometer is not identical with the curve which would have been obtained if measurements had been made directly in the blood stream *in vivo*. A distortion has been introduced by the sampling catheter, and we shall return to this problem later.

The type of flow-system with which we shall be concerned here is depicted in Fig. 34.1. It is the somewhat restricted system where blood enters via a single inlet or blood vessel and leaves via a single outlet. The capillary labyrinth which lies between is assumed to be impermeable to indicator[†] but is quite unrestricted with respect to geometry. The fundamental extrinsic variables in such systems are the flow rate and the volume of fluid within the system. Generally, we shall treat the case where indicator is infused just a little downstream to the fluid inlet, and the fluid which carries the indicator enters at a rate very much less than the flow rate of blood at the inlet. Fluid leaving the system is sampled, either directly or by means of a sampling catheter. Such a flow-system, of course, forms part of the entire organism, so that fluid will recirculate from the outflow back to the inflow (after a suitable time lag) an indefinite number of times.

Measurement of Constant Flow Rate by the Indicator Infusion Method

Suppose that we deal with the case where the flow is constant, so that fluid enters and leaves the system at the steady rate Q [vol. $\cdot$ time^{-1}]. Suppose, moreover, that we can establish experimentally the condition where indicator (tracer) is in a steady state by the criteria of section 13. The source strength of tracer, $R^{*}_{a,v}$, may be represented using the three-dimensional delta function as

$$R^{*}_{a,v} = R^{*}_{a} \, \delta(\vec{r} - \vec{r}_{in}) \tag{34.1}$$

where $\vec{r}_{in}$ is the position vector of the site of indicator infusion. Since there are no sites of indicator degradation, by assumption, within the capillary network we may set

$$R^{*}_{d,v} = 0 \tag{34.2}$$

[†]This impermeablility condition will be relaxed in our discussion of flow measurement by the continuous indicator infusion method.

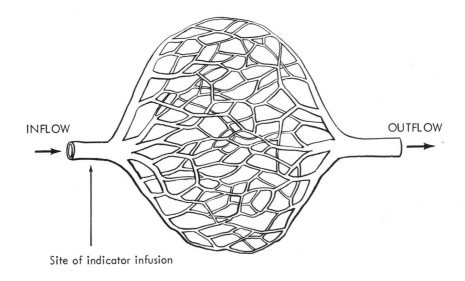

INFLOW   OUTFLOW

Site of indicator infusion

Fig. 34.1.   A Capillary labyrinth of arbitrary geometry, but impermeable to
             indicator, separates a single inflow from a single outflow vessel.
             Indicator infusions are applied at the inflow and blood is sampled
             at the outflow.   Blood leaving the outflow will eventually re-
             appear at the inflow.   The inlet should be taken to be a little
             upstream to the site of indicator infusion.

In our studies of metabolic transport, we have met and briefly discussed the
significance of the molecular diffusion coefficient, D.   In studies of intravascular transport the molecular diffusion coefficient is, generally, of
lesser importance than in metabolic studies.   However, other diffusion coefficients arise to take its place.   In the type of flow-system illustrated
in Fig. 34.1 it is possible to have a coefficient of diffusion associated
with turbulent flow of fluid.   Such a diffusion coefficient would have the
same mathematical properties as the molecular coefficient but would be about
10^6 times greater in magnitude.   Thus the equation of convective diffusion
will be unchanged in form, but it will now be understood that the "D" refers

to this *virtual* coefficient of diffusion associated with turbulent flow. The
interested reader is referred to the now-classical papers by Sir Geoffrey
Taylor (1953, 1954). As in the metabolic studies, we shall confine ourselves
to a single diffusion coefficient throughout.

Introducing (34.1) and (34.2) into (11.23), the equation of convective diffusion, and equating the derivative of time to zero (steady state) leaves

$$D\nabla^2 C* - \vec{u} \cdot \nabla C* + R_a^* \delta(\vec{r} - \vec{r}_{in}) = 0 \; ^\dagger \qquad (34.3)$$

Since the fluid is incompressible ((11.18), (11.19))

$$\vec{u} \cdot \nabla C* = \nabla \cdot (C*\vec{u}) \qquad (34.4)$$

Making this substitution in (34.3) and integrating over the entire internal
volume of the system between infusion and sampling sites,

$$D\int_V \nabla \cdot \nabla C* dV - \int_V \nabla \cdot (C*\vec{u}) dV + \int_V R_a^* \delta(\vec{r} - \vec{r}_{in}) dV = 0$$

Applying the divergence theorem,

$$D\int_S \nabla C* \cdot \vec{n} \, dS - \int_S C* \, \vec{u} \cdot \vec{n} \, dS + R_a^* = 0 \qquad (34.5)$$

We shall assume henceforth that after some time the indicator concentration
assumes the constant value, C_{out}^*, over the entire cross-section of the outflow
channel, and C_{in}^* over the entire cross-section of the inflow channel. (The
entire curve, $C_{out}^*(t)$, is called the *F curve* in the chemical engineering
literature, e.g. Levenspiel, 1962).

Boundary Conditions (i): Zero Gradient in C* at the Inflow and Outflow, Zero Gradient in C* over the Remainder of the Surface

These are the boundary conditions which are usually accepted. Expressed
explicitly,

$$[\nabla C*]_{C* \, = \, C_{out}^*} = 0 \qquad (34.6)$$

$\dagger$When using the constant infusion method, the single indicator inlet, $\vec{r}_{in}$, is
really overly restrictive. Indicator may enter anywhere between the fluid
inlet and outlet sites. For example, oxygen (used as an indicator) enters
over the entire pulmonary capillary bed. The important thing is that when
the distributed source is integrated over the total volume, i.e. $\int_V R_{a,v}^* dV$,

we obtain R_a^* in Eq. (34.5).

$$[\nabla C^*]_{C^* = C^*_{in}} = 0 \tag{34.7}$$

$$[\nabla C^*]_{surface} = 0 \tag{34.7a}$$

Thus (34.5) becomes, by virtue of (34.6), (34.7) and (34.7a),

$$-C^*_{in} \int_{S_{in}} \vec{u} \cdot \vec{n} \, dS - C^*_{out} \int_{S_{out}} \vec{u} \cdot \vec{n} \, dS + R^*_a = 0 \tag{34.8}$$

where the integration is carried out over cross-sectional areas of the inflow
and outflow channels, S_{in} and S_{out}, respectively. Since we have confined ourselves to the case of a single inlet and a single outlet for fluid, it must
be true that

$$Q = -\int_{S_{in}} \vec{u} \cdot \vec{n} \, dS = \int_{S_{out}} \vec{u} \cdot \vec{n} \, dS \tag{34.9}$$

which permits us to obtain from (34.8)

$$Q = R^*_a \Big/ (C^*_{out} - C^*_{in}) \tag{34.10}$$

That is, the constant flow rate, Q, is given as the ratio of the steady rate
of infusion of indicator to the difference between the inflow and outflow indicator concentrations. When no recirculation of indicator occurs,

$$C^*_{in} = 0$$

effectively, since the indicator infusion site is downstream from the inlet,
and (34.10) becomes simply

$$Q = R^*_a \Big/ C^*_{out} \tag{34.11}$$

Eq. (34.11) is analogous to Eq. (25.6) which gives the rate of appearance of
a metabolite in a steady state metabolic system. Equations (34.10) and
(34.11) define the "direct Fick" method for measuring steady flow.

Boundary Conditions (ii): Nonzero Gradients in C* at the Inflow and Outflow, Zero Gradient in C* over the Remainder of the Surface

Although all time derivatives must vanish using our definition of a steady
state system, spatial derivatives need not. Hence, for example, if indicator
is removed from the flowing stream at a point downstream from the sampling
site, a gradient in indicator concentration could be present at the outlet.
Because of the nature of the boundary conditions, the surface integrals of
Eq. (34.5) vanish at all points except across the inflow and outflow channels.
Hence (34.5) now assumes the form

$$D \int_{S_{in}} \nabla C^* \cdot \vec{n} \ dS + D \int_{S_{out}} \nabla C^* \cdot \vec{n} \ dS - C^*_{in} \int_{S_{in}} \vec{u} \cdot \vec{n} \ dS$$

$$-C^*_{out} \int_{S_{out}} \vec{u} \cdot \vec{n} \ dS + R^*_a = 0 \qquad (34.12)$$

If the flow is not turbulent, the only type of diffusion which could be implied by the first two integrals would be molecular diffusion, and transport by molecular diffusion will be orders of magnitude slower than transport by convection. Hence the first two integrals may be dropped, leaving us again with (34.8). If, however, the flow is turbulent, the first two integrals may represent diffusive transport governed by "virtual" coefficients of diffusion (Taylor, 1954), in which event the integrals do not necessarily represent negligible quantities. Representing $- D\nabla C^*$ as a diffusive flux, $\vec{J}^*_{diff}$, taking A as the sectional area of the inflow and outflow channels, and assuming that C^* is constant across the channel, we obtain from (34.12)

$$A \ J^{* \ in}_{diff} - A \ J^{* \ out}_{diff} + Q \cdot C^*_{in} - Q \cdot C^*_{out} + R^*_a = 0$$

or

$$Q = \frac{R^*_a}{C^*_{out} - C^*_{in}} - \frac{A \ J^{* \ out}_{diff} - A \ J^{* \ in}_{diff}}{C^*_{out} - C^*_{in}} \qquad (34.13)$$

The second term on the right-hand side may be taken to be a perturbation in the usual direct Fick equation. It would be very difficult to measure this quantity experimentally. The primary feature to remember here is that the very popular Eq. (34.10) is not universally valid.

Measurement of Constant Flow Rate by the Indicator Injection Method

Just as in the measurement of steady state rates of appearance in a metabolic system, the two fundamental methods are the constant infusion and the sudden injection methods, so too is the case in measurement of constant flow rates. Suppose that a sudden injection or impulse of indicator of mass M_o^* is applied at the inflow channel and the concentration-time curve is measured again in the outflow. Suppose, again, that C_o^* is constant over a cross-section of the outflow channel. The function $C_o^*(t)$ defines the *C curve* (Levenspiel, 1962). The resulting concentration-time curve, Fig. (32.2a), is called an *indicator-dilution curve*. Because we shall have occasion to use this term frequently,

it will be abbreviated to *IDC*. Since injected indicator may pass throught the
heart many times past the sampling site, a measured IDC may consist of a primary, a secondary, and even a tertiary peak (Fig. 34.2b). Although we can
deal with the recirculated indicator when using the steady infusion method
(see Problem 34.1) we must usually remove all but the first peak when using
the injection method. Hence we shall presume that the secondary and higher
order peaks have been removed for purposes of our present discussion. Methods
or correcting curves for recirculation of indicator will be described in a
later section, in conjunction with the modelling of an IDC.

a)

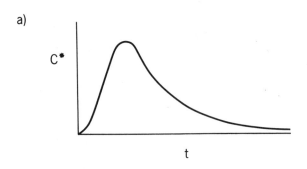

b)

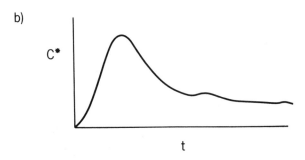

Fig. 34.2. (a) An indicator-dilution curve (IDC) obtained from a system
                 where indicator does not recirculate.
            (b) An IDC obtained from a system where indicator recirculates.

            These are schematic drawings, not measured curves.

The sole input to the system (when there is no recirculation) is just the impulse of indicator that the experimenter injects. Let the corrected IDC be represented by the function $C_0^*(t)$, and the unit impulse response function then by $C_0^*(t)/M_0^*$. Applying the convolution integral (12.25) then, the response of the system to a step input, R_a^*, is given by

$$C_{out}^*(t) = \frac{R_a^*}{M_0^*} \int_0^t C_0^*(\tau)d\tau \qquad (34.14)$$

The limiting value of the step response, $C_{out}^*(t)$, as $t \to \infty$ will be given by

$$C_{out}^* = \frac{R_a^*}{M_0^*} \int_0^\infty C_0^*(t)dt \qquad (34.15)$$

Case (i): Indicator Injection Method. Zero Gradients in C* at the Inflow and Outflow etc. *If a Steady Infusion were Maintained in the Same System*[+]
From (34.11)

$$C_{out}^* = R_a^*/Q \qquad (34.16)$$

By equating the right-hand sides of (34.15) and (34.16) we obtain

$$Q = \frac{M_0^*}{\displaystyle\int_0^\infty C_0^*(t)dt} \qquad (34.17)$$

Eq. (34.17) was developed through the work of G.N. Stewart (1897) and W.F. Hamilton (1932). This equation is analogous to (26.6), the equation for calculating steady state rate of appearance by the tracer injection method.

Although (34.17) has been developed for the case where the injection of indicator was sudden, it should be equally valid for the case where the indicator was infused over a finite interval 0 to T. For a fuller description of the

---

[+]We are concerned here with making calculations using the data of an injection experiment. But if we are to apply the simple Stewart-Hamilton formula for flow, Q, we must know something about the gradients which *would have occurred* had we performed a steady infusion experiment (cf practical summary at the end of section 26).

finite infusion theory the reader is referred to the paper by W. Perl (1971). However, one can get the gist from a simple conservation of mass argument. Suppose that a total mass M_0^* of indicator is infused at the inlet at any infusion rate, $R_a^*(t)$. In the absence of a significant contribution to the transport of indicator by diffusion, the amount of indicator passing the sampling site in time, Δt, is given by $QC_0^*(t)\Delta t$. After a long time (much greater than T) has passed, the total indicator mass, M_0^* will have passed the sampling site. That is,

$$\int_0^\infty R_a^*(t)dt = M_0^* = Q\int_0^\infty C_0^*(t)dt \qquad (34.18)$$

from which we immediately retrieve (34.17)

Case (ii): Indicator Injection Method. Non-Zero Gradients in C* Allowed at the Inflow and Outflow *If a Steady Infusion were Maintained in the Same System* Let us proceed from first principles.

Returning to (11.23) and introducing the arbitrary infusion of finite length at a point $R_a^*(t) \cdot \delta(\vec{r} - \vec{r}_{in})$,

$$D\nabla^2 C^* - \vec{u}\cdot\nabla C^* + R_a^*(t)\cdot\delta(\vec{r} - \vec{r}_{in}) = \frac{\partial C^*}{\partial t} \qquad (34.19)$$

For the case of the sudden injection, the final term on the left-hand side will be $M_0^* \cdot \delta(t) \cdot \delta(\vec{r} - \vec{r}_{in})$. As before

$$\int_0^T R_a^*(t)dt = M_0^* \qquad (34.20)$$

Introducing (34.4) and integrating over the total constant internal volume,

$$D\int_V \nabla\cdot\nabla C^* \, dV - \int_V \nabla\cdot(C^*\vec{u})dV + \int_V R_a^*(t)\cdot\delta(\vec{r}-\vec{r}_{in})dV = \frac{d}{dt}\int_V C^* dV$$

Applying the divergence theorem for the boundary conditions of the labyrinth, and letting M* be the total tracer mass present in the system at any time.

$$D\int_{S_{in}} \nabla C^*\cdot\vec{n} \, dS + D\int_{S_{out}} \nabla C\cdot\vec{n} \, dS - C_i^*(t)\int_{S_{in}} \vec{u}\cdot\vec{n} \, dS$$

$$-C_0^*(t)\int_{S_{out}} \vec{u}\cdot\vec{n} \, dS + R_a^*(t) = \frac{dM^*}{dt} \qquad (34.21)$$

or

$$-A \; j_{\mathrm{diff}}^{*\mathrm{out}} + A \; J_{\mathrm{diff}}^{*\mathrm{in}} + C_i^*(t) \cdot Q - C_o^*(t) \cdot Q + R_a^*(t) = \frac{dM^*}{dt} \qquad (34.22)$$

where C_i^* and C_o^* are the impulse response functions at the inflow and out-flow respectively. (Recall that the site of tracer injection is downstream from the fluid inflow.) Integrating now with respect to time,

$$-A \int_0^\infty J_{\mathrm{diff}}^{*\mathrm{out}}(t)dt + A \int_0^\infty J_{\mathrm{diff}}^{*\mathrm{in}}(t)dt - Q \int_0^\infty \left[C_o^*(t) - C_i^*(t) \right] dt$$

$$+ \int_{0+}^\infty R_a^*(t)dt = 0$$

or using (34.20),

$$Q = \frac{M_o^*}{\displaystyle\int_0^\infty \left[C_o^*(t) - C_i^*(t) \right] dt} - \frac{A \displaystyle\int_0^\infty J_{\mathrm{diff}}^{*\mathrm{out}}(t)dt - A \displaystyle\int_0^\infty J_{\mathrm{diff}}^{*\mathrm{in}}(t)dt}{\displaystyle\int_0^\infty \left[C_o^*(t) - C_i^*(t) \right] dt}$$

$$(34.23)$$

Eq. (34.23), applied for an impulse or finite infusion, corresponds to (34.13) applied for a steady, infinite infusion. If diffusion is negligible, the two final terms on the right-hand side of (34.23) vanish, and $C_i^*(t)$ becomes much smaller than $C_o^*(t)$, so that (34.23) reduces to (34.17). Eq. (34.23) is the generalization of the Stewart-Hamilton equation (34.17) for the case where diffusion (or virtual diffusion) is not negligible. In section 35 we shall solve the convective diffusion equation for the case of a *sudden* injection of indicator, and in this way obtain a simpler expression for Q than the one given by (34.23).

Example 1

For the case of no recirculation of indicator we may set $C_i^* = 0$. And when diffusion of indicator is of negligible proportions we may set $J_{\mathrm{diff}}^{*\mathrm{out}} = J_{\mathrm{diff}}^{*\mathrm{in}} = 0$. Making these changes in (34.23) leaves

$$Q = \frac{M_o^*}{\displaystyle\int_0^\infty C_o^*(t)dt}$$

which is the Stewart-Hamilton equation (34.17). This equation is the ana-logue of (26.6), the equation for calculating the steady state rate of appear-ance using the tracer injection method.

Example 2

Let us take the hypothetical example of measurement of flow in an infinite uniform cylinder, when the flow is actually zero. The experimenter does not know that the flow rate is zero and proceeds to measure it by indicator-dilution techniques. The simplified geometry is illustrated in Fig. 34.3. The "inflow" is taken a little to the left of the injection site. Sampling occurs at inflow and outflow positions. By symmetry we see that over all time, as much indicator must diffuse to the left of the injection site as to the right. That is,

$$A \int_0^\infty J^{*out}_{diff}(t)dt = -A \int_0^\infty J^{*in}_{diff}(t)dt = M_0^*/2 \qquad (34.24)$$

Making this substitution in (34.23), we obtain

$$Q = 0$$

as required. (See also problem 34.3).

Pulsatile Flow

Equations (34.10) and (34.17) are routinely applied in the measurement of blood flow which is pulsatile in nature rather than steady. Eq. (34.18) should more properly be written

$$M_0^* = \int_0^\infty Q(t)C_0^*(t)dt \qquad (34.25)$$

where $Q(t)$ describes the pulsatile flow. By removing $Q(t)$ from under the integral sign,

$$M_0^* = Q \int_0^\infty C_0^*(t)dt$$

we make the assumption that Q is some sort of mean or average flow. This assumption has been discussed in some depth by Cropp and Burton (1966) and the interested reader is referred to their paper. Aaron and Fine (1975) also give a very neat proof to show that (34.17) is valid for interrupted, "square-wave" pulsatile flow from a well mixed reservoir. We shall return later to a discussion of indicator methods in pulsatile blood flow.

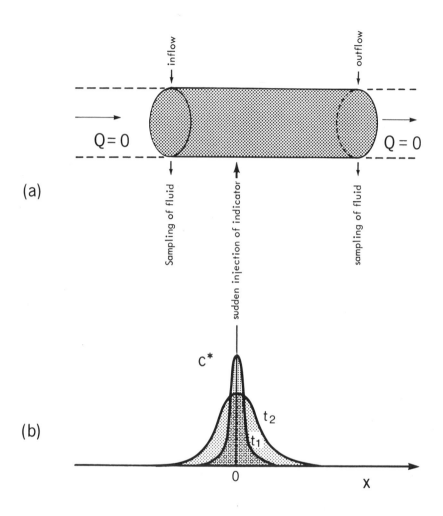

Fig. 34.3. (a) An infinite uniform cylinder containing static fluid.
            (b) After injection of indicator, dispersion is entirely by
                 diffusion. Concentration as a function of axial distance
                 from the injection site is shown for two times, $t_2 > t_1$.
                 The areas beneath the two curves are equal.

Multiple Outlets

In the derivation of the injection equation (34.17) we have assumed that the capillary labyrinth takes on the anatomical configuration of Fig. (34.1): namely a single fluid inlet and a single fluid outlet. This equation should also be valid for a system with multiple outlets, provided that the areas under the concentration-time curve obtained at all outlets are equal. This may be seen immediately from (34.21). Assuming that diffusion is negligible, the first two integrals vanish. If recirculation of indicator is prevented so that $C_i^* = 0$, the third term vanishes. Suppose now that there are two outlets rather than one, so that the fourth term is replaced by two new ones. We then have

$$-C_{01}^*(t) \int_{S_1} \vec{u} \cdot \vec{n} \ dS - C_{02}^*(t) \int_{S_2} \vec{u} \cdot \vec{n} \ dS + R_a^*(t) = \frac{dM^*}{dt} \qquad (34.26)$$

where C_{01}^* and C_{02}^* are the concentrations of indicator in the two outlets, etc.

$$-C_{01}^* \cdot Q_1 - C_{02}^* \cdot Q_2 + R_a^*(t) = \frac{dM^*}{dt}$$

where Q_1 and Q_2 are the flow rates through the two outflow channels. Integrating with respect to time,

$$-Q_1 \int_0^\infty C_{01}^*(t)dt - Q_2 \int_0^\infty C_{02}^*(t)dt + M_0^* = 0$$

If the two integrals are equal,

$$\int_0^\infty C_{01}^*(t)dt = \int_0^\infty C_{02}^*(t)dt = \int_0^\infty C_0^*(t)dt$$

and

$$Q = Q_1 + Q_2 = \frac{M_0^*}{\int_0^\infty C_0^*(t)dt} \qquad (34.27)$$

That is, total flow is given by dividing the injected mass of indicator by the area under either IDC.

By invoking this assumption of IDC's with equal areas one may estimate, for example, cardiac output (the rate at which the heart pumps blood) by injecting indicator into the main pulmonary artery and sampling in the brachial artery.

Mean Residence Time and Blood Volume

The theory of residence times of blood particles in a capillary labyrinth is
similar to but simpler than the theory of residence times of molecules of a
metabolic system. Since, by hypothesis, all fluid leaves the capillary labyrinth through a single outlet, we are assured that the sampling site in the
outflow channel *is* a representative site for particles leaving the system.
(In the case of the metabolic system this was a moot point.) We shall apply
(29.3) without further proof to give

$$\bar{t}_v = \bar{t} = \frac{\int_0^\infty t C_0{}^* dt}{\int_0^\infty C_0{}^* dt} \qquad (34.28)$$

Or we may just write (34.28) as a definition of $\bar{t}$. The distinction should be
drawn between a *mean residence time* as given by (34.28) and a *mean transit
time*. The *mean transit time* usually means *mean time of first transit*, or the
average time which elapses from the instant the particle enters the system at
the injection site until the first time it passes the sampling site. The *mean
residence* or *sojourn time* usually means the average time which elapses from
the instant the particle enters the system at the injection site until it
passes the sampling site forever. In the absence of diffusion in the region
of the sampling site, the mean transit and residence times should be equal.
In the presence of diffusion in the region of the sampling site, Brownian
motion of the indicator molecules will cause the same particle to pass the
sampling site several times, so that mean residence time will exceed mean
transit time. The calculation of mean first transit time is not simple. It
has been calculated, for example, for a drifting Maxwellian gas (where gas
molecules may be taken as analogous to indicator) by Schrödinger (1915), and
his arguments have been translated by Kennard (1938). In many biological
treatises, the two terms *mean transit* and *mean residence* are used synonymously
with loss of an important distinction.

Calculating the volume of the capillary labyrinth in a system where diffusion
at the exit is not significant is analogous to calculating the pool size of a
metabolic system in steady state, as was shown in section 29. The derivation
will be reproduced rather briefly. The impulse response, at the outlet, to a
sudden injection of tracer at the inlet will be designated, as before, by $C_0{}^*$.
The probability that a particle of indicator which was injected at t = 0 will

leave the system between the times t and t + dt is given by

$C_0^*(t)dt \Big/ \int_0^\infty C_0^*(t)dt$. Define $f(t)$ by

$$f(t) = \frac{\int_0^t C_0^*(\tau)d\tau}{\int_0^\infty C_0^*(\tau)d\tau} \qquad (34.29)$$

As required

$$\lim_{t\to\infty} f(t) = 1 \qquad (34.30)$$

Then $f(t)$ gives the probability that a particle which enters the system between times 0 and dt will leave by time, t. And $(Q\ dt) \cdot f(t)$ gives the volume
of fluid which enters the system between times 0 and dt which will leave by
time, t. Hence

$$\int_0^t (Q\ d\tau) \cdot f(\tau)$$

gives the volume of fluid which enters the system between the times 0 and t
which will leave by time, t. Therefore

$$Qt - \int_0^t Qf(\tau)d\tau \qquad (34.31)$$

expresses the difference between the total amount of fluid which entered during the time interval 0 to t and that which left during this period. Expression (34.31) then gives the amount of fluid which entered during the interval
0 to t which remains there at time, t. Therefore

$$V = \lim_{\tau\to\infty} Q \left[t - \int_0^t f(\tau)d\tau \right] \qquad (34.32)$$

must give the volume of the system, because for very large t, all fluid in the
system will be fluid which entered after zero time. If we now set

$$g(t) = \frac{C_0^*(t)}{\int_0^\infty C_0^*(t)dt} \qquad (34.33)$$

we can apply the arguments of (29.14) to (29.17) to obtain

$$V = \lim_{t\to\infty} Q \left[t - tf(t) + \int_0^t \tau g(\tau)d\tau \right] \qquad (34.34)$$

Applying (34.30) leaves

$$V = Q \int_0^\infty t\ g(t)dt$$

or

$$V = Q \frac{\int_0^\infty t C_0^*(t)dt}{\int_0^\infty C_0^*(t)dt} \qquad (34.35)$$

or, from (34.28)

$$V = Q \cdot \bar{t}$$

The volume of the capillary labyrinth is equal to the product of the steady
flow rate and the mean residence time, *in a system where random motion of
molecules is not significant.*

Speaking Practically

The two fundamental equations for calculating steady blood flow are the constant indicator infusion equation (34.10) and the indicator injection equation
(34.17). These equations are analogues of the two equations for calculating
the steady state rate of appearance of a metabolite by the tracer infusion and
injection methods. Similarly, the blood volume contained by a system such as
that of Fig. 34.1 is given by the product of blood flow rate by mean residence time, with mean residence time defined by (34.28). The volume equation
for blood is therefore, the analogue of the *mass* equation (29.20), for a metabolite. The analogues are,

specific activity   indicator concentration
rate of appearance of tracee   blood flow rate
mass of tracee   volume of system

The analogy, however, appears to hold with certainty only when diffusion or
virtual diffusion or quasi diffusion (section 35) does not contribute to the
transport of indicator. If diffusion is not negligible one should refer to
(34.23) rather than (34.17), etc., but this is not usually a practical
consideration.

Problems

34.1 In applying the Fick principle to the calculation of cardiac output,
the chief problem may be confusion with the units of measurement. Suppose
that oxygen concentration in mixed venous blood is 13.9 "volumes %" (13.9 ml
of oxygen in 100 ml of blood) and the oxygen concentration in arterial blood
is 19.1 volumes %. The oxygen uptake by the lungs is 300. ml/min. Calculate
the cardiac output in litres/minute.

34.2 A patient with a cardiomyopathy was investigated in the cardiovascular
unit of the Toronto General Hospital. An indicator-dilution curve was prepared by injecting 5.0 mg of indocyanine green into the main pulmonary artery,
and sampling at the root of the aorta. The IDC, corrected for catheter distortion and recirculation of indicator, is given below, Calculate the cardiac
output.

time (seconds from injection)	concentration (mg/litre)
0.0	0.0
1.0	0.0
2.0	0.0
2.4	0.0
2.6	0.179
2.8	0.405
3.0	0.787
3.2	1.35
3.4	2.11
3.6	3.01
3.8	4.00
4.0	5.01
4.2	5.94
4.4	6.74
4.6	7.35
4.8	7.74
5.0	7.90
5.2	7.85
5.4	7.62
5.6	7.23
5.8	6.74
6.0	6.17
6.2	5.55
6.4	4.93
6.6	4.32
6.8	3.75
7.0	3.21
7.2	2.72
7.4	2.29
7.6	1.91

7.8	1.58
8.0	1.30
8.2	1.06
8.4	0.859
8.6	0.693
8.8	0.555
9.0	0.443
9.2	0.352
9.4	0.278
9.6	0.219
9.8	0.172
10.0	0.152
10.4	0.105
10.8	0.0630

34.3 The solution of the equation of diffusion for the infinite cylinder of example 2 above is

$$C*(x,t) = \frac{M_0{}^*}{\sqrt{4\pi DA^2}} \, t^{-\frac{1}{2}} \, e^{-\frac{x^2}{4Dt}}$$

where x is the distance along the cylinder measured from the site of the sudden injection of indicator. Show that each of the integrals in Eq. (34.24) is equal to $M_0{}^*/2$ by evaluating the integrals explicitly.

34.4 Suppose we deal with flow in an infinite, uniform cylinder at a mean speed u. Indicator is injected suddenly at t = 0, x = 0. Let the sampling site be located at $x = x_0 > 0$ and define τ by the equation

$$\tau = x_0/u$$

Then the solution of the equation of convective diffusion for $x = x_0$ is

$$C*(x_0,t) \equiv C_0{}^*(t) = \frac{M_0{}^*}{\sqrt{4\pi DA^2}} \, t^{-\frac{1}{2}} \, \exp\left[\frac{-(uA)^2}{4DA^2}\frac{(t-\tau)^2}{t}\right]$$

Referring to the text by Kennard we see that τ is the mean (first) transit time of the system. Calculate $\bar{t}$, the mean residence time and show that it exceeds τ when diffusion is not negligible. Show that, in fact

$$\bar{t} = \tau + \frac{2D}{u^2}$$

Show that the volume of the cylinder between injection and sampling sites is *not* given by $Q \cdot \bar{t}$. Why?

34.5 Suppose you have two indicators, one of which remains within the vascular system and the other which passes freely out of the vascular system
into the interstitial fluid, and equally freely back into the vasuclar network. Devise a method for measuring the volume of interstitial fluid. Such
methods are used for measuring the volume of pulmonary oedema.

SECTION 35

MODELLING THE INDICATOR — DILUTION CURVE: CORRECTING FOR RECIRCULATION OF INDICATOR

It was demonstrated in section 34 that the determination of steady blood flow
by the indicator injection method, in a system where diffusion of indicator
is negligible, is most readily achieved when the indicator-dilution curve
(IDC) is free of the effects of recirculated indicator. In this case the
steady flow rate, Q, may be obtained using measurements only of $C_o*(t)$, the
impulse response function at the output, by employing (34.17), the Stewart-
Hamilton equation. If the IDC contains a secondary or recirculated peak, one
may still calculate Q but a knowledge of $C_i*(t)$, the impulse response function at the inlet, is required. For various reasons measurement of $C_i*(t)$ is
difficult. It necessitates using a second catheter or probe, which adds technical problems. And of greater importance, it is often difficult to position
the inflow catheter or probe so that it receives a sample of incoming blood
which has a uniform concentration of indicator over a cross-section — a condition required for the derivation of all the equations of section 34. Because recirculation cannot be eliminated easily by making measurements, it is
usually eliminated by modelling the IDC — which is the primary reason why we
shall now consider modelling. A model of the IDC will also be seen to be
helpful in removing the distortion due to the catheter sampling.

Functions without Definite Physical Basis

There are a number of mathematical functions which define curves having the general appearance of the primary IDC (sketched in Fig. 35.2a) without having any obvious physical reason for doing so. Perhaps the best known of these are the lognormal and gamma functions, the correspondence between which is shown clearly in tabular form by Wise (1966). Because of the absence of physical basis for these functions we shall simply note their existence and pass on to consider various physical models.

Single, Well-Mixed Tank

Models which represent the capillary labyrinth by a single, well-mixed chamber are, in effect, one-compartment models. The difference between tank and compartment models is as follows. In the former the volume of fluid is kept constant by means of a constant, non-zero rate of fluid inflow and outflow, and it is this convective flux that induces any changes in concentration within the tank. In the latter, we are not much concerned with the flow of fluid through the chamber, and the changes in concentration within are not necessarily related to any convective flux of the carrier fluid.

A single-tank model (Fig. 35.1) is not really capable of accounting for the observed shape of the entire IDC, but this model is widely used anyway as a means of correcting for recirculation of indicator. Fluid enters the tank at the constant rate Q [vol $\cdot$ time^{-1}] and leaves at the same rate. The volume of fluid within the tank remains constant at V. To simulate the IDC, we imagine a mass M_0^* of indicator injected suddenly into the tank at zero time. Since fluid is carried away at the rate Q, and no new indicator is added, the rate of change of indicator mass, M^*, within the tank is given by

$$\frac{dM^*}{dt} = 0 - Q \cdot C^* \tag{35.1}$$

where C^* is, as usual, the indicator concentration. Since the contents are well-mixed

$$C^* = M^*/V \tag{35.2}$$

so that (35.1) becomes

$$\frac{dC^*}{dt} = -\frac{Q}{V} C^* \tag{35.3}$$

Hence

$$C_0^*(t) = C_0^*(0)e^{-\frac{Q}{V}t} \tag{35.4}$$

where $C_o^*(t)$, the impulse response function at the outflow, is identical with $C^*(t)$, the impulse response function within the tank. Since the monoexponential decay function, (35.4), decreases monotonically with increasing t, it cannot fully account for the shape of the IDC (Fig. 35.2a).

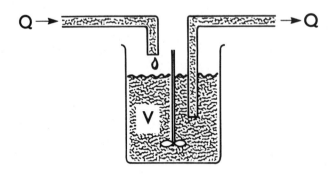

Fig. 35.1. Single, well-mixed tank. Fluid enters and leaves at the constant flow rate, Q, leaving a constant volume of fluid, V, inside the tank. The stirring rod within implies that the contents are well mixed.

The assumption is often made that the tail or descending limb of the IDC, if it were not obscured by recirculation, would fall monoexponentially. Thus (following W.F. Hamilton and his colleagues) the descending limb of the measured (uncorrected) IDC is plotted on semi-logarithmic paper. Part of the graph so-obtained is a straight line (Fig. 35.2b). This straight line is extrapolated downward on the semi-log paper. The antilogarithms of points which lie on the extrapolated portion of the straight line are then taken to define the portion of the original IDC which was obscured by recirculation. The area under the corrected IDC is then used to calculate the steady flow rate using (34.17). This is probably the commonest way currently used to correct for recirculation, for two reasons:

      (i) It is simple

     (ii) In the absence of hard-core experimental evidence to the contrary, there seems to be little reason to deviate from a method employed for nearly half a century.

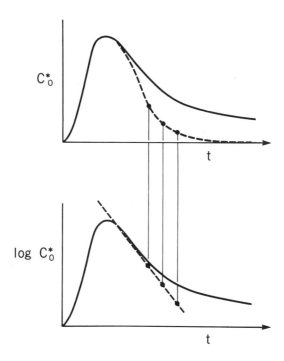

Fig. 35.2. Using a semi-logarithmic plot to correct an indicator-dilution
            curve for recirculation of indicator.
            (a) The original IDC plotted with standard linear ordinate.
            (b) The same IDC plotted on semi-logarithmic paper. A
                 straight line has been drawn along the descending limb and
                 extrapolated downwards. Antilogarithms of points on the
                 extrapolated limb are then used to correct the original IDC
                 for recirculation.
            These are schematic drawings.

Series of Well-Mixed Tanks

Let us now consider a catenary system comprising a number, n, of well-mixed
tanks (Fig. 35.3) each having the same volume, V_0. Then the total volume V
is given by

$$V = nV_0 \qquad\qquad (35.5)$$

The theory is well-treated by Denbigh (1965) whose development we shall follow here. Let the mass and concentration of indicator in the jth compartment be M_j^* and C_j^* respectively. The differential equation governing concentration change in the first compartment is (as in (35.1))

$$\frac{dM_1^*}{dt} = Q\,\frac{M_1^*}{V_0} \tag{35.6}$$

whose solution is again

$$M_1^*(t) = M_1^*(0)e^{-\frac{Q}{V_0}t} \tag{35.7}$$

The differential equation for the second tank is

$$\frac{dM_2^*}{dt} = Q\,\frac{M_1^*}{V_0} - Q\,\frac{M_2^*}{V_0} \tag{35.8}$$

Eliminating M_1^* by substituting from (35.7)

$$\frac{dM_2^*}{dt} + \frac{Q}{V_0}\,M_2^* = \frac{QM_1^*(0)}{V_0}\,e^{-\frac{Q}{V_0}t}$$

Multiplying by the integrating factor $e^{\frac{Q}{V_0}t}$ we obtain the solution

$$M_2^*(t) = \left[\frac{Qt}{V_0}\right] M_1^*(0)\,e^{-\frac{Q}{V_0}t} \tag{35.9}$$

The solution for M_2^* from (35.9) is the same as the solution for M_1^* from (35.7) except for the factor in square parentheses. Proceeding in the same way we obtain for M_3^*

$$M_3^*(t) = \left[\frac{Qt}{V_0}\right]^2 M_1^*(0)\,\frac{e^{-\frac{Q}{V_0}t}}{2!} \tag{35.10}$$

and for the jth tank

$$M_j^*(t) = \left[\frac{Qt}{V_0}\right]^{j-1} M_1^*(0)\,\frac{e^{-\frac{Q}{V_0}t}}{(j-1)!} \tag{35.11}$$

The solutions (35.7), (35.9) and (35.10) are shown in Fig. 35.4. One could, therefore, employ (35.11) to correct a measured IDC for the effects of recirculation of indicator. For example, one might select the curve corresponding to j = 3 as best and then curve-fit the initial portion of the measured IDC

to (35.10), thereby ascertaining the best values of Q, V_0 and $M_1^*(0)$. In this
way, one obtains Q explicitly from the curve-fitting process. Alternatively,
Q may be obtained by extrapolating the fitted function through the time period
obscured by recirculation to obtain a pure or primary curve. The area under
the primary curve should then give the same value of Q when (34.17) is applied
(see problem 35.1). The selection of an "initial portion" of the IDC — a
portion during which recirculation of indicator is believed to be zero or
negligible — always involves a certain arbitrary element.

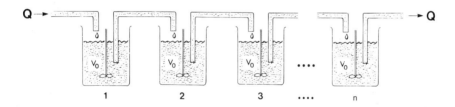

Fig. 35.3. A Series of n identical, well-stirred tanks, each of volume V_0.
          The flow through all conduits is Q.

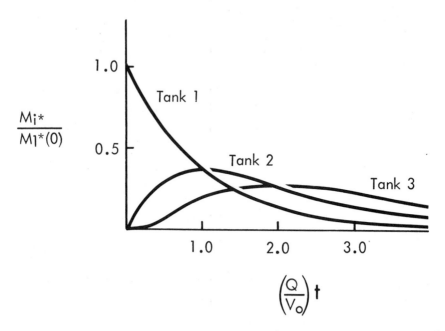

Fig. 35.4. Progressive movement of a tracer through a catenary system of well-stirred tanks (after Denbigh, 1965). Tracer mass and time are expressed in dimensionless units.

From Chemical Reactor Theory: An Introduction, p.82, 1965 by K. Denbigh. Reproduced by permission of Cambridge University Press.

Series of Well-Mixed Elastic Cylinders

The tank- or compartment-models with which we have been dealing hitherto have all depicted the capillary labyrinth as a structure with rigid walls. This constraint was relaxed by Beneken and Rideout (1968), who represented the labyrinth as an elastic tube consisting of a finite number of discrete elements — a number of "well-mixed" elastic cylinders (Fig. 35.5). To express their equations we must define a number of new variables.

$V_{j\ tot}$ is the total volume of the jth section.

V_{ju} is the unstressed volume of the jth section.

$V_j = V_{j\ tot} - V_{ju}$

p_j is the pressure in the jth section.

Q_j is the flow from the $j\text{-}1^{th}$ to the jth section.

R_j, L_j and C_j are the resistance to flow, the inertance of the fluid and the compliance of the jth section respectively.

It is seen that one pays for the new degree of freedom (elastic walls) by the creation of a plethora of variables. The haemodynamic equations for each section may now be set down in straightforward fashion. For the second section, for example,

$$V_{2tot} = \int_0^t (Q_2 - Q_3)dt + V_{2tot}(0) \tag{35.12}$$

$$p_2 = \frac{1}{C_2}(V_{2tot} - V_{2u}) = \frac{V_2}{C_2} \tag{35.13}$$

$$p_1 - p_2 = R_2 Q_2 - L_2 \frac{dQ_2}{dt} \tag{35.14}$$

M_j^* and C_j^* have their usual meaning as the mass and concentration respectively of indicator in the jth section of the tube, and are related by the equation

$$C_j^* = M_j^*/V_{j\ tot} \tag{35.15}$$

The equations governing indicator transport in the second section or segment of tube are then

$$M_2^*(t) = \int_0^t (C_{1,2}^* Q_2 - C_{2,3}^* Q_3)dt + M_2^*(0) \tag{35.16}$$

where

$$C_{1,2}^* = C_1^* \text{ if } Q_2 > 0$$

$$= C_2^* \text{ if } Q_2 < 0$$

$$C_{2,3}^* = C_2^* \text{ if } Q_3 > 0$$

$$= C_3^* \text{ if } Q_3 < 0$$

In this way a set of equations can be developed. The above equations were solved by Beneken and Rideout with an analog computer using perturbation techniques, and some of their results are reproduced in Fig. 35.6. Notice how the characteristic form of the IDC or C-curve develops as we move to the more distal arterial segments. The usefulness of this model transcends its application to IDC's, and the reader is referred to the original paper.

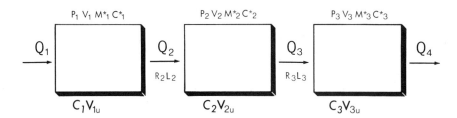

Fig. 35.5. Block diagram of three sections of an elastic tube, (after Beneken
            and Rideout, 1968). All symbols are defined in the text.
From IEEE Transactions on Bio-Medical Engineering, Vol BME-15, No. 4, Oct.
1968 by J.E.W. Beneken and V.C. Rideout. Reproduced by permission of the
Institute of Electrical and Electronic Engineers.

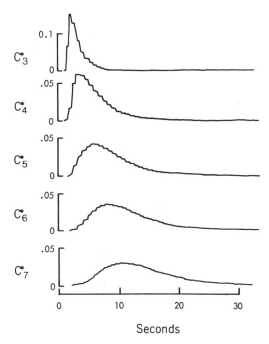

Seconds

Fig. 35.6.   Indicator-dilution curves in the 3rd to 7th segments of the
             elastic tube. Obtained by solving equations (35.12) to (35.16)
             using an analog computer.

From IEEE Transactions on Bio-Medical Engineering, Vol BME-15, No. 4, Oct.
1968 by J.E.W. Beneken and V.C. Rideout. Reproduced by permission of the
Institute of Electrical and Electronic Engineers.

Discrete Random Walk Models

(i) <u>Mathematical</u>. Stephenson (1958) suggested a model of the capillary labyrinth in which a particle after each interval of time, Δt, may take either a
forward step (from inlet in the direction of outlet) or a sideways step (which
does not advance it toward the outlet). Suppose that it takes i forward steps
for a particle to travel from inlet to outlet. Then the number of particles
emerging at time $i\Delta t$ is the number making i forward steps in i trials; the
number of particles emerging at time $(i+1)\Delta t$ is the number making i forward
steps in i+1 trials etc. If a particle emerges at the time $n\Delta t$, the nth step
must have been a forward step. Suppose that the probability of a forward step
is always p and that of a sideways step is q = 1-p. Therefore the probability
that a particle which enters the inlet at t = 0 emerges at time $n\Delta t$ is

(the probability that it made i-1 forward steps in the previous n-1 steps) ·
(the probability that the nth step was forward)
= (term in $p^{i-1} q^{n-i}$ in the binomial expansion of $[p+q]^{n-1}$) · p

$$= \frac{(n-1)!}{(i-1)!(n-i)!} p^i q^{n-i} \qquad\qquad (35.17)$$

The probability that a particle will eventually emerge ($i \leq n \leq \infty$) must be
unity. That is

$$\sum_{n=i}^{\infty} \frac{(n-1)!}{(i-1)!(n-i)!} p^i q^{n-i} = p^i \sum_{n=i}^{\infty} \frac{(n-1)!}{(i-1)!(n-1)!} q^{n-i}$$

should equal unity. Expanding the first few terms of the series

$$\sum_{n=i}^{\infty} \frac{(n-1)!}{(i-1)!(n-i)!} q^{n-i} = 1 + iq + \frac{i(i+1)}{2!} q^2 + \dots$$

Also

$$[1-q]^{-i} = 1 + iq + \frac{i(i+1)}{2!} q^2 + \dots$$

Hence the probability that a particle will ultimately emerge is

$$p^i [1-q]^{-i} = p^i [1 - (1-p)]^{-i} = 1$$

as required.

Curves generated by the probability function (35.17) bear some resemblance to
IDC's.

(ii) Mechanical. A very simple mechanical model which has some of the properties of a labyrinth of tubes may be easily constructed. A number of small
nails are hammered part-way into a smooth board (Fig. 35.7) in a more or less
random fashion, but such that the average distance between adjacent nails does
not change much from one place to another. Such a board is similar but not
necessarily identical to the well-known Galton board, where the nails are
arranged more uniformly. The board is inclined a few degrees to the hori-
zontal such that a marble or ball-bearing dropped at the starting point will
roll in an interrupted course to the finish line. One can time the passage of
a marble from start to finish, and when this process is repeated many times,
one obtains a frequency distribution of transit times.

START

FINISH LINE

Fig. 35.7. Modified Galton board. The board was inclined slightly to the
            horizontal and marbles rolled from the starting point to the fin-
            ish line. The transit times of 150 rolls are recorded in the
            following histogram. Those marbles that rolled off the side were
            discounted.

My wife, Barbara, and I, a few years ago, whiled away an interesting evening
timing the passage of marbles with a stop-watch. The board was first set up
as in Fig. 35.7, with a rectangular area covered by nearly-equally-spaced
nails. The distribution of transit times, as shown in Fig. 35.8, was nearly
symmetrical and hence not a very good representation of the IDC. However,
when a parabolic region of the board in the lower central region was cleared
of nails (Fig. 35.9), so that some marbles could reach the finish line more
directly and quickly, the resulting frequency distribution (Fig. 35.10) took
on more of the skewed appearance of the IDC. Such a board, with a "preferred
channel" in the centre is somewhat analogous to the hydrodynamic system shown
in Fig. 35.11.

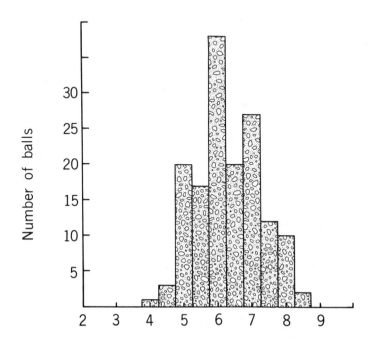

Fig. 35.8.   Histogram of transit times for the Galton board.   The histogram
            is roughly symmetrical, as opposed to an indicator-dilution curve
            which is skewed to the right.

START

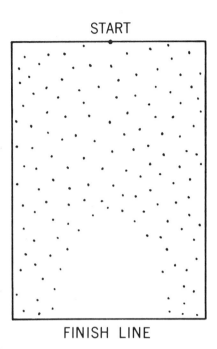

FINISH LINE

Fig. 35.9.   Modified Galton board with short paths permitted. The board was
             inclined slightly to the horizontal and marbles rolled from the
             starting point to the finish line. The transit times of 174 rolls
             are recorded in the following histogram. Those marbles that
             rolled off the sides of the board were discounted.

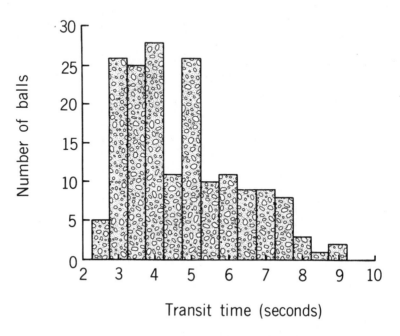

Fig. 35.10. Histogram of transit times for the short-path Galton board.
Notice that the histogram assumes the general shape of an indicator-dilution curve.

START

FINISH

Fig. 35.11. Capillary labyrinth drawn by analogy to the short-path histogram.
            A direct arterio-venous shunt passes through the centre. About
            as much blood travels through the direct shunt as travels through
            the longest path, which is about 4 times longer.

This whole experiment is good fun but should probably not be taken too seriously.

Distributed or Continuous Random Walk Models

Since the molecular basis of diffusion phenomena is the random motion or
Brownian motion of the molecules, models involving diffusion are essentially

random walk models[†]. Perl and Chinard (1968) presented a convective diffusion
model of the transport of indicator through an organ. The indicator was taken
to be one which follows the extravascular fluid (as opposed to one which remains inside the system of blood vessels). They assumed that indicator diffusion perpendicular to the inlet-outlet axis (see Fig. 34.1) is very rapid
in comparison with diffusion along the inlet-outlet axis. After a number of
simplifying assumptions they emerged with the differential equation

$$\frac{\partial C^*}{\partial t} = D \frac{\partial^2 C^*}{\partial x^2} - \frac{QL}{\lambda V} \frac{\partial C^*}{\partial x}$$ (35.18)

where the x-axis is taken along the inlet-outlet axis.

L is the length and V the volume of a unit tissue slab, and λ is a partition
coefficient.

Solutions of (35.18) were shown to be compatible with the observed C^* for a
number of indicators such as ^{14}C-creatine and ^{22}N$_a$. That is, the IDC's produced by injecting a number of extravascular indicators (and even Evans Blue
which is largely confined to the vascular space) could be expressed as solutions to the convective diffusion equation (35.18) with appropriate values
of the constant coefficients.

The authors corrected their experimental IDC's for recirculation of indicator
by means of the standard semilogarithmic plot of the right-hand limb of the
curve; and they corrected their curve for catheter distortion by a technique
which will be discussed generally in the next section.

A somewhat simpler but parallel treatment of the transport of an intravascular
indicator by a diffusion or random walk model was given by Norwich and Zelin

---

[†]The molecular basis of diffusion was first described by Einstein (1905).
The nucleus of his demonstration consists of a single page of equations which
are so clearly and simply set forth that any attempt to simplify it further
is probably futile. The reader is heartily encouraged to read the English
translation of this classical paper.

(1970). Essentially it represents the capillary labyrinth by a uniform tube of cross-sectional area A through which fluid flows at the mean speed u. Hence

$$Q = uA$$

The dispersion of indicator is taken to occur under the influence of a *quasi* diffusion coefficient, D. Thus we can apply the equation of convective diffusion in one dimension,

$$\frac{\partial C^*}{\partial t} = D \frac{\partial^2 C^*}{\partial x^2} - u \frac{\partial C^*}{\partial x} \qquad (35.19)$$

where x is the distance measured along the axis of the tube. Eq. (35.19) is identical in form with (35.18), the velocity u replacing the coefficient $\frac{QL}{\lambda V}$.

While Eq. (35.19) should be perfectly comprehensible to the reader who has followed the derivation given in section 11, it is, nevertheless, instructive to examine it from a slightly different point of view. The *substantial derivative* (10.12) gives the rate of change of a dynamic variable as seen by an observer moving with velocity $\vec{u}$. That is in general

$$\frac{d}{dt} = \frac{\partial}{\partial t} + \vec{u} \cdot \text{grad}$$

and in one dimension

$$\frac{d}{dt} = \frac{\partial}{\partial t} + u \frac{\partial}{\partial x} \qquad (35.20)$$

Making this substitution in (35.19), we see that the rate of change of concentration with time as seen by an observer moving with speed u, the mean speed of fluid in the tube, is given by

$$\frac{dC^*}{dt} = D \frac{\partial^2 C^*}{\partial x^2} \qquad (35.21)$$

Suppose we take a new distance axis, 0'x', which translates along the axis of the tube with speed u. Then, making a transformation of variables (Galilean transformation)

$$x' = x - ut \qquad (35.22)$$

so that when t is held constant

$$\frac{\partial^2 C^*(x',t)}{\partial x'^2} = \frac{\partial^2 C^*(x,t)}{\partial x^2}$$

and (35.21) becomes (since the moving observer will write $\frac{\partial}{\partial t}$ for $\frac{d}{dt}$)

$$\frac{\partial C*}{\partial t} = D \frac{\partial^2 C*}{\partial x'^2} \tag{35.23}$$

for an observer on the moving frame. The effect has been to transform away the convective term; i.e. Eq. (35.23) is just Eq. (35.19) without the convective term.

The solution to (35.19) for a point source of indicator applied as an impulse at the origin (delta function in space and in time, $M_0^* \delta(x)\delta(t)$) in an infinite tube is

$$C*(x,t) = \frac{M_0^*}{\sqrt{4\pi DA^2}} \; t^{-\frac{1}{2}} \; \exp\left[-\frac{(x - ut)^2}{4Dt}\right] \tag{35.24}$$

and the solution to (35.23) for the same input and boundary conditions is

$$C*(x',t) = \frac{M_0^*}{\sqrt{4\pi DA^2}} \; t^{-\frac{1}{2}} \; \exp\left[\frac{-x'^2}{4Dt}\right] \tag{35.25}$$

For a derivation of the point source solution to the diffusion equation, the reader is referred to any text on partial differential equations. To an observer moving with the stream and using the x'-system of coordinates and Eq. (35.25), the dispersion of indicator will appear to be symmetrical about the point of indicator injection (Fig. 35.12). On the other hand, to an observer in the stationary reference frame using the x-system of coordinates and Eq. (35.24), the dispersion of indicator will appear to be non-symmetrical and the IDC will have the characteristic skewed shape (Fig. 35.13). If the observer on the stationary frame measures the concentration, C*, at the point $x = x_0$, then from (35.24)

$$C*(x_0,t) = \frac{M_0^*}{\sqrt{4\pi DA^2}} \; t^{-\frac{1}{2}} \; \exp\left[-\frac{(uA)^2}{4DA^2} \frac{(t - \tau)^2}{t}\right] \tag{35.26}$$

where

$$\tau = x_0/u \tag{35.27}$$

Hence a symmetrical distribution on the moving reference frame gives rise to a non-symmetrical distribution on the stationary reference frame.

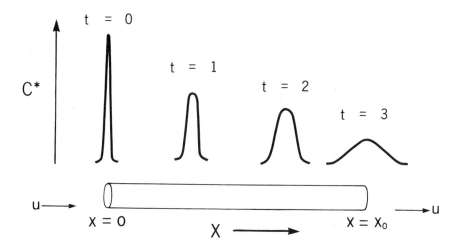

Fig. 35.12.   Dispersion as seen by an observer moving with the speed u. Fluid
               is flowing through a tube of uniform bore at the constant speed
               u. Indicator is injected as an impulse at t = 0 at x = 0. An
               observer moving with the fluid would record a (symmetrical)
               Gaussian curve which was flattening progressively as time in-
               creased and as the bolus of fluid moved to the right (Eq.
               (35.25)).

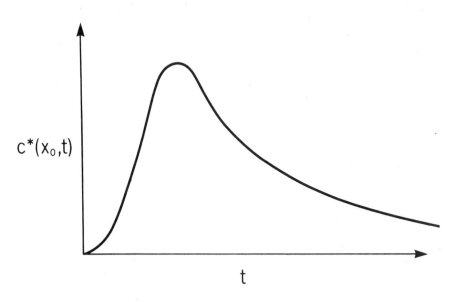

$c^*(x_0, t)$

t

Fig. 35.13. Dispersion as seen by an observer in the stationary frame. This
             observer records concentrations at the point $x = x_0 = u\tau$ and he
             obtains a non-symmetrical curve (Eq. (35.26)).

Experimentally-measured IDC's can easily be fitted to a function of the form
(35.26) by plotting $t^{\frac{1}{2}} C^*$ against t to obtain an estimate of τ ($t^{\frac{1}{2}} C^*$ takes
its maximum value at $t = \tau$), and then plotting the logarithm of $t^{\frac{1}{2}} C^*$ against
$(t-\tau)^2/t$ to obtain an estimate of uA and $A^2 D$. By using only the early portion
of the IDC, that part which extends from the first appearance of the indicator
until the point where recirculation is believed to contribute substantially to
the shape of the curve (that arbitrary decision again), the IDC can be corrected for the effects of recirculation of indicator. These calculations are
all discussed in greater detail in the original paper. In my own work with
indicator-dilution curves I usually correct for recirculation both by the
Hamilton semi-log method (Eq. (35.4)) *and* by the random walk method (Eq.
(35.26)). The former method usually leaves a corrected IDC of greater area
than the latter. I feel that the best correction probably lies between those
extremes, but as expressed earlier, this is hard to prove experimentally.

The effect of random noise in the IDC might be expected to produce a greater error with the semi-log than with the random walk method, since the semi-log method uses only the descending limb of the curve while the random walk method uses both the ascending and descending limbs.

With the random walk model, the steady flow rate Q is still given by the Stewart-Hamilton equation (34.17)

$$Q = \frac{M^\star_0}{\int_0^\infty C^\star dt} \qquad (35.28)$$

with C* evaluated from the right-hand side of (35.26). (See problem 35.4).

One final note before leaving the subject of random walk models: a few remarks on first passage times. The indicator concentration, $C^\star(x,t)$, as given by (35.24) takes into account both indicator particles which have reached the sampling site, $x = x_0$, for the first time as well as those which by virtue of the random motion of the diffusion process have passed the sampling site several times. If the nature of the sampling site were such that no particle could traverse it more than once, we should then be interested in calculating the mass of indicator which crosses the plane $x = x_0$ per unit time for the first time at time t in a system where transport is governed by (35.19). Kennard (1938) shows very clearly that this rate is given by

$$J'A = \frac{x_0 M^\star_0}{\sqrt{4\pi D}}\, t^{-3/2}\, \exp\left[-\frac{(x_0 - ut)^2}{4Dt}\right] \qquad (35.29)$$

where J' is the "first time flux" in units of mass $\cdot$ area^{-1} $\cdot$ time^{-1}. We could now go on to construct a model of the IDC based on this type of boundary condition (e.g. Sheppard, 1962).

Speaking Practically

Models of indicator-dilution curves are useful for correcting these curve for the presence of recirculated indicator. Just about *any* model you might select will give an acceptable correction. In the absence of a model which is clearly superior, you might just as well use the time-honoured Stewart-Hamilton, monoexponential method. Or, as a recent paper (Eichling et al., 1975) recommends, plotting log C* against log t results in a nearly straight line for at least one type of indicator, and the line can easily be extrapolated to correct for recirculation. My own preference is for a random walk function such as (35.26) because it has at least *some* physical basis and has been tested experimentally for a number of indicators. No model-correction is as good as a direct measurement of the recirculated indicator. If you can measure recirculation directly, do so.

Problems

35.1 Show that when (35.10) is used to evaluate $C_o*(t)$, the correct value for flow rate is given by Eq. (34.17), the Stewart-Hamilton equation.

35.2 For the tanks-in-series model, the fraction F of the original indicator mass, $M_1^*(0)$, which has escaped from the system by time t is given (Denbigh, 1965) by

$$F = \frac{M_1^*(0) - (M_1^* + M_2^* + M_3^* + \ldots + M_n^*)}{M_1^*(0)}$$

Taking the mean residence time $\bar{t}$ as

$$\bar{t} = \int_0^1 t dF$$

show that once again

$$V = Q\bar{t}$$

35.3 Explore $C* = J'A/Q$ as a model of the IDC. J'A is given by Eq. (35.29). Show by integrating that the Stewart-Hamilton relationship

$$Q = \frac{M_o^*}{\int_0^\infty C*(t)dt}$$

is obeyed. Was this obvious from the physics of the problem?

35.4 Demonstrate (35.28) as follows.
First, by substituting for C* from (35.26), show that

$$uA = \frac{M_o^*}{\int_0^\infty C*dt}$$

Second, by evaluating the right-hand side of (34.23) using (35.26) show that

$$Q = uA$$

Combining these equations gives (35.28). This result was obtained in another manner in the original paper.

REMOVAL OF CATHETER DISTORTIONS:
NUMERICAL DECONVOLUTION

As we have discussed earlier, it is quite common to sample the fluid in a
flowing stream by withdrawing a small quantity of fluid continuously through a
catheter, which is a long, flexible tube of fine bore. For example, to determine the optical density of indocyanine green in blood flowing through the
ascending aorta, a catheter may be introduced through a small incision made
in the femoral artery, and the tip of the catheter advanced to the aortic
root. Blood is then withdrawn through this catheter by means of a withdrawal
pump, at a rate which is negligible (say 20.ml/min) in comparison with blood
flow in the aorta (say 5000.ml/min), and is led to a cuvette-densitometer system. The signal from the densitometer is often displayed on an oscilloscope,
while a permanent record of the indicator-dilution curve is obtained on a
chart recorder (Fig. 36.1). The problem with a catheter sampling system is
that the indicator concentration-time curve obtained at the cuvette is not
exactly the same as the curve which would have been obtained had the concentration been measured directly in the blood stream. We shall be concerned
largely with indicator-dilution curves (IDC's) obtained downstream after a
sudden injection of indicator upstream. The IDC has a characteristic shape
(illustrated in the two preceding sections). The effect of the sampling catheter is to "smear" the curve; that is the IDC at the cuvette has a peak of
lower amplitude and is more spread out in time than the IDC in the blood
stream (Fig. 36.2). It is also shifted to the right along the time-axis. The
mean residence time, and hence the blood volume, cannot be calculated from the
the catheter-distorted curve, which is our incentive for correcting the curve.

Let us see what can be said about the transport of indicator in a catheter,
without knowing too much about the hydrodynamics. Let us first make the
assumption that the interior surface of the catheter is inert with respect to
the indicator so that no indicator reacts with or adsorbs to a component of

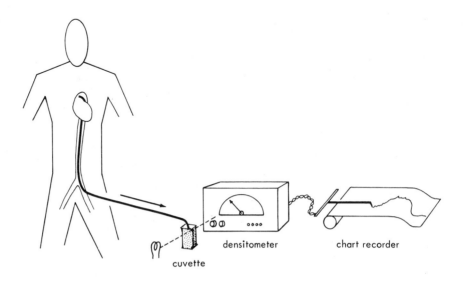

Fig. 36.1. Sampling system for the extracorporeal measurement of optical den-
            sity. A catheter has been inserted through an incision in the
            femoral artery and the tip advanced to the aortic root. The opti-
            cal density is obtained using a cuvette-densitometer system.

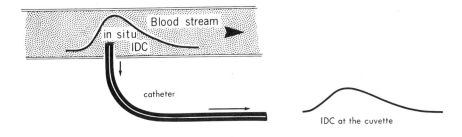

Fig. 36.2. The effect of a sampling catheter on an IDC. The IDC at the
            cuvette is lower and wider than the IDC in the blood stream. It
            is also shifted to the right along the time-axis, but this cannot
            be seen from the diagram.

the catheter wall[†]. Thus all the indicator which enters the catheter will
eventually leave. Suppose we make the additional assumption that transport
of indicator through the catheter is largely convective in nature. Let us retain the symbol $C_o^*(t)$ to represent the actual concentration of indicator in
the outflow channel of the labyrinth following the impulse injection at the

---

[†]This is not as innocent an assumption as it may look. We have observed in
our laboratory how Cardio-Green[(R)] in *water* (not in whole blood) will adhere
to the interior surface of a catheter commonly used in the clinic.

inflow. Hence $C_0*(t)$ also represents the concentration of indicator at the
tip of the sampling catheter. Represent the concentration at the cuvette end
of the sampling catheter by $C_d*(t)$ (d for *distorted*). The mass of indicator,
dM*, which passes through the tip of the catheter in time dt is given by

$$dM* = F\ C_0*(t)dt \qquad (36.1)$$

where F is the steady flow rate through the catheter [volume · time^{-1}]. Hence
the total mass of indicator to pass through the tip of the catheter during
the recording of one IDC which is free of recirculation is given by

$$M^*_{TOT} = F \int_0^\infty C_0*(t)dt \qquad (36.2)$$

for convective flow (cf. (34.18)). Since no mass is lost within the catheter
we must have

$$M^*_{TOT} = F \int_0^\infty C_d*(t)dt \qquad (36.3)$$

Hence

$$\int_0^\infty C_0*(t)dt = \int_0^\infty C_d*(t)dt \qquad (36.4)$$

Eq. (36.4) has an important implication: the distortion due to the sampling
catheter may have no effect on the calculation of blood flow rate by (34.17).
Only the area under the IDC (free of recirculation) is important in the application of (34.17), and the area under the distorted curve, $C_d*(t)$, is equal
to the area under the pure curve, $C_0*(t)$, by (36.4).

Linearity of the Catheter System

The transport of indicators through the catheter could be described rather
generally by a convective diffusion equation of the form

$$\frac{\partial C*}{\partial t} = D(\vec{r})\nabla^2\ C* - \vec{u}\cdot\nabla C* + R_a^*(t)\cdot\delta(\vec{r} - \vec{r}_{in}) \qquad (36.5)$$

where indicator enters at the point designated by position vector, $\vec{r}_{in}$, and
the coefficient of diffusion may vary with position. One might write even
more generally

$$\frac{\partial C*}{\partial t} = (\text{Linear operator}) \cdot C* + R_a^*(t) \cdot \delta(\vec{r} - \vec{r}_{in}) \qquad (36.6)$$

In either of the latter two equations C* enters only to the first power, so
that following the arguments of section 13, equations (13.4) to (13.9), one
can say that the system whose input is the rate of infusion of indicator at
$\vec{r}_{in}$ and whose output is the indicator concentration at the cuvette $C_d*(t)$, is

a linear one. The catheter system then possesses a unit impulse response
function which we may designate by $h_c(t)$. In keeping with the approach initi-
ated earlier, (Eq. (13.12) and Eq. (26.13)), we shall regard $h_c(t)$ as a tracer
concentration per unit mass of tracer injected: units [volume^{-1}]. As in
(26.13)

$$C_d^*(t) = \int_0^t R_a^*(\tau)h_c(t - \tau)d\tau \tag{36.7}$$

But $R_a^*(t)$, the rate at which indicator enters the catheter, is related to the
flow, F, through the catheter and C_0^*, the indicator concentration entering
the catheter, by the equation

$$R_a^*(t) = F \cdot C_0^*(t) \tag{36.8}$$

convection predominating, so that (36.7) becomes

$$C_d^*(t) = \int_0^t C_0^*(\tau)Fh_c(t - \tau)d\tau \tag{36.9}$$

Let us define the function $h_c'(t)$ [time^{-1}] by

$$h_c'(t) = Fh_c(t) \tag{36.10}$$

so that (36.9) may be written simply

$$C_d^*(t) = \int_0^t C_0^*(\tau)h_c'(t - \tau)d\tau \tag{36.11}$$

From Eq. (26.12) we know that

$$\int_0^\infty C_d^*(t)dt = \left[\int_0^\infty C_0^*(t)dt\right]\left[\int_0^\infty h_c'(t)dt\right] \tag{36.12}$$

so that by introducing (36.4) we obtain

$$\int_0^\infty h_c'(t)dt = 1 \tag{36.13}$$

Equivalently, using (36.10),

$$\int_0^\infty h_c(t)dt = 1/F \tag{36.14}$$

Equations (36.13) and (36.14) are valid only when dealing with a stationary,
linear system in which mass is conserved.

Equations (36.11) and (36.13) are the ones which are found useful in removing
catheter distortions. In order to apply these equations we must first be able
to determine $h_c'(t)$ numerically.

320   Indicators and Blood Flow

Experimental Determination of the Unit Impulse Response Function of the Catheter

One might think that the simplest way to obtain $h_c(t)$ is just to apply an impulse at the "vascular" end of the catheter, and record the response at the cuvette. However, such an experiment is somewhat difficult to conduct. It is simpler to measure the *step* response of the system and to make use of the theorem which states that the unit impulse response function is the time derivative of the unit step response function (Cheng, 1959). This may be seen using a fundamental property of the Riemann integral, viz that

$$\frac{d}{dt} \int_a^t f(x)dx = f(t) \qquad (36.15)$$

The step response to the catheter system is given by (36.7) using (12.25) and (12.26):

$$\text{step response} = R_a^* \int_0^t h_c(\tau)d\tau \qquad (36.16)$$

where R_a^* is a constant infusion of indicator beginning at zero time. Differentiating the step response with respect to t using (36.15),

$$\frac{d(\text{step response})}{dt} = R_a^* h_c(t) \qquad (36.17)$$

from which the desired result follows. From (36.10) and (36.17) we see that

$$\frac{d(\text{step response})}{dt} \propto h_c'(t); \qquad (36.18)$$

that is the time derivative of the step response is proportional to the primed form of the unit impulse response function. We shall see that the existence of proportionality is all that matters; we really don't have to know the proportionality constant.

The step response can be measured using the method of Fox et al. (1957), Glassman et al. (1969) or Wong (1972). Wong's apparatus is illustrated in Fig. 36.3. Two open glass cylinders are connected by means of flexible tubing to a three-way stopcock which leads to the "vascular" end of the catheter. The other end of the catheter leads to the cuvette-densitometer system. One cylinder is filled with undyed blood; the other is filled with blood containing an optical indicator. Air bubbles are carefully removed from the tubing. At zero time, the stopcock is turned so that undyed blood flows through the catheter, and a base-line is obtained by the densitometer. After a few seconds the stopcock is suddenly turned so that the flow of undyed blood is cut off, and dyed blood of uniform concentration enters the "vascular" end of the

catheter. The densitometer then records the step response function (not necessarily the unit step function) of the catheter.

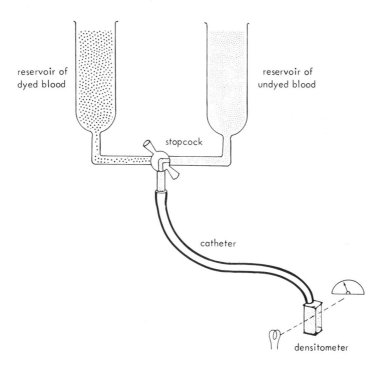

Fig. 36.3. Apparatus for measuring the step response of a sampling catheter
(Wong 1972). Undyed blood flows from the right-hand reservoir
through the sampling catheter to the cuvette-densitometer. At
some instant in time the stopcock is turned, the flow of undyed
blood is stopped suddenly and replaced by the flow of dyed blood
from the left-hand reservoir. The step response is recorded at
the densitometer.

It is really not even necessary to calibrate the densitometer's output, since
only the shape of the step response curve is important. We must, however,
have the optical density plotted on a linear, rather than a logarithmic scale.
The curve is smoothed, if necessary, and the graph of its derivative against
time is obtained numerically. From (36.18) we see that the amplitude of the
derivative is proportional to $h_c'(t)$. The area contained under the curve

d(step response)/dt versus t, is then calculated, and the ordinate of each point on the curve is divided by the area. The result is a curve whose ordinate is given by area^{-1} · d(step response)/dt; the area under this curve has then been normalized to unity. Suppose we write (36.18) in the form

$$k \cdot \frac{d(step\ response)}{dt} = h_c'(t) \qquad\qquad (36.19)$$

where k is the proportionality constant.   Then

$$k \cdot area^{-1} \cdot \frac{d(step\ response)}{dt} = area^{-1} \cdot h_c'(t)$$

and

$$k \int_0^\infty area^{-1} \cdot \frac{d(step\ response)}{dt} \cdot dt = area^{-1} \int_0^\infty h_c'(t)dt$$

$$k \cdot 1 = area^{-1} \int_0^\infty h_c'(t)dt \qquad\qquad (36.20)$$

Using (36.13) we can see that

$$k = area^{-1} \qquad\qquad (36.21)$$

so that from (36.19)

$$h_c'(t) = area^{-1} \cdot \frac{d(step\ response)}{dt} \qquad\qquad (36.22)$$

Thus the normalized curve, in fact, defines the function $h_c'(t)$ numerically.

Deconvolution

In equation (36.11), two of the three functions are known in numerical or tabular form:

$C_d*(t)$, the measured (catheter-distorted) IDC, and

$h_c'(t)$, the normalized impulse reponse function.

The problem is then to solve the integral equation (36.11) numerically for

$C_o*(t)$, the undistorted IDC.

Because (36.11) expresses the convolution of C_o* with h_c', the solution for C_o* is sometimes called *deconvolution* (although I am never sure whether there exists an infinitive "to deconvolve"). There is little doubt that a book the size of this one could be written purely on the subject of deconvolution. There is also little doubt in my mind that I shall not be the one to write it. A pragmatic, two-statement summary of the state of the art of deconvolution would go something like this:

(1) You can always deconvolve if you spend enough time at it.

(2) Deconvolution to correct for catheter distortion can be carried out routinely — to a good approximation.

Broadly speaking, there are two approaches to the problem of deconvolution: analog techniques and digital techniques. In the former an analog computer is used, and in the latter a digital computer. Scores of papers have been written to describe various analog circuits which have been used successfully in the solution of this problem. But analog computers are not suitable for routine deconvolution in hospital or clinic, and so we shall not review these papers here. There are, however, two fine papers which provide three general approaches to deconvolution using the analog computer, and I recommend them to the interested reader: Parrish, Gibbons and Bell, 1962; Goresky and Silverman, 1964. Digital computers are more readily accessible and more suitable for routine deconvolution, so let us focus our attention upon digital techniques.

The most obvious digital approach to the solution of (36.11) seems to be to convert the integral to a summation (Eq. (12.33)):

$$C_d^*(t) = \sum_{\tau=0}^{t} C_o^*(\tau) h_c'(t - \tau) \Delta t$$

Let us take

$$\tau = n \cdot \Delta t, \quad n = 0, 1, 2, \ldots t/\Delta t$$

so that

$$C_d^*(t) = \sum_{n=0}^{t/\Delta t} C_o^*(n \cdot \Delta t) h_c'(t - n \cdot \Delta t) \cdot \Delta t \qquad (36.23)$$

We can now evaluate $C_d^*(t)$ for any t (greater than zero):

From the initial conditions or (36.11)

$$C_d^*(0) = 0 \qquad (36.24)$$

and

$$C_o^*(0) = 0 \qquad (36.25)$$

From (36.23) for $t = \Delta t$

$$C_d^*(\Delta t) = \sum_{n=0}^{1} C_o^*(n \cdot \Delta t) h_c'(\Delta t - n \cdot \Delta t) \Delta t$$

$$= C_o^*(0) h_c'(\Delta t) \Delta t + C_o^*(\Delta t) h_c'(0) \Delta t$$

or

$$C_d^*(\Delta t) = C_0^*(\Delta t)h_c'(0)\Delta t \qquad (36.26)$$

because of (36.25).   For $t = 2\Delta t$

$$C_d^*(2\Delta t) = \sum_{n=0}^{2} C_0^*(n \cdot \Delta t)h_c'(2\Delta t - n \cdot \Delta t)\Delta t$$

$$C_d^*(2\Delta t) = C_0^*(\Delta t)h_c'(\Delta t)\Delta t + C_0^*(2\Delta t)h_c'(0)\Delta t \qquad (36.27)$$

etc.

Representing $\dfrac{C_d^*(i \cdot \Delta t)}{\Delta t}$, $C_0^*(j \cdot \Delta t)$ and $h_c'(k \cdot \Delta t)$ by C_{di}^*, C_{0j}^* and h_{ck}' respectively, we have

$$C_{d1}^* = C_{01}^* \, h_{c0}'$$

$$C_{d2}^* = C_{01}^* \, h_{c1}' + C_{02}^* \, h_{c0}'$$

$$C_{d3}^* = C_{01}^* \, h_{c2}' + C_{02}^* \, h_{c1}' + C_{03}^* \, h_{c0}'$$

$$\cdot \qquad \cdot \qquad \cdot \qquad \cdot$$

or

$$\begin{pmatrix} C_{d1}^* \\ C_{d2}^* \\ C_{d3}^* \\ \cdot \\ \cdot \\ \cdot \end{pmatrix} = \begin{pmatrix} h_{c0}' & 0 & 0 & \cdots \\ h_{c1}' & h_{c0}' & 0 & \cdots \\ h_{c2}' & h_{c1}' & h_{c0}' & \cdots \\ \cdot & \cdot & \cdot & \cdot \end{pmatrix} \begin{pmatrix} C_{01}^* \\ C_{02}^* \\ C_{03}^* \\ \cdot \\ \cdot \\ \cdot \end{pmatrix} \qquad (36.28)$$

Since the C_d^* - vector is known from the measured IDC and the h_c' - matrix is known from the measured step response of the catheter, it should be a simple matter to invert the triangular matrix to obtain the unknown C_0^* - vector:

$$(C_0^*) = (h_c')^{-1}(C_d^*) \qquad (36.29)$$

This method works well — with simulated, noise-free data. When using real data which contain noise it is my experience that (36.29) does not give a satisfactory solution: the C_0^* - vector bears little resemblance to its expected form and may even contain negative elements. By all means try it for yourself, because the alternatives to this method are harder by several degrees.

What digital methods, then, work well? In my experience, at least four.

(1) Deconvolution by Multiexponential Curve-Fitting

Suppose that both the distorted curve, C_d^*, and the catheter impulse response function, h_c' were fitted to multiexponential functions. That is

$$C_d^*(t) = \sum_{i=1}^{m} A_i \, e^{-\beta_i t} \qquad (36.30)$$

and

$$h_c'(t) = \sum_{i=1}^{p} B_i \, e^{-\gamma_i t} \qquad (36.31)$$

Taking the Laplace transformation of both sides of (36.11) (see Eq. (12.36)),

$$C_d^*(s) = C_0^*(s) \cdot h_c'(s)$$

or

$$C_0^*(s) = C_d^*(s)/h_c'(s) \qquad (36.32)$$

From (36.30) and (36.31), the Laplace transforms of $C_d^*(t)$ and $h_c'(t)$ are respectively

$$C_d^*(s) = \sum_{i=1}^{m} \frac{A_i}{s + \beta_i} \qquad (36.33)$$

and

$$h_c'(s) = \sum_{i=1}^{p} \frac{B_i}{s + \gamma_i} \qquad (36.34)$$

Combining the latter three equations

$$C_0^*(s) = \frac{\sum_{i=1}^{m} \dfrac{A_i}{s + \beta_i}}{\sum_{i=1}^{p} \dfrac{B_i}{s + \gamma_i}} \qquad (36.35)$$

The right-hand side of (36.35) may then be simplified, a process usually involving decomposition of the expression into partial fractions, and the inverse Laplace transform found. This technique for deconvolution has been used successfully by several authors (e.g. Wraight, 1969; Norwich, 1972) but not usually to remove catheter distortion.

There are several disadvantages to this method: curve-fitting to an "optimal" number of exponential functions is a cumbersome process. It is difficult to conform a typical IDC to a sum of exponentials, a minimum of four or five

being necessary in our experience. And the algebraic labour involved in simplifying the right-hand side of (36.35) is considerable.

(2) Deconvolution by Polynomial Curve-Fitting

This method is similar in principle to the above. Instead of fitting to an
optimal number of exponentials one fits to an optimal number of polynomials of
optimal order. Moreover, the various polynomial functions and their first
derivatives are continuous at the points of meeting. It sounds much more difficult than the previous method; and it is. What makes this technique worth
considering is that a fairly general computer program for its implementation
has already been written in Fortran IV by J. Radziuk (1974), and will, hope-
fully, be made available by its author upon request. This method has been
used to correct IDC's for catheter distortion (Norwich et al., 1974).

(3) Deconvolution by Fourier Transformation

Use of the Fourier transformation to deconvolve seems to have a number of pro-
ponents. The methods of Phillips (1962), Twomey (1965) and Hunt (1970) were
combined and modified by Wong (1972), who used the Fast Fourier Transformation
as a means of removing the distortion from IDC's. A review of the relevant
papers can be found in Wong's work. Fourier transform methods have been applied in other investigations of the circulation involving convolution (Coulam
et al., 1966) and in determining the response of an oxygen electrode (Ellis et
al., 1974). We have used Fourier transform methods successfully in our studies
of the circulation. However, we find that the deconvolution proceeds somewhat
more routinely and with lower computer costs when we use the "hill-climbing"
approach to be described next.

(4) Deconvolution by Hill-Climbing

Hill-climbing is the term applied rather loosely to the process of searching
for the optimal value of a function using a digital computer. The particular
hill-climber we are concerned with utilizes the principle that both the distorted and the corrected IDC are well-fitted by a function of the form (35.26)

$$C*(t) = \frac{M_0{}^*}{\sqrt{4\pi DA^2}}\, t^{-\frac{1}{2}}\, \exp\left[\frac{-Q^2}{4DA^2}\,\frac{(t-\tau)^2}{t}\right] \qquad (36.36)$$

where we have set

$$Q = uA$$

This function, although it contains only three parameters which can be estimated from the data, does conform fairly well to the measured-distorted IDC up to the point where recirculation is believed to begin. It is perhaps reasonable to assume, then, that with appropriate re-evaluation of the three parameters, it would also describe the original or pure IDC. This assumption is, in fact, born out.

The hill-climbing process can be carried out by a method similar to that described in Appendix B. Let us represent the three parameters of (36.36) which characterize the measured-distorted curve by τ^d, A^2D^d and Q^d, and the three parameters of (36.36) which characterize the original-correct curve by τ^c, A^2D^c and Q^c. We are assured by (36.4) and (34.17) that

$$Q^c = Q^d \tag{36.37}$$

By the process described in section 35 we can fit the measured curve to function (36.36) and hence determine τ^d, A^2D^d, and Q^d.

The hill-climbing or searching process begins by guessing nominal values for τ^c and A^2D^c. A *trial value* of C_o^* is then constructed by introducing these values of τ^c and A^2D^c in (36.36). This trial value of C_o^* is then convolved with the known function h_c' to give a trial match for C_d^*. The correspondence of the trial C_d^* with the measured C_d^* is determined by the square of the area contained between the curves: the smaller the area, the better the correspondence. Slightly different values for τ^c and A^2D^c are then selected, and a new trial function C_o^* is constructed. The new trial function is convolved with h_c' to give a new trial C_d^*, whose correspondence with the measured C_d^* is then determined. The better the correspondence between the trial C_d^* and the measured C_d^*, the closer are the values of τ^c and A^2D^c to optimal. In this way the search continues for the best values of τ^c and A^2D^c, and hence for the "true" deconvoluted or pure $C_o^*(t)$

This method, based on a simple hill-climbing procedure suggested by Hazelrig et al., leads to a rapid convergence upon the deconvoluted curve. It is described in some detail by Norwich et al. (1974), and is being applied quite routinely in the analysis of IDC's made at the Toronto General Hospital. We find this approach useful because it is simple, can be handled completely by the technical staff, and seldom goes awry. However, it is not a general method for deconvolution, but rather one which is confined to removal of catheter distortions.

A fifth digital technique for correcting IDC's has been proposed by Oleksy et al., (1969) based upon a tanks-in-series model of the IDC. There is every reason to believe that it will function as accurately and reliably as any of the four preceding methods.

Speaking Practically

The usual reason why people want to correct their curves for catheter distortion is that they wish to measure the blood volume of a labyrinth. If you have some facility with methods of numerical analysis (not much is needed) and have access to a digital computer, you can correct indicator-dilution curves for the distorting effects of catheter sampling by a simple, straightforward method as described in the text above. Once corrected, $\bar{t}$ can be obtained in the usual manner, and volume obtained as the product of Q with $\bar{t}$. However, if you are content with a first approximation of volume, you can proceed as follows. Measure the impulse response function of the catheter and record the time interval between the instant of injection and the arrival of the peak of the response curve. Call this time interval $\bar{t}_{cath}$ (an approximation). Obtain a standard indicator-dilution curve using this catheter and measure the time interval between the instant of injection and the arrival of the peak of the IDC. Call this time interval $\bar{t}_d$. Then

$$\bar{t} \doteq \bar{t}_d - \bar{t}_{cath}$$

SECTION 37

INDICATOR-DILUTION TECHNIQUE IN NONSTEADY FLOW

We have discussed the theory of measurement of steady flow rates by indicator-dilution techniques in section 34. The use of the Fick (steady indicator infusion) and Stewart-Hamilton (indicator injection) methods for the measurement of steady flow rates has now become quite commonplace. However, to my knowledge, no papers have yet emerged dealing with the use of indicators in the

resolution of nonsteady flow rates. In order to resolve the pulsatile nature
of the flow of blood in arteries, people have turned to cineangiography,
electromagnetic flowmeters, Doppler flowmeters and various correlating time-
variables such as aortic pressure. It is, perhaps, reasonable to conclude our
discussion of indicators with a brief exploration of the potential of indicators in this area.

It was observed in section 34 that the steady indicator infusion equation[†]

$$Q = R_a^*/C_{out}^* \tag{34.11}$$

is analogous to the steady tracer infusion equation

$$R_a = R_a^*/\alpha \tag{25.6}$$

It was also observed that the Stewart-Hamilton equation

$$Q = \frac{M_0 ^*}{\displaystyle\int_0^\infty C_0^*(t)dt} \tag{34.17}$$

is analogous to the tracer injection equation

$$R_a = \frac{M_0 ^*C}{\displaystyle\int_0^\infty C^*(t)dt}$$

One cannot but speculate, then, that there exist an indicator equation,

$$Q(t) = R_a^*(t)/C_0^* \tag{37.1}$$

for determining the nonsteady, pulsatile flow rate $Q(t)$, analogous to the
tracer equation

$$R_a(t) = R_a^*(t)/\alpha \tag{32.2}$$

for determining nonsteady rate of appearance, $R_a(t)$. Eq. (37.1) expresses
such a simple relationship that one would expect a very simple demonstration
to exist — perhaps along thermodynamic lines. However, I have not been able

---

[†]It was necessary, in section 34 to distinguish between the step response at
the outlet, C_{out}^*, and the impulse response at the outlet, C_0^*. In this section, all outlet concentrations will be represented simply by C_0^*.

to find a demonstration which is really simple and still completely convincing.
So until one emerges, I offer the following proof for (37.1).

Consider a system of exactly the same geometry and boundary conditions as the
one described in section 34 and depicted in Fig. 34.1. However, instead of
restricting ourselves to steady flow, we shall permit the flow to vary with
time. Let us represent the inflow by the time-varying function $Q_i(t)$, and the
outflow by the time-varying function $Q_o(t)$. The labyrinth together with its
inlet and outlet may be elastic so that the total volume of the labyrinth may
vary with time. We shall assume the flow to be pulsatile with period p, and
we shall assume, moreover, the existence of a *cyclical steady state for fluid*
as defined by the relationship

$$\int_p Q_i(t)dt = \int_p Q_o(t)dt \qquad (37.2)$$

where $\int_p$ implies the integral over one period. Finally, we shall assume the

existence of a cyclical *steady state for indicator*, as expressed by the
equation

$$\int_p C_i^*(t)Q_i(t)dt = \int_p C_o^*(t)Q_o(t)dt \qquad (37.3)$$

This equation expresses a balance of indicator in the system; as much indicator enters in one period as leaves. Conservation of fluid volume is expressed by the equation (e.g. Slattery, 1972)

$$\int_S \vec{u} \cdot \vec{n} \, dS = Q_i(t) - Q_o(t) \qquad (37.4)$$

The surface of the labyrinth is, in general, expanding and contracting with
each pulse of fluid. Rather than infusing indicator at a constant rate as we
did in section 34, we now infuse indicator at the time-varying rate $R_a^*(t)$.
We must again assume that indicator distributes itself rapidly over a cross-
section of the inlet, so that there exists a uniform indicator concentration,
$C_i^*(t)$, over the section and

$$R_a^*(t) = C_i^*(t) \cdot Q_i(t) \qquad (37.5)$$

Similarly, the total rate at which indicator leaves the system is given by
$R_d^*(t)$, where

$$R_d^*(t) = C_o^*(t) \cdot Q_o(t) \qquad (37.6)$$

The equation of convective diffusion for the nonsteady system is then

$$\frac{\partial C^*}{\partial t} + \nabla \cdot (C^* \vec{u}) - D\nabla^2 C^* = C_i^*(t) \cdot Q_i(t) \cdot \delta(\vec{r} - \vec{r}_{in}) - C_o^*(t) \cdot Q_o(t) \cdot \delta(\vec{r} - \vec{r}_{out}) \quad (37.7)$$

which upon integration over the interior volume of the labyrinth becomes

$$\int_V \frac{\partial C^*}{\partial t} dV + \int_V \nabla \cdot (C^* \vec{u} - D\nabla C^*) dV = C_i^*(t) \cdot Q_i(t) - C_o^*(t) \cdot Q_o(t)$$

Applying the divergence theorem with D constant throughout

$$\int_V \frac{\partial C^*}{\partial t} dV + \int_S (C^* \vec{u} - D\nabla C^*) \cdot \vec{n} \, dS = C_i^*(t) \cdot Q_i(t) - C_o^*(t) \cdot Q_o(t) \quad (37.8)$$

Let us now try as a solution to (37.8)

$$C^* = \overline{C_o^*} = \text{a constant in space and in time} \quad (37.9)$$

That is

$$\frac{\partial \overline{C_o^*}}{\partial t} = 0$$

and

$$\nabla \overline{C_o^*} = 0$$

Let the rate of infusion of indicator associated with the solution (37.9) be $R_{ao}^*(t)$. Substituting for C^* in (37.8) from (37.9) we obtain

$$\int_S \vec{u} \cdot \vec{n} \, dS = Q_i(t) - Q_o(t)$$

which just retrieves the expression (37.4) for conservation of fluid. Therefore

$$C^* = \overline{C_o^*} = \text{a constant}$$

is a valid solution to (37.8). The result which we are seeking, (37.1), follows directly from two simple lemmata.

Lemma 1. *In the cyclical steady state, no constant concentration of indicator at the outlet except $\overline{C_o^*}$ is possible in response to the indicator infusion rate, $R_{ao}^*(t)$.*

Proof: We have already seen that an outlet concentration, $\overline{C_o^*}$, is possible in response to an infusion rate, $R_{ao}^*(t)$. Therefore, in a cyclical steady state, an outlet concentration of $\overline{C_o^*}$ will conserve tracer mass in the system: as much indicator leaves as enters during one cycle. Since the outflow rate, $Q_o(t)$, is fixed by the system (not affected by the investigator), an increase or decrease in the outlet concentration above or below the value $\overline{C_o^*}$ will remove too much or too little indicator to maintain a cyclical steady state in indicator. Therefore $\overline{C_0}^*$ is the only possible constant outlet concentration.

A mathematical derivation of this lemma is offered in the appendix to the associated paper (Norwich, 1975).

<u>Lemma 2</u>. *In the cyclical steady state, no constant concentration of indicator at the <u>inlet</u> except $\overline{C_0}^*$ is possible in response to the indicator infusion rate $R_{ao}^*(t)$.*

Proof: Substituting $R_a^*(t) = R_{ao}^*(t)$ in (37.5)

$$C_i^*(t) = R_{ao}^*(t)/Q_i(t) \tag{37.10}$$

The inflow rate, $Q_i(t)$, is fixed by the system (i.e. not affected by the investigator), so that $C_i^*(t)$ is determined solely by $R_{ao}^*(t)$. But we have already seen that $C_i^*(t) = \overline{C_0}^*$ is a solution corresponding to $R_a^*(t) = R_{ao}^*(t)$. Therefore $\overline{C_0}^*$ is the only possible inlet concentration and

$$\overline{C_0}^* = R_{ao}^*(t)/Q_i(t) \tag{37.11}$$

Eq. (37.1) may now be readily demonstrated. Let us express it in the form of a theorem: <u>Theorem</u>. *If a system is in cyclical steady state with respect to both flow and indicator, and indicator is infused at the inlet at a time-varying rate, $R_{ao}^*(t)$, such that the concentration of indicator at the outlet assumes a constant value $\overline{C_0}^*$, then the time-varying flow at the inlet is given by*

$$Q_i(t) = R_{ao}^*(t)/\overline{C_0}^* \tag{37.11}$$

Proof: From Lemma 1, $\overline{C_0}^*$ is the only possible outlet concentration corresponding to the indicator infusion rate $R_{ao}^*(t)$.

From Lemma 2, then, (37.11) emerges as the more precise form of (37.1).

Thus the nonsteady flow rate is obtained by infusing indicator at the inlet at such a rate that the concentration at the outlet remains constant.

I sincerely encourage the reader to look for a simpler and more general demonstration.

For a discussion of the problems anticipated in the experimental implementation of (37.11) I refer you to the original paper on the subject. [For example, why bother with outlet concentrations at all? Why not infuse indicator
at any rate, $R_a^*(t)$, and calculate $Q_i(t)$ by directly measuring $C_i^*(t)$ and applying (37.5) to give

$$Q_i(t) = R_a^*(t)/C_i^*(t) \ ?]$$

At the time of writing, this method had not been tested experimentally.

References

Aaron, A. and Fine, S., Simulation and analysis of indicator dilution methods for measuring mean flow from a pulsatile ventricle, *Proceedings of the 28th Annual Conference on Engineering in Medicine and Biology*, 232 (1975).

Benekin, J.E.W. and Rideout, V.C., The use of multiple models in cardiovascular system studies: transport and perturbation methods, *IEEE Transactions on Bio-Medical Engineering* BME-15, 281-289 (1968).

Cheng, D.K. *Analysis of Linear Systems*, Chapter 8, Addison Wesley, Reading, Mass., 1959.

Coulam, C.M., Warner, H.R., Wood, E.H. and Bassingthwaighte, J.B., Transfer function analysis of coronary and renal circulation calculated from upstream and downstream indicator-dilution curves, *Circ. Res.* 19, 879 (1966).

Cropp, G.J.A. and Burton, A.C., Theoretical considerations and model experiments on the validity of indicator dilution methods for measurements of variable flow, *Circ. Res.* 18, 26-48 (1966).

Denbigh, K.G., *Chemical Reactor Theory: An Introduction*, Cambridge University Press, 1965.

Eichling, J.O., Raichle, M.E., Grubb, R.L., Larson, K.B. and Ter-Pogossian, M., In vivo determination of cerebral blood volume with radioactive oxygen-15 in the monkey, *Circ. Res.* 37, 707-714 (1975).

Einstein, A., *Investigations on the Theory of Brownian Movement*, Dover, New York, 1956.

Ellis, C.G., Goldstick, T.K., Caprini, J.A. and Zuckerman, L., Analysis of renal oxygen transport using an oxygen impulse, *Proceedings of the 5th Canadian Medical and Biological Engineering Conf.*, 12.5 (1974).

Fox, I.J., Sutterer, W.F. and Wood, E.H., Dynamic response characteristics of systems for continuous recording of concentration changes in a flowing liquid (for example, indicator-dilution curves), *J. Appl. Physiol.* 11, 390-404 (1957).

Glassman, E., Blesser, W. and Mitzner, W., Correction of distortion in dye dilution curves due to sampling systems, *Cardiovasc. Res.* 3, 92-99 (1969).

Goresky, C.A. and Silverman, M., Effect of correction of catheter distortion on calculated liver sinusoidal volumes, *Am. J. Physiol.* 207, 883-892 (1964).

Hamilton, W.F., Moore, J.W., Kinsman, J.M. and Spurling, R.G., Studies on the circulation: IV. Further analysis of the injection method, and of changes in hemodynamics under physiological and pathological conditions, *Am. J. Physiol.* 99, 534-551 (1932).

Hazelrig, J.B., Ackerman, E. and Rosevear, J.W., An iterative technique for conforming mathematical models to biomedical data, *Proceedings of the 16th Annual Conference on Engineering in Medicine and Biology* 5, 8-9 (1963).

Hunt, B.R., The inverse problem of radiography, *Math. Biosci.* 8, 161-179 (1970).

Kennard, E.H., *Kinetic Theory of Gases*, Sec. 164, McGraw Hill, 1938.

Levenspiel, O., *Chemical Reaction Engineering*, Wiley, 1962.

Norwich, K.H. and Zelin, S., The dispersion of indicator in the cardio-pulmonary system, *Bull. Math. Biophys.* 32, 25-41 (1970).

Norwich, K.H., Rates of plasma protein synthesis by deconvolution, *Biochem. J.* 126, 1124-1126 (1972).

Norwich, K.H., Pinto, C., Morch, J.E. and Zelin, S., A practical method for removing catheter distortions from indicator-dilution curves, *Cardiovasc. Res.* 8, 430-438 (1974).

Norwich, K.H., Determination of pulsatile flow rates by indicator-dilution methods, *J. Theor. Biol.* 50, 353-361 (1975).

Oleksy, S.J., Weinstein, H. and Shaffer, A.B., Correction of dye-curve distortion with the use of a perfect mixers in series model, *J. Appl. Physiol.* 26, 227-232 (1969).

Parrish, D., Gibbons, G.E. and Bell, J.W., A method for reducing the distortions produced by catheter sampling systems, *J. Appl. Physiol.* 17, 369-371 (1962).

Perl, W. and Chinard, F.P., A convection-diffusion model of indicator transport through an organ, *Circ. Res.* 22, 273-298 (1968).

Perl, W., Stimulus-response method for flows and volumes in slightly perturbed constant parameter systems, *Bull. Math. Biophys.* 33, 225-233 (1971).

Phillips, D.L., A technique for the numerical solution of certain integral equations of the first kind, *J. Assoc. Comp. Mach.* 9, 84-97 (1962).

Radziuk, J.M., Ph.D. Thesis, Department of Physics, University of Toronto 1974.

Schrödinger, E., Physik. Zeits. 16, 289 (1915).

Sheppard, C.W., *Basic Principles of the Tracer Method*, Wiley, 1962.

Slattery, J.C., *Momentum, Energy and Mass Transfer in Continua*, pp. 479 and 664, McGraw Hill, 1972.

Stephenson, J.L., Theory of measurement of blood flow by dye dilution technique, *IRE Transactions on Medical Electronics* PGME-12, 82-88 (1958).

Stewart, G.N., Researches on the circulation time and on the influences which affect it: IV. The output of the heart, *J. Physiol. London* 22, 159 (1897).

Taylor, G., Dispersion of soluble matter in solvent flowing slowly through a tube, *Proc. Roy. Soc. (London)* A, 219, 186-203 (1953).

Taylor, G., The dispersion of matter in turbulent flow through a pipe, *Proc. Roy. Soc. (London)* A, 223, 446-468 (1954).

Twomey, S., The application of numerical filtering to the solution of integral equations encountered in indirect sensing measurements, *Journal of the Franklin Institute* 279, 95-109 (1965).

Wise, M.E., Tracer dilution curves in cardiology and random walk and lognormal distributions, *Acta Physiol. Pharmacol.* Neerl 14, 175-204 (1966).

Wong, W. K-K., M.A.Sc. Thesis, Department of Electrical Engineering, University of Toronto, 1972.

Wraight, E.P., The place of deconvolution analysis in plasma protein turnover studies, *Phys. Med. Biol.* 14, 463-470 (1969).

APPENDIX A

SOLVING DIFFERENTIAL EQUATIONS BY THE METHOD OF FINITE ELEMENTS

This method for the numerical solution of differential equations is sometimes referred to as "Euler's method". It is not a very sophisticated method. In fact it is the infant of the species of numerical solutions. But it is commended by its stark simplicity and the ease with which it may be implemented on a digital computer. Time and again I have been impressed by the degree of accuracy obtained by the method of finite elements with an inconsequential expenditure of computer time. This method certainly seems adequate in the "systems" approach to biology where accuracy beyond the third digit remains a dream.

The technique is easy to explain, especially for the first degree, ordinary differential equations — which are the type usually encountered in this book. If the differential equation can be expressed in the form

$$\frac{dy}{dx} = f(x,y) \tag{A.1}$$

then, representing the differentials dx and dy by their finite elements Δx and Δy respectively,

$$\Delta y = f(x,y)\Delta x . \tag{A.2}$$

Suppose we are given the initial condition

$$f_0 = f(x_0,y_0) \tag{A.3}$$

Suppose moreover, that we adopt some "small" value for Δx: Then from (A.2)

$$\Delta y = f_0 \Delta x \tag{A.4}$$

And the value of y corresponding to $x_0 + \Delta x$ is

$$y_1 = y_0 + \Delta y = y_0 + f_0 \Delta x . \tag{A.5}$$

Now we set

$$x_1 = x_0 + \Delta x \tag{A.6}$$

and repeat the process:

$$\Delta y = f_1 \Delta x \qquad\qquad (A.7)$$

$$y_2 = y_1 + \Delta y$$

$$x_2 = x_1 + \Delta x$$

etc. By repeating this process often enough we can evaluate $f(x,y)$ for any value of its arguments. It is clear, however, that (A.4) is an approximation and hence (A.5) gives an approximate value for y_1. The error in y_1 is propagated through (A.7) etc., so that the estimated value for $f(x,y)$ may diverge from the "true" value. If Δx is kept small enough, the divergence may not be great. Therefore small increments Δx and Δy and numerous iterations on the computer are desiderata.

Example

Solve $\dfrac{dy}{dx} = xy$ given: $y = 1$ when $x = 1$ $\qquad\qquad$ (A.8)

Suppose we take $\Delta x = .25$. Following (A.2) we write

$$\Delta y = xy\ \Delta x \qquad\qquad (A.9)$$

The first calculation gives

$$\Delta y = (1)(1)(.25) = .25$$

so that

$$y_1 = 1 + .25 = 1.25$$

$$x_1 = 1 + .25 = 1.25$$

etc. The process can be summarized in tabular form:

k	x_k	y_k	$x_k y_k \Delta x$
0	1.00	1.00	0.250
1	1.25	1.25	0.391
2	1.50	1.64	0.615
3	1.75	2.26	0.999
4	2.00	3.26	1.13
5	2.25	4.39	2.46
6	2.50	6.85	4.28
7	2.75	11.13	7.65
8	3.00	18.75	

From the final row of the Table we are led to believe that y = 18.8 when
x = 3.

In this particular example, the differential equation may be easily solved by
analytical means, and so we can check the accuracy of the numerical technique.
Separating the variables,

$$\frac{dy}{y} = xdx$$

$$\log y - \log y_0 = \frac{1}{2}(x^2 - x_0^2)$$

$$\log y = \frac{1}{2}(x^2 - 1)$$

$$y = e^{\frac{1}{2}(x^2 - 1)} \qquad \qquad \text{(A.10)}$$

Evaluating y(3) from (A.10) gives

$$y(3) = 54.598 \qquad (?) \qquad \qquad \text{(A.11)}$$

The method of finite elements has given us 18.8, which is barely one-third of
the correct value of 54.6. We have been caught in the quagmire of finite in-
finitesimals.

Let us begin again, this time taking Δx = .001 rather than .25. Calculating
by hand or slide rule is now out of the question, so let us write a brief
program in Fortran IV:

```
C     SOLVE DY/DX = XY : (X = 1, Y = 1)
C     USE FORWARD DIFFERENCES
      X = 1.0
      Y = 1.0
      DELX = 0.001
2     DELY = X*Y*DELX
      Y = Y + DELY
      X = X + DELX
      IF (X-3.)2,3,3
3     PRINT, X,Y
      STOP
      END
```

The output obtained from this simple program is:

$$3.00 \qquad 0.5455...E\ 02 \qquad .$$

That is, $y(3) = 54.55$, which gives an error of .05 (see (A.11)). If Δt can be reduced by a factor of 10, without affecting $y(t)$ appreciably, then one has probably made a reasonable estimate.

The method of finite elements is also of use in solving sets of simultaneous differential equations as seen in section 31. Suppose we wish to solve the following set of simultaneous, linear differential equations:

$$\frac{du}{dt} = a_1 u + a_2 v + w(t) \tag{A.12}$$

$$\frac{dv}{dt} = b_1 u + b_2 v \tag{A.13}$$

where $w(t)$ is given and a_1, a_2, b_1, b_2, $u(0)$, $v(0)$ are known. The corresponding finite element equations are

$$\Delta u = (a_1 \Delta t)u + (a_2 \Delta t)v + w(t)\Delta t \tag{A.14}$$

$$\Delta v = (b_1 \Delta t)u + (b_2 \Delta t)v \quad . \tag{A.15}$$

By selecting a "small" value for Δt, and starting with the known values $u(0)$, $v(0)$ we can solve for $u(t)$, $v(t)$ as before.

The finite element method converts a differential equation (like (A.8)) to an algebraic equation (like (A.9)), and as such can be used not only for solving or integrating the differential equation but also for estimation of parameters (e.g. sections 30 and 31). Put quite simply, *if the various coefficients of the differential equation are known, we can then integrate or solve it. If the solutions are known (i.e. experimental values have been measured), then we can estimate parameters.* As an example of parameter estimation, consider the problem of section (30):

$$R_a(t) = \frac{R_a^\star(t)}{a(t)} - \frac{pVC}{a(t)} \frac{da}{dt} \tag{30.7}$$

The variable a is obtained as a discrete set of measured values. By a process of interpolation, curve fitting, or manual smoothing the function $a(t)$ is obtained in tabular form for any desired "granularity" — i.e. any desired increment in t: $a(t)$, $a(t+\Delta t)$, $a(t+2\Delta t)$ ... $a(t_{max})$. Then (30.7) can be solved for the parameter $R_a(t)$ using the finite element representation

$$R_a(t) = \frac{R_a^\star(t)}{a(t)} - \frac{pVC}{a(t)} \frac{\Delta a(t)}{\Delta t} \tag{A.16}$$

where
$$\Delta a(t) = a(t + \Delta t) - a(t) \qquad\qquad (A.17)$$
Eq. (A.17) uses a type of forward "difference". A better estimate of the differential da may be given by $\frac{1}{2}[a(t + \Delta t) - a(t - \Delta t)]$, and we are into the niceties of numerical analysis. The crux of the matter is that acceptable results are obtainable using the simple representation (A.17) with small enough values of Δt.

The reader should not be left with the impression that the method of finite elements is a distillate of all that's fine in twentieth century analysis. *Runge-Kutta* and *predictor-corrector* methods are much more delicate tools for the precise craft of numerical solution of differential equations[†]. But finite elements is a simple and versatile procedure which can, perhaps, be commended in the current state of development of systems biology.

[†]For example, the HP-25 hand calculator may be used to solve a first order differential equation by a modified-Euler (predictor-corrector) technique by introducing a supplied program (Hewlett-Packard, 1975). The differential equation (A.8) may be solved very simply be entering the function xy into the supplied program, as follows:

Line	Code	Key Entry
18	61	X
19 13	31	GTO 31

Taking $\Delta x = .1$ and entering the initial conditions (1,1), one obtains after only 20 iterations
$$y(3) = 53.0868,$$
a much more rapid convergence than we obtained with the unmodified technique.

References

Hewlett-Packard HP-25 Application Programs, Hewlett-Packard Co. Publication, 83, 1975.

La Fara, R.L., *Computer Methods for Science and Engineering,* Hayden Book Co., 1973.

Moore, R., Davids, N. and Berger, R.L., Finite element methods in cell dynamics: fundamentals, *Currents in Modern Biology* 3, 95-109 (1969).

Pennington, R.H., *Computer Methods and Numerical Analysis,* 2nd Ed., Collier-MacMillan Ltd., 1970.

APPENDIX B
CURVE FITTING, PARAMETER ESTIMATION, AND THINGS LIKE THAT

This appendix has an element in common with the previous one. They both deal with the subject: How to Do a Reasonably Good Job Yourself Even If an Expert Could Do it Better. It is fair to assume that most readers will have had some dealings with the general problem of curve fitting, but let's review some of the fundamentals anyway.

Suppose we have a set of N data points (N pairs of measured values), (y_i, t_i), $i = 1, 2, ..., N$. We wish to find a function of the form $y = f(A,B,C, ... t)$ which "best fits" the data points; i.e. we wish to find values for the parameters A, B, C ... such that $y = f(A,B,C, ... t)$ "best fits" the data points. First we plot the data points and take a good look at them. If they seem to lie on a straight line, for example, we may seek a function of the form

$$Y = A + Bt \qquad \qquad (B.1)$$

If we feel that the data points should be conformed to a parabola, we may seek a function of the form

$$Y = A + Bt + Ct^2 \qquad \qquad (B.2)$$

The expression *best fits* usually means *fits with a minimum sum of squares of errors*. It is usually assumed that values of t_i (time) are accurate, and that scatter of the data around some smooth function is due to scatter in the measured values y_i. Hence one looks for a function $f(A, ... t)$ which minimizes the sum of squares of the deviations of the ordinates (Fig. B.1)

The most practical approach for any investigator with a curve fitting problem is to visit his computing centre and enquire regarding the existence of library tape subprograms for curve fitting. Some remarkably versatile programs now exist. The user need only read in his data, specify the form of the function to which he requires a fit, provide reasonable starting points and criteria for satisfactory results, and the library subprogram does the rest.

343

If such programs exist and operate satisfactorily, one need go no further. But, upon frequent occasions, one runs into difficulty with these packages or desires to modify the method of fitting. Sometimes one is using a small computer and library programs are not available. Under these circumstances he must write his own program. So let's continue with some fundamental concepts of curve fitting.

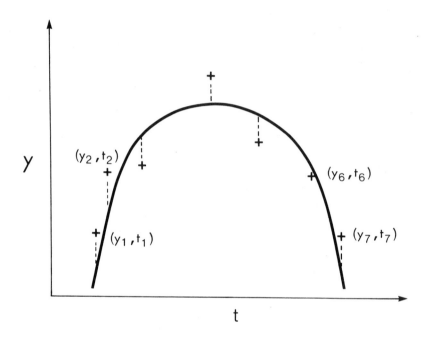

Fig. B.1   The crosses indicate measured data points (y_i, t_i), i = 1, 2, ... 7. In curve fitting we try to select a function of the form f(A,B,C, ... t) such that the sum of squares of the vertical deviations (in dashed lines) is minimized. The problem is to estimate the optimal set of parameters (A,B,C, ...).

The sum of squares of deviations may be represented by the expression

$$S = \sum_{i=1}^{N} \left[y_i - f(A,B,C, \dots t_i) \right]^2 .$$ (B.3)

The manner by which extrema of the function S may be obtained is by solving simultaneously the set of algebraic equations

$$\frac{\partial S}{\partial A} = 0, \ \frac{\partial S}{\partial B} = 0, \ \frac{\partial S}{\partial C} = 0 \ \dots \ , \tag{B.4}$$

one equation for each of the parameters which are sought. Equations (B.4) are in general *neither a necessary nor a sufficient condition for an extremum* and the reader is referred to a standard text on mathematical analysis for a complete discussion. In this abbreviated approach we shall simply assume their sufficiency.

Linear Functions of the Parameters A, B, C ...

Suppose the desired function is of the form

$$f(A,B,C, \ \dots \ t) = a(t)A + b(t)B + c(t)C + \ \dots \tag{B.5}$$

Such functions may be nonlinear in the coefficients $a(t)$, $b(t)$, $c(t)$ ... but are linear in the parameters A, B, C ... — which are the quantities we are trying to determine. Eq. (B.3) then gives

$$S = \sum_{i=1}^{n} \left[y_i - (aA + bB + cC + \ \dots) \right]^2$$

and (B.4) gives rise to a series of equations of the form

$$-\frac{1}{2} \frac{\partial S}{\partial A} = \sum_i a(t_i) \left[y_i - a(t_i)A - b(t_i)B - c(t_i)C - \ \dots \right] = 0$$

$$\left[\sum_i a(t_i)^2 \right] A + \left[\sum_i a(t_i)b(t_i) \right] B + \left[\sum_i a(t_i)c(t_i) \right] C + \ \dots = \sum_i y_i a(t_i).$$

$$\tag{B.6}$$

By permuting the coefficients we obtain one equation of the form (B.6) for each of the parameters A, B, C ... . These equations may then be solved simultaneously to give values for A, B, C ... which optimize the fit. The coefficients $a(t_i)$, $b(t_i)$, $c(t_i)$ ... are quantities which can be evaluated from the data.

Example: Suppose we are fitting to the straight line (B.1). Then

$$S = \sum_i \left[y_i - (A + Bt_i) \right]^2$$

$$-\frac{1}{2} \frac{\partial S}{\partial A} = \sum_i \left[y_i - A - Bt_i \right] = 0$$

$$-\frac{1}{2}\frac{\partial S}{\partial B} = \sum_i \left[y_i - A - Bt_i \right] t_i = 0$$

which becomes

$$NA + \left[\sum_i t_i \right] B = \sum_i y_i$$

$$\left[\sum_i t_i \right] A + \left[\sum_i t_i^2 \right] B = \sum_i y_i t_i \qquad\qquad (B.7)$$

These two linear equations may then be solved simultaneously for A and B, which permits us to place the best straight line, y = A + Bt, through a set of data points.

Example: Find the best fit to the function $y = Ae^{Bt}$ by the least squares method.

Taking logarithms of both sides of this equation,

$$\ln Y = \ln A + Bt \qquad\qquad (B.8)$$

(B.8) is now a linear function of the form (B.5) and we can proceed as in the previous example.

Similarly, power functions of the form (21.5), $y = At^{-\alpha} e^{-\beta t}$, should pose no problem. Taking logarithms of both sides leaves a linear equation.

Non-Linear Functions of the Parameters A, B, C ...

e.g.   Y = A cos Bt

Curve fitting to functions of this type is considerably harder than the fitting to linear functions, and a number of approaches may be taken (see, for example, La Fara). An important non-linear function with which we deal in this book is the multiexponential function:

$$Y = Ae^{-Bt} + Ce^{-Dt} + \ldots \qquad\qquad (B.9)$$

For purposes of example, let us consider the simplest multiexponential function, the double exponential; and to maintain a nomenclature consistent with the main text, let us express it in the form

$$Y = Ae^{-\beta_1 t} + Be^{-\beta_2 t} \qquad\qquad (B.10)$$

We are primarily concerned with the case where β_1 and β_2 are greater than zero, and we shall take $\beta_1 > \beta_2$.

A great number of papers have been written on the subject of multiexponential curve fitting, and most of them, I believe, are quite good. I shall draw attention here to only three of them. The paper by Perl (1960) is noteworthy because it demonstrates the fundamental technique of curve fitting manually, using a pencil, paper and brain. (In the age of the "747" it is hard to remember that people used to fly by the seat of their pants.) The paper by Worseley and Lax (1962) is noteworthy because it was a pioneer effort in non-linear curve fitting on the computer. The method was developed by Dr. Lax together with the late Dr. Worseley on an IBM 650 computer in Toronto. And finally the paper by Hazelrig et al. (1963) is noteworthy because of its simplicity and versatility. Much of the remainder of this appendix is de-voted to a discussion of the latter approach. The matter of multiexponential curve fitting is not really "resolved"; new methods are continually being proposed. Even as I am drafting this review (1974) I see a group of papers recently abstracted which deal with the application of Fourier transform methods to multiexponential curve fitting.

Reduction of the Non-Linear Multiexponential Function to Linear Form

We have seen how curve fitting to linear functions may be achieved by the least squares criterion. The method of Hazelrig et al. is one for reducing the multiexponential function (e.g. (B.10)) to the linear form (B.5). Sup-pose we were to guess at the best values for β_1 and β_2 in (B.10). Then $e^{-\beta_1 t}$ and $e^{-\beta_2 t}$ become known functions of t, say a(t) and b(t) respectively, and

$$Y = a(t)A + b(t)B$$

which is the linear form (B.5). We can then complete the optimization pro-cess using (B.6), giving us the best values for A and B *provided that* β_1 and β_2 take on the values which were guessed. The idea is, then, to carry out the linearization and optimization process for a great number of guesses for the value of β_1 and β_2. The procedure for searching for values for β_1 and β_2 ("hill-climbing") need not be a random one. If for a fixed value of β_2, a value† $\beta_1{}^2$ gives a better curve fit (smaller sum of squares) than $\beta_1{}^1$ when $\beta_1{}^2 > \beta_1{}^1$, then it is a fair guess that $\beta_1{}^3$ will give a better curve fit than $\beta_1{}^2$ when $\beta_1{}^3 > \beta_1{}^2$. Not always, but often.

Recalling that β_1 and β_2 are always real, positive numbers, let us suppose that $\beta_1{}^1 < \beta_1{}^2 < \beta_1{}^3$ and that $\beta_2{}^1 > \beta_2{}^2 > \beta_2{}^3$. Suppose that we make an initial guess at the best values for β_1 and β_2, and call this pair of numbers $(\beta_2{}^2, \beta_1{}^2)$. Suppose moreover, that we calculate the numbers

$$\left.\begin{array}{ll} \beta_1{}^1 = 0.7\ \beta_1{}^2 & \beta_2{}^3 = 0.7\ \beta_2{}^2 \\[2mm] \beta_1{}^3 = 1.3\ \beta_1{}^2 & \beta_2{}^1 = 1.3\ \beta_2{}^2 \end{array}\right\} \qquad \text{(B.11)}$$

We then have 3 values for β_1 and 3 values for β_2 which, when combined in all possible ways, gives 9 pairs of values for (β_2, β_1): $(\beta_2{}^1, \beta_1{}^1)$, $(\beta_2{}^1, \beta_1{}^2)$ etc. These 9 pairs of values may be arranged to form a matrix, as shown in Fig. B.2. Then for each pair of numbers we can convert (B.10) into a linear function, and hence evaluate the best A and B. The pair of numbers, (β_2, β_1), which give the smallest sum of squares of errors can then serve as the centre of a new matrix. Thus, if the smallest sum of squares is given by $(\beta_2{}^3, \beta_1{}^3)$, this pair of numbers will then assume the new name $(\beta_2{}^2, \beta_1{}^2)$ and a new matrix will be evaluated using (B.11). The search matrix has "shifted" diagonally, downwards to the right to give a new search matrix. If the smallest sum of squares is obtained from the central entry, $(\beta_2{}^2, \beta_1{}^2)$, then the peripheral entries are computed from

$$\left.\begin{array}{ll} \beta_1{}^1 = (1 - .3^k)\beta_1{}^2 & \beta_2{}^3 = (1 - .3^k)\beta_2{}^2 \\[2mm] \beta_1{}^3 = (1 + .3^k)\beta_1{}^2 & \beta_2{}^1 = (1 + .3^k)\beta_2{}^2 \end{array}\right\} \qquad \text{(B.12)}$$

with $k = 2$. (When $k = 1$, (B.12) is identical with (B.11).) That is, the

---

†The subscript of a β indicates which exponential it belongs to (see Eq. B.10), and the superscript is not a "power" but rather indicates a particular value for β.

search is conducted over a smaller square in the β_1 - β_2 plane. Each time the smallest sum of squares is obtained at the $(\beta_2{}^2, \beta_1{}^2)$ - position, k is increased by unity, until the search square becomes "small enough". The resolution of the curve fitting process is determined, in part, by the largest value which k may assume. Any searching (or hill-climbing) algorithm may become entrapped in a "local minimum" (as opposed to an absolute minimum) of the sum of squares of errors. The only hope of avoiding this pitfall lies in good luck and skilful selection of the initial pair of values, $(\beta_2{}^2, \beta_1{}^2)$.

Parameter Estimation with the Linearized Multiexponential Function

The double exponential function, (B.10), has now been reduced to the linear form (B.5), by the process of selecting numerical values for (β_2, β_1). For the general pair $(\beta_2{}^m, \beta_1{}^j)$, (B.5) takes on the form

$$f(A,B) = Ae^{-\beta_1{}^j t} + Be^{-\beta_2{}^m t} \qquad (B.13)$$

and the sum of squares of errors, S is given by[†]

$$S = \sum_{i=1}^{N} \left[y_i - \left(Ae^{-\beta_1{}^j t_i} + Be^{-\beta_2{}^m t_i} \right) \right]^2$$

Differentiating partially with respect to A and then to B gives (B.6)

$$\left[\sum_{i=1}^{N} e^{-2\beta_1{}^j t_i} \right] A + \left[\sum_{i=1}^{N} e^{-(\beta_1{}^j + \beta_2{}^m) t_i} \right] B = \sum_{i=1}^{N} y_i e^{-\beta_1{}^j t_i}$$

$$\left[\sum_{i=1}^{N} e^{-(\beta_1{}^j + \beta_2{}^m) t_i} \right] A + \left[\sum_{i=1}^{N} e^{-2\beta_2{}^m t_i} \right] B = \sum_{i=1}^{N} y_i e^{-\beta_2{}^m t_i}$$

$$(B.14)$$

[†] If we were fitting two functions simultaneously to a linear combination of the same two exponentials (as in section 17), so that in addition to Eq. (B.10) we had $z = De^{-\beta_1 t} + Ee^{-\beta_2 t}$, the sum of squares would just be

$$S = \sum_{i=1}^{N} \left[y_i - \left(Ae^{-\beta_1{}^j t_i} \dots \right) \right]^2 + \sum_{i=1}^{N} \left[z_i - \left(De^{-\beta_1{}^j t_i} \dots \right) \right]^2$$

and we would proceed in much the same manner.

Notice that the summation is only with respect to i; β^j and $\beta_2{}^m$ are constant throughout Eq. (B.14). The various sums are easily obtained from the data using a computer. Let us rewrite (B.14) more simply as

$$S_5 A + S_4 B \; = \; S_1$$

$$S_4 A + S_2 B \; = \; S_3 \tag{B.15}$$

where $S_1 \ldots S_5$ can be defined by comparing (B.15) with (B.14). Equations (B.15) are linear in A and B, and may be solved by Cramer's rule. Let

$$\text{DENOM} = S_5 \cdot S_2 - (S_4)^2 \tag{B.16}$$

Then

$$A = (S_1 S_2 - S_3 S_4)/\text{DENOM}$$

$$B = (S_3 S_5 - S_1 S_4)/\text{DENOM} \tag{B.17}$$

The A and B values so-obtained are the optimal values for the pair $(\beta_2{}^m, \beta_1{}^j)$, and so we should really replace A and B by the symbols

$$A^{m,j} = A$$

$$B^{m,j} = B \tag{B.18}$$

Thus it is clearly seen that we obtain 9 pairs of values $(A^{m,j}, B^{m,j})$, corresponding to the 9 pairs of values $(\beta_2{}^m, \beta_1{}^j)$, by solving (B.17) 9 times. One pair of values $(A^{m,j}, B^{m,j})$ will give the smallest sum of squares. The search matrix will now shift, as shown in Fig. B.4, in the direction of this best position, designated by m and j.

Computer Program for Parameter Estimation with the Linearized Multiexponential Function

It is usual to presume at this point, that the reader can assemble his own computer program — and often he can. But to write a program which converges fairly rapidly on the best values of the parameters requires a quantity of time and a degree of dedication often in excess of that available. And since I feel that any worker in the field of biokinetics should possess his own double exponential package (at least), I offer here a complete program in Fortran IV. It may not be the swiftest or the newest, but it works, and works well. It has toiled faithfully in the service of about a dozen people since 1966. The reader who follows the various annotations will be able to alter the program to fit his own specifications.

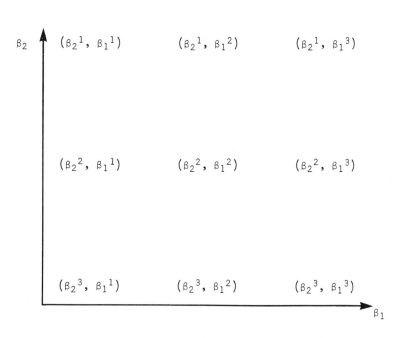

Fig. B.2. Matrix representation of the 9 pairs of values (β_2, β_1). β_1 increases as we move to the right along the abscissa, and β_2 in-
creases as we move upwards along the ordinate. The central value
$(\beta_2{}^2, \beta_1{}^2)$ is the most probable guess. The superscripts give the
standard representation of position of an element in the matrix;
i.e. $(\beta_2{}^3, \beta_1{}^1)$ is the element in the 3rd row, 1st column. The
matrix functions primarily as a mnemonic; a glance at it tells the
relative values of the various entries. This matrix divides naturally into two single matrices, which are shown in the next
figure.

First let us split up the double matrix of Fig. B.2 into two simple matrices,
one for β_1 and one for β_2 as shown in Fig. B.3. The evaluation of A and B
for one complete (β_2, β_1)-matrix will be called an *iteration*. When the matrix
shifts position, the values of some, but not all, of the elements have
changed. On the next iteration, it is desirable to avoid re-calculating A
and B for those (β_1, β_2) which have already been searched (see, for example,

Fig. B.4). This is achieved by the "guard ring" of zeros, discussed in the legends to Figs. B.5a and B.5b. Without further ado, let us to the program.

$$\begin{pmatrix} \beta_2{}^1 & \beta_2{}^1 & \beta_2{}^1 \\ \beta_2{}^2 & \beta_2{}^2 & \beta_2{}^2 \\ \beta_2{}^3 & \beta_2{}^3 & \beta_2{}^3 \end{pmatrix}$$

$$\begin{pmatrix} \beta_1{}^1 & \beta_1{}^2 & \beta_1{}^3 \\ \beta_1{}^1 & \beta_1{}^2 & \beta_1{}^3 \\ \beta_1{}^1 & \beta_1{}^2 & \beta_1{}^3 \end{pmatrix}$$

Fig. B.3. The figurative matrix of Fig. B.2 breaks up into two standard matrices. The upper β_2-matrix has equal column vectors; the lower β_1-matrix has equal row vectors.

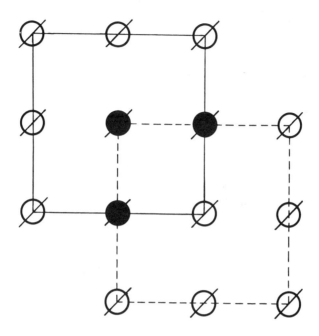

Fig. B.4. A schematic representation of the (β_2, β_1)-matrix of Fig. B.2,
           evaluated over two successive iterations. The first iteration
           evaluated the region encompassed by the solid line square, and the
           minimum sum of squares was found at the lower right-hand corner.
           Hence, the second iteration evaluated the region encompassed by
           the dashed line square. But it would be desirable to avoid re-cal-
           culation of A and B for the positions indicated by the solid cir-
           cles, because they have already been done and found to be not-opti-
           mal. A way to avoid re-evaluation of the solid circles is shown
           in Fig. B.5.

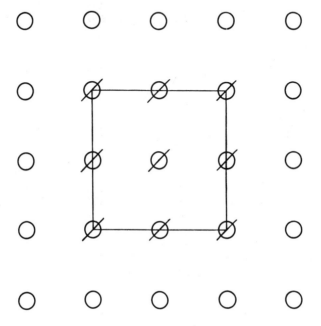

Fig. B.5a. The original (β_2, β_1)-matrix is represented by the 9 circles Ø.
These 9 pairs of values are surrounded by a "guard ring" of
zeros, O. This means that each of the standard matrices of Fig.
B.3 is really a 5 x 5 matrix, with an outer ring of zeros. Only
the inner, 3 x 3 matrix contains nonzero elements.

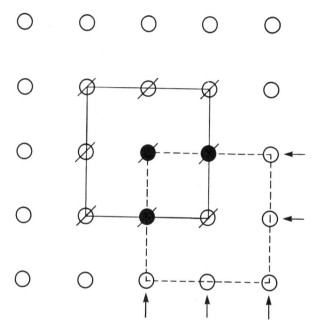

Fig. B.5b.   In the following iteration the search area has shifted to the
dashed square. But rather than searching all 9 positions of the
dashed square, the computer searches only those positions (marked
by arrows) which encroached upon the guard ring of the previous
iteration. A new guard ring of zeros is then set up etc.

```
C      PROGRAM TO FIT SET OF DATA POINTS TO A LINEAR COMBINATION OF
C      TWO EXPONENTIALS.  K.H. NORWICH.
1             DIMENSION T(100), Y(100), P(100), U(100), B1(5,5), B2(5,5),
       1 SUM1(3,3), SUM2(3,3), SUM3(3,3), SUM4(3,3), SUM5(3,3),
       2 C1(3,3), C2(3,3), S(3,3), BTR(5,5), DENOM(3,3),
       3 YSQ(100), WEIGHT(100)
```

T and P are the data (to be read in); Y = P - FBS, where FBS is a known constant to be subtracted before curve fitting. B1 and B2 are the matrices of Fig. B.3, surrounded by a ring of zeros. SUM1, ... SUM5 are S_1, ... S_5 of Eq. (B.15). C1 = A, C2 = B. S is the sum of squares of erros. DENOM is defined by Eq. (B.16). WEIGHT is the statistical weighting of each Y-value. Other arrays will be defined later.

```
2      103    READ (5,2) N, FBS
       C    MEANS "READ FROM TAPE 5 ACCORDING TO FORMAT 2".  OTHER COMPILERS
       C    MAY USE DIFFERENT INPUT CODES.
3             DEX1 = 0.1
4             DEX2 = 0.01
5             MP = 50
```

N is the number of data points.

DEX1 and DEX2 are the initial estimates for β_1 and β_2.

MP is the total allowable number of iterations.

```
6             A = 5.0E+37
7             MM = 0
8             KN = 1
9             DO  300  I = 1,5
10            DO  300  J = 1,5
11            B1(I,J) = 0.0
12            B2(I,J) = 0.0
13     300    CONTINUE
```

Statements 6 - 13 provide initial conditions for the program.

KN is k of Eq. (B.12).

```
14     2      FORMAT (I5, F6.1)
15            READ (5,4) (T(K), K = 1,N)
16            READ (5,4) (P(K), K = 1,N)
17     4      FORMAT (8F10.4)
18            DO 5 I=1,N
19     5      Y(I) = P(I) - FBS
       C    THE WEIGHT GIVEN TO A DATA POINT Y(K) IN THE COMPUTATION OF THE
       C    SUM OF SQUARES OF ERRORS IS EQUAL TO YMAX**2/Y(K)**2, WHERE YMAX
       C    IS THE LARGEST Y-VALUE
20            DO 400 K=1,N
21     400    YSQ(K) = Y(K)**2
22            YMAXSQ = AMAX(YSQ,N)
23            DO 401 K=1,N
24     401    WEIGHT(K) = YMAXSQ/YSQ(K)
25            WRITE (6,6)
```

```
      C  MEANS "WRITE ON TAPE 6 ACCORDING TO FORMAT 6". OTHER COMPILERS MAY
         E.G. ALLOW FREE OUTPUT:  PRINT, ...
26          FORMAT(1H1,6X,2HC1,12X,2HC2,12X,2HB1,12X,2HB2,6X,14HSUM OF
                                                                   SQUARES)
```

Statements 15-24 are concerned with reading in the data, subtracting a constant if necessary, and weighting the Y-value.

```
27          MARK = 0
28          B1(3,3) = DEX1
29          B2(3,3) = DEX2
```

Further initial conditions. Marker set to zero. The centre of arrays set at the initial estimates for β_1 and β_2, DEX1 and DEX2 respectively.

```
      C  IF MATRIX ELEMENT IS ZERO, EVALUATE IT.  OTHERWISE EQUATE TO TEN.
30    29    DO 8 I = 2,4
31          DO 8 J = 2,4
32          IF (MARK-1) 111,10,111
33   111    IF (B1(I,J)) 9,10,9
34     9    IF (I-3) 101,105,101
35   105    IF (J-3) 101,10,101
36   101    B1(I,J) = 10.0
37          GO TO 8
38    10    B1(I,J) = EXP((1.0+FLOAT(J-3)*.3**KN)*ALOG(B1(3,3)))
39     8    CONTINUE
40          DO 13 I = 2,4
41          DO 13 J = 2,4
42          IF (MARK-1) 112,12,112
43   112    IF (B2(I,J)) 11,12,11
44    11    IF (I-3) 102,106,102
45   106    IF (J-3) 102,12,102
46   102    B2 (I,J) = 10.0
47          GO TO 13
48    12    B2(I,J) = EXP((1.0+FLOAT(3-I)*.3**KN)*ALOG(B2(3,3)))
49    13    CONTINUE
50          MARK = 0
```

Statements 30-50 evaluate all matrix elements which are initially equal to zero (see Fig. 5.5) according to the equations (B.12). The logarithms of the β's are incremented, rather than the β's themselves. The k of (B.12) is KN in the program, and KN is initialized at unity (statement 8). B1 and B2 will be set equal to 10.0 (an arbitrary number) for positions which have been searched before. The sum of squares is not re-calculated for these positions (line 61 below).

```
51          DO 14 I=1,3
52          DO 14 J=1,3
53          SUM1(I,J) = 0.0
54          SUM2(I,J) = 0.0
55          SUM3(I,J) = 0.0
56          SUM4(I,J) = 0.0
57          SUM5(I,J) = 0.0
58    14    CONTINUE
```

```
59              DO 15 I=2,4
60              DO 15 J=2,4
61              IF (B1(I,J)-10.0) 16,15,16
62      16      DO 201 K=1,N
63              SUM1(I-1,J-1) = WEIGHT(K)*
        1                       Y(K)*EXP(-B1(I,J)*T(K)) + SUM1(I-1,J-1)
64              SUM2(I-1,J-1) = WEIGHT(K)*
        1                       EXP(-2.0*B2(I,J)*T(K)) + SUM2(I-1,J-1)
65              SUM3(I-1,J-1) = WEIGHT(K)*
        1                       Y(K)*EXP(-B2(I,J)*T(K)) + SUM3(I-1,J-1)
66              SUM4(I-1,J-1) = WEIGHT(K)*
        1                       EXP(-(B2(I,J)+B1(I,J))*T(K)) + SUM4(I-1,J-1)
67              SUM5(I-1,J-1) = WEIGHT(K)*
        1                       EXP(-2.0*B1(I,J)*T(K)) + SUM5(I-1,J-1)
68      201     CONTINUE
69      15      CONTINUE
```

DO 14 I = 1,3:   DO 14 J = 1,3 ... i.e. we evaluate the sums S_1 ... S_5 a maximum of 3 x 3 times: once for each of the 9 potential nonzero pairs (β_2, β_1). But (statement 61): if a pair (β_2, β_1) did not overlie the guard ring of the previous iteration do not bother to evaluate it; it has already been done.

```
70              DO 17 I=1,3
71              DO 17 J=1,3
72              IF (B1(I+1,J+1)-10.0) 18,17,18
73      18      DENOM(I,J) = SUM5(I,J)*SUM2(I,J) - SUM4(I,J)**2
74              IF (DENOM(I,J)) 202,203,202
75      203     C1(I,J) = 0.0
76              C2(I,J) = 0.0
77              GO TO 17
78      202     C1(I,J)=(SUM1(I,J)*SUM2(I,J)-SUM3(I,J)*SUM4(I,J))/DENOM(I,J)
79              C2(I,J)=(SUM5(I,J)*SUM3(I,J)-SUM4(I,J)*SUM1(I,J))/DENOM(I,J)
80      17      CONTINUE
81              DO 19 I=1,3
82              DO 19 J=1,3
83      19      S(I,J) = 0.0
84              DO 20 I=2,4
85              DO 20 J=2,4
86              IF (B1(I,J)-10.0) 22,21,22
87      21      S(I-1,J-1) = 1.0E+38
88              GO TO 20
89      22      DO 200 K-1,N
90              S(I-1,J-1)=WEIGHT(K)*(Y(K)-C1(I-1,J-1)*EXP(-B1(I,J)*T(K))
        1 -C2(I-1,J-1)*EXP(-B2(I,J)*T(K)))**2 + S(I-1,J-1)
91      200     CONTINUE
92      20      CONTINUE
```

Statements 70-92 evaluate C1 and C2 (A and B) using (B.16) and (B.17), as well as S (sum of squares of errors) only for positions which previously overlay the guard ring. Division by zero is excluded.

```
93              D = AMUM(S,3,3)
94              DO 23 II = 2,4
95              DO 23 JJ = 2,4
96              I = II
97              J = JJ
```

```
98              IF (D-S(II-1,JJ-1)) 23,24,23
99      23      CONTINUE
100     24      WRITE(6,25)C1(I-1,J-1),C2(I-1,J-1),B1(I,J),B2(I,J),S(I-1,J-1)
101     25      FORMAT (1X, 5E14.7)
102             WRITE(6,110) MM, KN
103     110     FORMAT(1X, 2I10)
104             MM = MM+1
```

D is equated to the minimum of the 9 sums of squares. AMUM is a subprogram for obtaining the minimum element of an array. The best values of A, B, β_1 and β_2 are printed. MM, the number of iterations made, is increased by 1.

```
105             IF (MM-MP) 7,7,35
106     7       IF (ABS(A-D) - 1.0E-04) 26,26,27
107     26      KN=KN+1
108             MARK = 1
109             IF (KN-10) 29,35,35
110     27      A=D
```

If the number of iterations has reached the maximum allowable number, MP, stop calculating. If the minimum sum of squares for this iteration is found for the same number pair (β_2,β_1) as before, make the search square smaller by increasing KN (k). If the square has reached an arbitrary degree of smallness (KN = 10), stop calculating.

```
111             KTRAN = I-3
112             MTRAN = J-3
113             DO 107 K=1,5
114             DO 107 M=1,5
115             BTR(K,M) = B1(K,M)
116     107     CONTINUE
117             DO 28 K=2,4
118             DO 28 M=2,4
119             KT = K + KTRAN
120             MT = M + MTRAN
121             B1(K,M) = BTR(KT,MT)
122     28      CONTINUE
123             DO 108 K=1,5
124             DO 108 M=1,5
125             BTR(K,M) = B2(K,M)
126     108     CONTINUE
127             DO 109 K=2,4
128             DO 109 M=2,4
129             KT = K + KTRAN
130             MT = M + MTRAN
131             B2(K,M) = BTR(KT,MT)
132     109     CONTINUE
133             GO TO 29
```

If searching is to continue, TRANslate the search square over the guard ring of zeros in the direction of the (β_2,β_1) pair which gave the minimum sum of squares. BTR is just an array to hold intermediary values while the arrays B1 and B2 are redefined.

```
134     35      DO 36 L=1,N
135             U(L)=C1(I-1,J-1)*EXP(-B1(I,J)*T(L))+C2(I-1,J-1)*EXP(-B2(I,J)
                1*T(L))+FBS
136     36      CONTINUE
137             WRITE(6,37)
138     37      FORMAT (1H0,6X,4HTIME,3X,7HY(CALC),3X,7HY(MEAS))
139             WRITE(6,38) (T(K),U(K),P(K), K=1,N)
140     38      FORMAT (2X,3F10.3)
```

$U(t)$ is evaluated from $U(t) = A_{best} e^{-\beta_{1_{best}} \cdot t} + B_{best} \cdot e^{-\beta_{2_{best}} \cdot t}$. For each time, T, the datum, P, and fitted value, U, are printed. The remainder of the program calculates zeroth and first moments about the y-axis and $\bar{t}$. There are two, short subprograms.

```
141             WRITE(6,39) FBS
142     39      FORMAT (1H0, 19HEQUILIBRIUM LEVEL= ,  F6.2)
143             WRITE(6,402)
144     402     FORMAT (1H0, 5X, 4HTIME, 6X, 10HDERIVATIVE)
145             DO 543 L=1,N
146             U(L) = -B1(I,J)*C1(I-1,J-1)*EXP(-B1(I,J)*T(L))
                1 -B2(I,J)*C2(I-1,J-1)*EXP(-B2(I,J)*T(L))
147             WRITE(6,544)T(L), U(L)
148     543     CONTINUE
149     544     FORMAT (1X, F10.3, 3X, E14.7)
150             AREA = C1(I-1,J-1)/B1(I,J) + C2(I-1,J-1)/B2(I,J)
151             WRITE(6,403) AREA
152     403     FORMAT (1H0, 6HAREA= ,  E16.8)
153             FIRMOM = C1(I-1,J-1)/B1(I,J)**2 + C2(I-1,J-1)/B2(I,J)**2
154             WRITE(6,104) FIRMOM
155     104     FORMAT (1H0,15HFIRST MOMENT = , E16.8)
156             TBAR = FIRMOM/AREA
157             WRITE(6,115) TBAR
158     115     FORMAT (1H0,22HMEAN RESIDENCE TIME = , E16.8)
159             MM = 0
160             KN = 1
161             GO TO 103
162             END




163             FUNCTION AMUM(S,N,M)
        C       FINDS MINIMUM OF ARRAY S(N,M)
164             DIMENSION S(N,M)
165             AMUM = S(1,1)
166             DO 2 I=1, N
```

```
167          DO 2 J=1,M
168          IF (AMUM-S(I,J)) 2,2,3
169    3     AMUM = S(I,J)
170    2     CONTINUE
171          RETURN
172          END

173          FUNCTION AMAX(S,N)
       C     FINDS THE MAXIMUM OF THE ARRAY S(N)
174          DIMENSION S(N)
175          AMAX = S(1)
176          DO 2 K=1,N
177          IF (AMAX-S(K)) 3,2,2
178    3     AMAX = S(K)
179    2     CONTINUE
130          RETURN
181          END
```

Figure B.6 is a sample computer output

The reader should have little or no difficulty adapting the program to his
needs. The number of iterations permitted (MP) may be increased or decreased,
as may be the number of times the search square is reduced (KN). The initial
estimates for β_1 and β_2 (DEX1 and DEX2) may be changed. The statistical
weighting may be altered and the statistical calculations at the end modified.
The fitted curve may be constrained to pass through the origin by setting
B = -A and optimizing

$$f(A,B) = A\left[e^{-\beta_1^j t} - e^{-\beta_2^m t}\right] \tag{B.19}$$

instead of (B.13). Take control of the program; make it yours.

A rather extensive review of the techniques of optimization is given by
Johnson (1974). In the work in which I am currently engaged at the Univer-
sities of Toronto and Ottawa we are using a variant of the method of Powell,
which is discussed in Johnson's review. Powell's method is more sophisticated
than the method followed by the Fortran program given above. For a comparable sum of squares of residuals, Powell's method will generally require less
execution time on the computer, which firmly establishes its superiority.
Either method may be easily adapted to curve fit to many other functions,
rather than just sums of exponentials. Powell's method is particularly versatile in this regard.

C1	C2	B1	B2	SUM OF SQUARES
0.3962766E+03 0	0.1715238E+03 1	0.5011887E-01	0.9999998E-02	0.7342931E+05
0.3781768E+03 1	0.2342021E+02 1	0.2041752E-01	0.2511897E-02	0.3555210E+05
0.3781768E+03 2	0.2342021E+02 1	0.2041752E-01	0.2511897E-02	0.3555210E+05
0.3665205E+03 3	0.3098923E+02 2	0.2041752E-01	0.4305284E-02	0.3522140E+05
0.3665205E+03 4	0.3098923E+02 2	0.2041752E-01	0.4305284E-02	0.3522140E+05
0.3709543E+03 5	0.4648680E+02 3	0.2267947E-01	0.4987512E-02	0.3466377E+05
0.3621746E+03 6	0.5229068E+02 3	0.2267947E-01	0.5754944E-02	0.3392596E+05
0.3513315E+03 7	0.5973351E+02 3	0.2267947E-01	0.6614856E-02	0.3335773E+05
0.3377634E+03 8	0.6943027E+02 3	0.2267947E-01	0.7574718E-02	0.3305724E+05
0.3232695E+03 9	0.9699281E+02 3	0.2512065E-01	0.8642185E-02	0.3176859E+05
0.3012187E+03 10	0.1287524E+03 3	0.2774791E-01	0.9825051E-02	0.3052927E+05
0.2687988E+03 11	0.1660948E+03 3	0.3056774E-01	0.1113119E-01	0.2858829E+05
0.2227446E+03 12	0.2106890E+03 3	0.3358625E-01	0.1256855E-01	0.2649239E+05
0.2312492E+03	0.2168323E+03	0.3680909E-01	0.1256855E-01	0.2513845E+05
.	.		.	
.	.		.	
.	.		.	
0.5167971E+03 41	0.3245486E+03 4	0.1731047E+00	0.1534488E-01	0.5178147E+03
0.5085254E+03 42	0.3244602E+03 4	0.1706632E+00	0.1534488E-01	0.4428999E+03
0.5003401E+03 43	0.3243704E+03 4	0.1682368E+00	0.1534488E-01	0.3765696E+03
0.4922344E+03 44	0.3242788E+03 4	0.1658256E+00	0.1534488E-01	0.3191553E+03
0.4842136E+03 45	0.3241858E+03 4	0.1634298E+00	0.1534488E-01	0.2710242E+03
0.4762656E+03 46	0.3240913E+03 4	0.1610497E+00	0.1534488E-01	0.2325276E+03
0.4684021E+03 47	0.3239951E+03 4	0.1586854E+00	0.1534488E-01	0.2040270E+03
0.4606182E+03 48	0.3238975E+03 4	0.1563370E+00	0.1534488E-01	0.1858875E+03
0.4529082E+03 49	0.3237981E+03 4	0.1540048E+00	0.1534488E-01	0.1784597E+03
0.4529082E+03 50	0.3237981E+03 4	0.1540048E+00	0.1534488E-01	0.1784597E+03

Fig. B.6a. Legend follows Fig. B.6b.

TIME	Y(CALC)	Y(MEAS)	TIME	DERIVATIVE
5.000	509.581	510.862	5.000	-0.3689606E+02
10.000	374.826	373.355	10.000	-0.1921419E+02
15.000	302.177	301.290	15.000	-0.1087005E+02
20.000	259.040	259.148	20.000	-0.6860909E+01
25.000	230.269	231.071	25.000	-0.4869661E+01
30.000	208.799	209.946	30.000	-0.3822670E+01
35.000	191.312	192.558	35.000	-0.3222105E+01
40.000	176.226	177.428	40.000	-0.2836790E+01
45.000	162.768	163.854	45.000	-0.2559057E+01
50.000	150.542	151.482	50.000	-0.2338472E+01
55.000	139.328	140.118	55.000	-0.2151139E+01
60.000	128.994	129.639	60.000	-0.1985496E+01
65.000	119.447	119.958	65.000	-0.1835722E+01
70.000	110.616	111.006	70.000	-0.1698693E+01
75.000	102.442	102.726	75.000	-0.1572564E+01
80.000	94.874	95.064	80.000	-0.1456113E+01
85.000	87.866	87.974	85.000	-0.1348427E+01
90.000	81.377	81.413	90.000	-0.1248772E+01
95.000	75.366	75.342	95.000	-0.1156513E+01
100.000	69.800	69.724	100.000	-0.1071084E+01
105.000	64.645	64.524	105.000	-0.9919738E+00
110.000	59.870	59.712	110.000	-0.9187081E+00
115.000	55.449	55.259	115.000	-0.8508858E+00
120.000	51.354	51.138	120.000	-0.7880155E+00
125.000	47.561	47.325	125.000	-0.7298159E+00
130.000	44.048	43.796	130.000	-0.6759154E+00
135.000	40.795	40.530	135.000	-0.6259959E+00
140.000	37.732	37.507	140.000	-0.5797626E+00
145.000	34.999	34.710	145.000	-0.5369446E+00
150.000	32.407	32.122	150.000	-0.4972881E+00

EQUILIBRIUM LEVEL = 0.0
AREA = 0.24042242E+05
FIRST MOMENT = 0.13942360E+07
MEAN RESIDENCE TIME = 0.57991089E+02

Fig. B.6b.

Fig. B.6. Sample computer output. The computer was given 30 data points generated from the function

$$Y_{meas} = 465.e^{-.162t} + 328.5e^{-.0155t}$$

and required to retrieve the 4 constants. The maximum number of
interations permitted (MP) was 50. The final estimate is given by
the last line of Fig. B.6a:

$$Y_{fitted} = 452.9e^{-.154t} + 323.8e^{-.0153t}$$

The correspondence between fitted and measured y's may be seen
in Fig. B.6b. The remainder of the output should be self-evident.

AREA $= \displaystyle\int_{0}^{\infty} y(t)dt$

FIRST MOMENT $= \displaystyle\int_{0}^{\infty} t\, y(t)dt$

MEAN RESIDENCE TIME $= \bar{t}$ = FIRST MOMENT/AREA

References

Hazelrig, J.B., Ackerman, E. and Rosevear, J.W., An iterative technique for conforming mathematical models to biomedical data, *Proceedings of the 16th Annual Conference on Engineering in Medicine and Biology* 5, 8-9 (1963).

Johnson, L.E., Techniques of Optimization, *CRC Critical Reviews in Bioengineering* Vol. 2, Issue 1 (1974).

La Fara, R.L., *Computer Methods for Science and Engineering*, Hayden Book Co., 1973.

Perl, W., A method for curve-fitting by exponential functions, *Int. J. Appl. Radiation and Isotopes* 8, 211-222 (1960).

Worseley, B.H. and Lax, L.C., Selection of a numerical technique for analyzing experimental data of the decay type, with special reference the use of tracers in biological systems, *Biochim Biophys. Acta.* 59, 1-24 (1962).

APPENDIX C
INTRODUCTION TO VECTOR ALGEBRA AND VECTOR CALCULUS

Whole books have been devoted to the algebra and calculus of vectors and there is no way whatever that a comprehensive introduction can be given in the few pages allocated to this appendix. It is, however, possible to introduce some of the functions and properties of vectors which are used in this book and in other works on kinetics, while omitting discussion of other vector functions which are more important in electromagnetic theory and hydrodymanics.

We are concerned, essentially, with vectors in 3-dimensional space. A vector in 3-space is an ordered triplet of numbers, e.g. (a_1, a_2, a_3). Geometrically it may be represented as an arrow, tethered to the origin of coordinates, with the tip of the arrow situated at the point whose spatial coordinates are (a_1, a_2, a_3). We have little need, in this book, to deviate from the rectangular or cartesian system of coordinates, so that the vector $\vec{A}$, whose components are (a_1, a_2, a_3), will have the projection a_1 on the x-axis, a_2 on the y-axis and a_3 on the z-axis (Fig. C.1).

Vectors can be added and subtracted by adding or subtracting their components. Thus

$$\vec{A}(a_1, a_2, a_3) + \vec{B}(b_1, b_2, b_3) = \vec{C}(a_1+b_1,\ a_2+b_2,\ c_1+c_2) \qquad (C.1)$$

Vectors can be multiplied by pure numbers (scalars) giving a new vector. For example, the product of the vector $\vec{A}$ by the scalar k is written $k\vec{A}$ and has the components (ka_1, ka_2, ka_3) — a new vector having the same direction as $\vec{A}$ but k times as large. Multiplication by a scalar is distributive so that

$$k(\vec{A} + \vec{B}) = k\vec{A} + k\vec{B} \qquad (C.2)$$

367

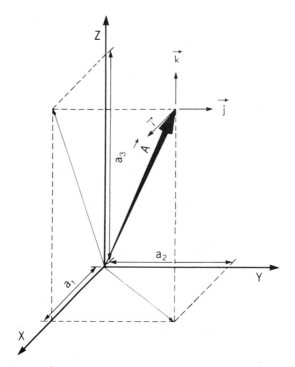

Fig. C.1. The vector $\vec{A}(a_1,a_2,a_3)$ when projected perpendicularly onto the x-,
          y- and z-axes will cut off lengths of a_1, a_2 and a_3 respectively.
          $\vec{i}$, $\vec{j}$ and $\vec{k}$ are unit vectors parallel to the x-, y- and z-axis re-
          spectively (an *orthonormal set*).

Certain physical quantities such as velocity or force may be represented by
vectors. These are quantities which involve both magnitude and direction.
If the vector $\vec{A}$ of Fig. C.1 represents a velocity, then the length of the
vector is proportional to the magnitude of the velocity and the direction of
the vector indicates the direction of the motion. If two forces act on a
particle, then the net force which acts on the particle is given by the vector sum of the two forces, as illustrated in Fig. 6.1.

The vectors $\vec{i}$, $\vec{j}$ and $\vec{k}$ are vectors of unit length, parallel to the x-, y- and
z-axis respectively. Thus $a_1\vec{i}$ is a vector of magnitude a_1 pointing along the
x-axis etc. This leads to two ways of representing a vector. We may write
          $\vec{A}(a_1,a_2,a_3)$

or equivalently, in terms of the *basis vectors* $\vec{i}$, $\vec{j}$ and $\vec{k}$

$$\vec{A} = a_1\vec{i} + a_2\vec{j} + a_3\vec{k}$$

The *scalar product* or *dot product* of two vectors is written $\vec{A}\cdot\vec{B}$. This product is *not* a vector but represents a scalar or pure number. $\vec{A}\cdot\vec{B}$ is defined by

$$\vec{A}\cdot\vec{B} = a_1b_1 + a_2b_2 + a_3b_3 \qquad (C.3)$$

The only scalar products used to any extent in this book are the scalar products of flux vectors with unit normal vectors. If $\vec{F}$ is some vector (say representing a flux) and $\vec{n}$ a unit vector (usually in the direction of a normal to a surface), then the scalar product $\vec{F}\cdot\vec{n}$ is a scalar quantity giving the length of the *projection* of $\vec{F}$ on $\vec{n}$ (Fig. C.2) *(normal scalar resolute)*. This result follows from simple trigonometry, and the reader is referred to any text on vector algebra. From the definition, (C.3), it is easily shown that the scalar product is commutative, so that

$$\vec{A}\cdot\vec{B} = \vec{B}\cdot\vec{A} \qquad (C.4)$$

and distributive, so that

$$\vec{A}\cdot(\vec{B} + \vec{C}) = \vec{A}\cdot\vec{B} + \vec{A}\cdot\vec{C} \qquad (C.5)$$

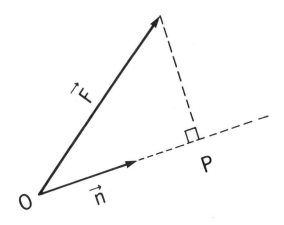

Fig. C.2. $\vec{F}$ is any vector; $\vec{n}$ is any vector of unit magnitude. $\vec{F}\cdot\vec{n}$ is a scalar quantity giving the length OP, the length of the projection of $\vec{F}$ on $\vec{n}$.

The derivative of a vector is another vector. The derivative with respect to time of the vector with components (a_1, a_2, a_3) is the vector with components $\left(\dfrac{da_1}{dt}, \dfrac{da_2}{dt}, \dfrac{da_3}{dt}\right)$.

Just as a general point may be represented by the coordinates (x,y,z), so too a general position vector is often represented by the vector $\vec{r}(x,y,z)$. Or equivalently

$$\vec{r} = x\vec{i} + y\vec{j} + z\vec{k} \tag{C.6}$$

Scalar and vector *point functions* are defined in section 5. We frequently find occasion to integrate scalar point functions over volumes and over surfaces. The *volume integral* of a scalar point function, $U(x,y,z)$,

$$\int_V U(\vec{r})dV, \text{ or } \int_V U(x,y,z)dxdydz \tag{C.7}$$

may be thought of as the limit of a summation involving the product of the scalar with an element of volume. Thus the product $U_i \Delta V_i$ is summated over all elements of volume contained by some surface or surfaces to give (C.7). If we take $U = 1$, (C.7) represents just the sum of all elements of volume. That is

$$\int_V dV = V \tag{C.8}$$

The *surface integral* of a scalar point function, $U(x,y,z)$,

$$\int_S U(\vec{r})dS \tag{C.9}$$

may also be thought of as a summation (Eq. (9.2)). The product of the scalar with an element of surface area is taken: $U_i \Delta S_i$. This product is then summated over all such elements lying on this surface or surfaces to give (C.9).

The operator ∇ (pronounced *del*), when applied to a scalar point function, denotes the *gradient* of that function. The gradient of a scalar U is given by

$$\nabla U \equiv \frac{\partial U}{\partial x}\vec{i} + \frac{\partial U}{\partial y}\vec{j} + \frac{\partial U}{\partial z}\vec{k} \tag{C.10}$$

Thus the gradient of a scalar is a vector. The gradient of U, ∇U, gives the maximum rate of change of $U(x,y,z)$ at the point (x,y,z) (see any text on vector calculus). Sometimes the operator *grad* is used in place of ∇; these operators are equivalent.

$$\text{grad } U \equiv \nabla U \qquad (C.11)$$

The gradient of a sum of point functions is the sum of the gradients:

$$\nabla(U + V) = \nabla U + \nabla V \qquad (C.12)$$

The gradient of a product of point functions is given by a familiar product rule:

$$\nabla(UV) = U\nabla V + V\nabla U \qquad (C.13)$$

The identities (C.12) and (C.13) follow directly from the definition (C.10) and may be easily demonstrated. Since

$$\frac{\partial(kU)}{\partial x} = k\frac{\partial U}{\partial x} \quad \text{etc.}$$

where k is a scalar constant, it will also be clear that

$$\nabla(kU) = k\nabla U \qquad (C.14)$$

In section 14 we dealt with the equation

$$\nabla C^* = \lambda\nabla C \qquad (14.36)$$

where λ is a constant scalar. This equation is interpreted to mean that the vectors ∇C and ∇C^* have the same direction but differ in magnitude.

When the operator ∇ is followed by a dot it may operate on a vector, and gives the *divergence* of that vector. Thus $\nabla \cdot \vec{G}$ (pronounced *del dot G*) denotes the divergence of the vector $\vec{G}$. If $\vec{G}$ has the three components g_1, g_2, g_3, then

$$\nabla \cdot \vec{G} \equiv \frac{\partial g_1}{\partial x} + \frac{\partial g_2}{\partial y} + \frac{\partial g_3}{\partial z} \qquad (C.15)$$

Hence, the divergence of a vector is a scalar. It is not really necessary to assign a geometric or physical significance to the divergence operator in order to use it. However physical meaning may be imported to it by Eq. (10.27), which may be generalized a little:

$$\nabla \cdot (\rho\vec{v}) = -\frac{\partial \rho}{\partial t} \qquad (C.16)$$

ρ is the density of some quantity (e.g. the number of molecules per unit volume, or the number of points in phase space per unit volume), and $\vec{v}(\vec{r}, t)$ is the velocity of a material point at $(\vec{r}, t)$. From (C.16) one might interpret $\nabla \cdot$, the operator consisting of spatial derivatives $\left(\frac{\partial}{\partial x}, \frac{\partial}{\partial y}\text{ and }\frac{\partial}{\partial z}\right)$ as that operator which corresponds to the "temporal" operator, $\frac{\partial}{\partial t}$. Roughly speaking, the divergence operator $\nabla \cdot$ gives the net quantity of material "diverging" from a given volume per unit time. Notice also Eq. (10.23), which shows that the divergence of the velocity vector $\nabla \cdot \vec{v}$ is equal to the material rate of change (*substantial derivative*) of an element of volume per unit of time.

Sometimes the operator *div* is used in place of $\nabla \cdot$; these operators are equivalent. Thus

$$\text{div } \vec{G} \equiv \nabla \cdot \vec{G} \qquad\qquad\qquad (C.17)$$

Since the gradient of a scalar, U, is the vector ∇U, we may find its divergence: $\nabla \cdot \nabla U$. The combination of operators $\nabla \cdot \nabla$ is encountered frequently and is usually abbreviated in some way. Thus

$$\nabla \cdot \nabla U \equiv \nabla^2 U \equiv \Delta U \equiv \text{Lap } U \equiv \text{div grad } U$$

are all equivalent representations. In this book we have used $\nabla^2 U$. This double operator is pronounced *del square U*, and is often called *Laplace's operator*.

$$\nabla^2 U = 0 \qquad\qquad\qquad (C.18)$$

is Laplace's equation.

The divergence of the product of a scalar with a vector point function is given by the identity (10.26):

$$\nabla \cdot (C\vec{v}) = \vec{v} \cdot \nabla C + C\nabla \cdot \vec{v} \qquad\qquad\qquad (C.19)$$

This identity has been applied frequently throughout the book, and so we shall prove it. Suppose that $\vec{v}$ has the components (v_1, v_2, v_3). Then from (C.15)

$$\nabla \cdot (C\vec{v}) = \frac{\partial(Cv_1)}{\partial x} + \frac{\partial(Cv_2)}{\partial y} + \frac{\partial(Cv_3)}{\partial z}$$

$$= \left[v_1\frac{\partial C}{\partial x} + v_2\frac{\partial C}{\partial y} + v_3\frac{\partial C}{\partial z} \right] + C\left(\frac{\partial v_1}{\partial x} + \frac{\partial v_2}{\partial y} + \frac{\partial v_3}{\partial z} \right)$$

$$= \vec{v} \cdot \nabla C + C\nabla \cdot \vec{v} \quad \text{q.e.d.}$$

One of the most important theorems involving the divergence of a vector is Gauss' Theorem (the divergence theorem) which states that for a bounded region and an arbitrary vector point function, $\vec{G}$,

$$\int_S \vec{G} \cdot \vec{n} \, dS = \int_V \nabla \cdot \vec{G} \, dV \qquad\qquad\qquad (9.4)$$

This theorem is important enough that the whole of section 9 was devoted to its demonstration.

In section 14, the third fundamental vector operator (following grad and div) was introduced: the *curl* or $\nabla \times$ operator. In this section use was also made of the *line integral* and the theorem of Stokes. However, I feel that a discussion of these matters does not properly belong in this brief introduction, and the interested reader is referred to any standard text on vector calculus.

Green's Theorem

In section 25 we had occasion to introduce Green's Theorem in the form

$$\int_V U\nabla^2 W \, dV = \int_S U\nabla W \cdot \vec{n} \, dS - \int_V \nabla U \cdot \nabla W \, dV \qquad (25.16)$$

where U and V are (uniform) scalar point functions which, together with their derivatives are continuous within some bounded region. Let us now demonstrate it.

In (C.19), replacing the scalar C by U and the vector $\vec{v}$ by ∇W,

$$\nabla \cdot (U\nabla W) = \nabla W \cdot \nabla U + U\nabla \cdot \nabla W$$
$$= \nabla W \cdot \nabla U + U\nabla^2 W$$

Integrating over the volume, V,

$$\int_V \nabla \cdot (U\nabla W)dV = \int_V \nabla W \cdot \nabla U \, dV + \int_V U\nabla^2 W \, dV \qquad (C.20)$$

From the divergence theorem (9.4), setting the vector $\vec{G}$ equal to the vector $U\nabla W$,

$$\int_V \nabla \cdot (U\nabla W)dV = \int_S U\nabla W \cdot \vec{n} \, dS$$

Substituting this relationship into (C.20),

$$\int_S U\nabla W \cdot \vec{n} \, dS = \int_V \nabla W \cdot \nabla U \, dV + \int_V U\nabla^2 W \, dV,$$

which is identical to (25.16) (bearing in mind the commutative property (C.4)).

Conclusion

The above has been a "crash course" in vector methods. Such courses are seldom completely satisfactory, but it is hoped that the reader who is uninitiated into the world of vectors will get the gist of their application in the study of biomolecular dynamics. This appendix might serve as a guide for all parts of the book except the somewhat speculative section 14, which will require a little more background reading. It is also most sincerely hoped that all vectorial neophytes will find a few evenings available to study one of the basic texts on vector analysis.

References

Hague, B. and Martin, D., *An Introduction to Vector Analysis for Physicists and Engineers*, 6th Ed. Methuen and Co. Ltd., 1970.

Weatherburn, C.E., *Advanced Vector Analysis*, Open Court Publishing Co., La Salle, Illinois, 1948.

SOLUTIONS TO SELECTED PROBLEMS

4.1

Let us first determine the relationship between the decay constant, λ, and the half life, $t_{\frac{1}{2}}$. From the definition of half life, we obtain, using (4.2),

$$\tfrac{1}{2}\, P^*_0 = P^*_0\, e^{-\lambda t_{\frac{1}{2}}}$$

or

$$\lambda = \ln 2/t_{\frac{1}{2}}$$

Suppose the physical decay of F^*_i is governed by $\lambda_p = \ln 2/H$ and the biological disappearance of F^*_i is governed by $\lambda_b = \ln 2/h$. Then the required differential equation is

$$\frac{d}{dt}[F^*_i\ PO_4] = -\,(\lambda_p + \lambda_b)\,[F^*_i\ PO_4]$$

$$= -\,(\ln 2/H + \ln 2/h)\,[F^*_i\ PO_4]$$

6.1

$$a \equiv \frac{C^*}{C + C^*} \qquad \text{(an alternative definition for this problem)}$$

$$\frac{\partial a}{\partial t}\Big|_D = \frac{\partial}{\partial t}\Big|_D \left[\frac{C^*}{C + C^*}\right] = 0$$

i.e.

$$\frac{(C + C^*)\,\dfrac{\partial C^*}{\partial t}\Big|_D - C^*\,\dfrac{\partial}{\partial t}\Big|_D(C + C^*)}{(C + C^*)^2} = 0$$

$$\frac{\partial C^*}{\partial t}\Big|_D = \frac{C^*}{C + C^*}\,\frac{\partial}{\partial t}\Big|_D(C + C^*)$$

or

$$R^*_{d,v} = a R_{d,v} \qquad \text{as before.}$$

377

7.1

My solution is

$$(a_1 \ a_2 \ a_3 \ a_4 \ a_5 \ a_6 \ a_7 \ a_8 \ a_9 \ a_{10} \ a_{11} \ a_{12})$$

$$= (2 \ \ 2 \ \ 1 \ \ 2 \ \ 1 \ \ 2 \ \ 1 \ \ 1 \ \ 2 \ \ 2 \ \ 1 \ \ 2 \ )$$

The reasoning involved is based upon simple collision theory, and may be demonstrated by reference to Eq. (7.22).

(i)   Every "successful" collision (x) of the form $[NO]^2$ x $[Br_2**]$
         will produce 2 NOBr* molecules. Therefore a_1 = 2.

(ii)   Every successful collision of the form [NOBr*] x [NOBr*]
         will destroy 2 NOBr* molecules. Therefore a_2 = 2.

(iii)   Every successful collision of the form $[Br_2*]$ x $[NO]^2$
         will produce 1 NOBr* molecule. Therefore a_3 = 1.

(iv)   Employing the binomial theorem it is seen that
         if the probability of a collision of the form [NOBr*] x [NOBr*]
                                           is [NOBr*][NOBr*],
         then the probability of a collision of the form [NOBr*] x [NOBr]
                                           is 2[NOBr*][NOBr].
   Every successful collision of the form [NOBr*] x [NOBr]
         will destroy only one NOBr* molecule, but such collisions are twice
         as probable as the collisions of case (ii). Therefore a_4 = 2.

The remaining a's may be evaluated using similar reasoning.

12.2

In (12.25) make the substitution

$$u = t - \tau$$
$$du = -d\tau$$

Then

$$\int_0^t h(\tau)X(t - \tau)d\tau = \int_t^0 h(t - u)X(u)(-du)$$

$$= \int_0^t h(t - u)X(u)du$$

Changing the dummy variable from u to τ gives (12.26)

15.2

Using Eq. (15.14)

$$\frac{dM_1^*}{dt} = .2M_2^* - \frac{.2}{2}M_1^* = .2M_2^* - .1M_1^*$$

$$\frac{dM_2^*}{dt} = \frac{.2}{2}M_1^* - .2M_2^* = .1M_1^* - .2M_2^*$$

For $\frac{dM_1^*}{dt} = \frac{dM_2^*}{dt} = 0$, we have

$$M_1^* = 2M_2^*$$

At t = 0, $\quad C_1^*(0) = M_1^*(0)/V_1 = 0$

$$C_2^*(0) = M_2^*(0)/V_2 = 3.0\mu C_i \text{ per unit volume}$$

At $t \to \infty$ $\quad C_1^*(\infty) = 2/1 = 2\mu C_i$ per unit volume

$$C_2^*(\infty) = 1/1 = 1\mu C_i \text{ per unit volume}$$

The tracer "gradient" has reversed direction.

19.1

$$F_{11}(D) \cdot F_{22}(D) \cdot C_2^* \equiv \left[-\frac{d}{dt} - K_{11}\right] \cdot \left[-\frac{d}{dt} - K_{22}\right] \cdot C_2^*$$

$$= \left(\frac{d^2}{dt^2} + K_{11}\frac{d}{dt} + K_{22}\frac{d}{dt} + K_{11}K_{22}\right) \cdot C_2^*$$

i.e. $\quad \frac{d}{dt} \cdot \frac{d}{dt} \cdot C_2^* \equiv \frac{d^2C_2^*}{dt^2}$ and does *not* equal $\left(\frac{dC_2^*}{dt}\right)^2$.

22.1

$$\frac{dC^*}{dt} + KC^* = \frac{R_a^*(t)}{V} \tag{22.11}$$

Multiplying by the integrating factor e^{Kt}

$$e^{Kt}\left[\frac{dC^*}{dt} + KC^*\right] = \frac{1}{V}e^{Kt}R_a^*(t)$$

$$\frac{d}{dt}\left[e^{Kt}C^*\right] = \frac{1}{V}e^{Kt}R_a^*(t)$$

$$\left[C^*e^{Kt}\right]_0^t = \frac{1}{V}\int_0^t e^{K\tau}R_a^*(\tau)d\tau$$

$$C^*(t) = C^*(0)e^{-Kt} + \frac{1}{V}e^{-Kt}\int_0^t e^{K\tau}R_a^*(\tau)d\tau$$

For $R_a^* = 0$ and $C^*(0) = M_o^*/V \equiv A$

$$R_d = KVC = \frac{KM_0^*C}{A}$$

which is Eq. (22.9).

For $R_a^* =$ a constant we obtain in the limit as $t \to \infty$

$$C^*(\infty) = \frac{R_a^*}{KV}, \quad \text{or} \quad R_d^* = R_a^* = KVC^*(\infty)$$

Since $R_d = KVC$ we have $R_d = R_a = R_a^*/a$ which is Eq. (22.10).

<u>24.4(a)</u>

The step response is given by $R_a^* \int_0^t A\tau^{-\frac{1}{2}} \, d\tau$, R_a^* constant

$$= 2R_a^* \, A\left[\tau^{\frac{1}{2}}\right]_0^t$$

The integral does *not* converge for $t \to \infty$. But, we know from experiment that the plasma levels of all metabolites tend to plateau (for nominal values of R_a^*). Hence $At^{-\frac{1}{2}}$ is not an acceptable function for representing the impulse response of a real system.

If an infusion to the hypothesized system is given at the rate $Bt^{-\frac{1}{2}}$, the response, by convolution is

$$AB \int_0^t \tau^{-\frac{1}{2}} \, (t - \tau)^{-\frac{1}{2}} \, d\tau$$

Making the substitution $u = \tau/t$, this integral becomes

$$AB \int_0^1 u^{-\frac{1}{2}} \, (1 - u)^{-\frac{1}{2}} du = AB\beta(\tfrac{1}{2}, \tfrac{1}{2}), \quad \text{(the beta function)}$$

$$= AB \, \frac{\Gamma(\tfrac{1}{2}) \cdot \Gamma(\tfrac{1}{2})}{\Gamma(1)} = \pi AB$$

(The integration can be carried out without the explicit use of beta and gamma functions.)

Therefore, in this hypothetical system, a diminishing infusion rate, $Bt^{-\frac{1}{2}}$, rather than a constant infusion rate, gives rise to a constant plasma level. (Not in Solarians, but in Barnardians perhaps?)

24.4b

$$\sqrt{.1} \int_0^\infty t^{-.5} e^{-.1t} \, dt = \Gamma(.5) = \sqrt{\pi}$$

Thus, in the limit of large t, the step response is constant as required.

29.2

Comparing the unit impulse response function

$$C^*(t) = B_1 e^{-\varepsilon t} + B_2 e^{-\beta_2 t}$$

with Eq. (20.5a) we can make the identification of B_1, the coefficient of the very slowly falling exponential, with E/V_1. That is

$$B_1 = E/V_1$$

Since we deal with a case where $\nabla C = 0$ and $\nabla R_{d,v} = 0$ we can calculate V_{TOT} from (28.20):

$$V_{TOT} = M_0^*/B_1 = 1 / \frac{E}{V_1}$$

From Table 20.3 E/V_1 has the value .10 and .05 for the smaller and larger Cylindrium respectively. Therefore V_{TOT} has the values 10. and 20. which are the values given in Table 20.1. However, the above is all rather academic. Since $R_{d,v} = 0$ everywhere and C^* approaches B_1 for large t, $V_{TOT} = M_0^*/B_1$ by the dilution principle, quite simply.

34.2

Using (34.17) I obtained 9.6 litres/min as the cardiac output.

34.3

$$C^*(x,t) = \frac{M_0^*}{\sqrt{4\pi A^2 D}} \, t^{-\frac{1}{2}} \exp\left(-\frac{x^2}{4Dt}\right)$$

$$J^{*out} = -D \frac{\partial C^*}{\partial x} \vec{i} \cdot \vec{n} = -D \frac{\partial C^*}{\partial x} = \frac{x}{2t} C^*$$

Thus

$$A \int_0^\infty J^{*out} \, dt = \frac{Ax}{2} \int_0^\infty \frac{C^*}{t} \, dt = \frac{M_0^* x}{4\sqrt{\pi D}} \int_0^\infty t^{-3/2} \exp\left(-\frac{x^2}{4Dt}\right) dt$$

Making the substitution $w^2 = \frac{x^2}{4Dt}$, we obtain after simplifying

$$A \int_0^\infty J^{*out} \, dt = \frac{M_o^*}{\sqrt{\pi}} \int_0^\infty e^{-w^2} \, dw = \frac{M_o^*}{\sqrt{\pi}} \cdot \frac{\sqrt{\pi}}{2} = \frac{M_o^*}{2}$$

Similarly $J^{*in} = -D \frac{\partial C^*}{\partial x} (-1) = D \frac{\partial C^*}{\partial x}$

since the unit outward normal vector $\vec{n}$ points in the direction on the negative x-axis. Etc.

34.4

The derivation of $V = Q\bar{t}$ was based upon the interpretation of the function g(t) (34.33). It was stated quite glibly on pages 287-8: "The probability that a particle of indicator which was injected at $t = 0$ will leave the system between the times t and t+dt is given by ..." g(t)dt. Now, while this may be the case when every indicator particle leaves the system after passing the sampling site once, it is not necessarily true for a diffusive system, where the same particle may pass the sampling site several times. Hence $V \neq Q\bar{t}$.

35.4

(i) First let us demonstrate that

$$uA = \frac{M_o^*}{\int_0^\infty C^* dt} \qquad \qquad ...(1)$$

In (35.26) set $P = \frac{M_o^*}{\sqrt{4\pi DA^2}}$, $R = \frac{u^2}{4D}$

and make the transformation of variables

$$u = (Rt)^{\frac{1}{2}}$$

Then

$$\int_0^\infty C^* dt = 2PR^{-\frac{1}{2}} e^{2R\tau} \int_0^\infty e^{-u^2 - \frac{(R\tau)^2}{u^2}} \, du$$

From a table of definite integrals it is found that

$$\int_0^\infty e^{-x^2 - a^2/x^2} \, dx = \frac{1}{2} e^{-2a} \sqrt{\pi}, \qquad a > 0 \quad ...(2)$$

Hence $\int_0^\infty C^* dt = \frac{M_o^*}{uA}$, which is the required result.

(ii) Evaluate the right-hand side of (34.23).

Since C_i^* is taken upstream, $\tau < 0$

$$\int_0^\infty C_i^* \, dt = \frac{M_0^*}{uA} e^{\frac{u^2}{4D} \tau} \qquad \qquad \ldots(3)$$

using (2). Therefore

$$\int_0^\infty \left[C_0^*(t) - C_i^*(t) \right] dt = \frac{M_0^*}{uA} \left(1 - e^{\frac{u^2}{4D} \tau} \right) \quad \ldots(4)$$

Now

$$\int_0^\infty J_{diff}^* \, dt = -D \int_0^\infty \frac{\partial C^*}{\partial x} \, dt$$

Using (2) it may then be shown that

$$\int_0^\infty J_{diff}^{*out} \, dt = 0, \quad \text{using } \tau > 0 \quad \text{and}$$

$$\int_0^\infty J_{diff}^{*in} \, dt = -M_0^* e^{\frac{u^2}{4D} \tau}, \quad \text{using } \tau < 0$$

Hence

$$Q = \frac{M_0^* \left[1 - e^{\frac{u^2}{4D} \tau} \right]}{\frac{M_0^*}{uA} \left(1 - e^{\frac{u^2}{4D} \tau} \right)} = uA \qquad \ldots(5)$$

Therefore, from (1) and (5)

$$Q = \frac{M_0^*}{\displaystyle\int_0^\infty C^* dt}$$

even though transport occurs by both convection and diffusion.

GLOSSARY

The following is a list of the symbols which are used frequently. All functions have the arguments $(\vec{r}, t)$, position and time, unless otherwise indicated or the function is specified as constant; eg. $R_{d,v}$ is $R_{d,v}(\vec{r}, t)$. The dpm, or equivalently the curie, may be regarded as having the dimensions of mass.

Symbol	Designation	Dimensions
α or a	specific activity	dimensionless or $dpm \cdot mass^{-1}$
β or b	positive constant: used in $e^{-\beta t}$	$time^{-1}$
C	concentration of tracee	$mass \cdot volume^{-1}$
C*	concentration of tracer	$dpm \cdot volume^{-1}$
D	coefficient of diffusion	$length^2 \cdot time^{-1}$
D*	coefficient of self-diffusion	$length^2 \cdot time^{-1}$
$\delta(\vec{r})$ or $\delta(t)$	Dirac delta function (distribution)	$argument^{-1}$
h	unit impulse response function	$volume^{-1}$
H(t)	Heaviside function (distribution)	dimensionless
$\vec{J}$	flux of tracee	$mass \cdot area^{-1} \cdot time^{-1}$
$\vec{J}*$	flux of tracer	$dpm \cdot area^{-1} \cdot time^{-1}$
k or K	rate constant	$time^{-1}$
$\vec{i}, \vec{j}, \vec{k}$	unit basis vectors parallel to the x-, y- and z-axis respectively	dimensionless
M(t)	quantity of tracee	mass
M*(t)	quantity of tracer	dpm
$\vec{n}$	unit outward-directed normal vector at a surface	dimensionless

385

Q	rate of flow of fluid	volume $\cdot$ time^{-1}
$\vec{r}(x,y,z)$	position vector	length
$R_a(t)$	rate of appearance of tracee	mass $\cdot$ time^{-1}
$R_a^*(t)$	rate of appearance of tracer	dpm $\cdot$ time^{-1}
$R_d(t)$	rate of disappearance of tracee	mass $\cdot$ time^{-1}
$R_d^*(t)$	rate of disappearance of tracer	dpm $\cdot$ time^{-1}
$R_{a,v}$	rate of appearance of tracee per unit volume	mass $\cdot$ volume^{-1} $\cdot$ time^{-1}
$R_{a,v}^*$	rate of appearance of tracer per unit volume	dpm $\cdot$ volume^{-1} $\cdot$ time^{-1}
R_{ai}^*	rate of appearance of tracer into the ith compartment of a system	dpm $\cdot$ time^{-1}
$R_{d,v}$	rate of disappearance of tracee per unit volume	mass $\cdot$ volume^{-1} $\cdot$ time^{-1}
$R_{d,v}^*$	rate of disappearance of tracer per unit volume	dpm $\cdot$ volume^{-1} $\cdot$ time^{-1}
s	†Laplace variable: e^{-st}	t^{-1}
S	designates a surface integral: $\int_S$	area
S	$^{\dagger\Delta}$symbol used to refer to tracee substance	——
S*	†symbol used to refer to tracer substance	——
t	time	time

---

†not regarded as a function of position and time.

$^{\Delta}$in actual usage, there is no possibility of confusing the two symbols, S.

$\bar{t}$	mean residence or sojourn time of a particle within its distribution space	time
$\vec{u}$	convective velocity	length · time^{-1}
v	[†]velocity of a material point	length · time^{-1}
V	designates a volume integral: $\int_V$, also designates a constant volume of distribution of a substance	volume
x,y,z	Cartesian coordinates	length

[†] $\vec{v} = \vec{u}$ if convection is the sole mode of transport.

AUTHOR INDEX

SUBJECT INDEX